Nanostructure Science and Technology

Series editor

David J. Lockwood, FRSC
National Research Council of Canada
Ottawa, Ontario, Canada

More information about this series at http://www.springer.com/series/6331

Nicolas Alonso-Vante

Chalcogenide Materials for Energy Conversion

Pathways to Oxygen and Hydrogen Reactions

Springer

Nicolas Alonso-Vante
University of Poitiers
Poitiers
France

ISSN 1571-5744 ISSN 2197-7976 (electronic)
Nanostructure Science and Technology
ISBN 978-3-030-07815-7 ISBN 978-3-319-89612-0 (eBook)
https://doi.org/10.1007/978-3-319-89612-0

Printed on acid-free paper

This Springer imprint is published by the registered company Springer International Publishing AG part of Springer Nature
The registered company address is: Gewerbestrasse 11, 6330 Cham, Switzerland

For Cristina, Pablo Alexander
and Gweni-Viani

Preface

This book is a review which focuses on electrocatalysis based on chalcogenides particularly in the nanoscale domain. The interest will be placed on the focus of two important reactions of technological interest, related to the electrochemistry of hydrogen and oxygen reactions. More particularly, the book will provide an introduction to materials synthesis, basic principles of electrochemistry and electrocatalysis, associating precious versus non-precious catalytic center chalcogenides as well as different supports. The role of such supports to stabilize the catalytic centers will be stressed. In short, it covers the properties of this class of electrocatalysts, from the bottom-up approach to the application in low-temperature fuel cell systems, such as the microfluidic fuel cells for portable devices. The book will illustrate the basic ideas behind materials design from cluster to cluster-like compounds providing current research progress. It will be ideal for students, professionals, and researchers in the search of non-noble metal centers for chemical energy conversion interested in the field of electrochemistry, renewable energy, and electrocatalysis.

Last but not least, this work reflects the product of many fruitful collaboration with a number of talented students and researchers in France and abroad, whose contributions are mentioned in the list of references in every chapter. Therefore, I heartily thank everyone.

Poitiers, France
February 2018

Nicolas Alonso-Vante

Contents

About the Author

Dr. Nicolas Alonso-Vante is Professor of Chemistry, doing research at the IC2MP laboratory at the University of Poitiers. He holds a Docteur Troisième Cycle, in Electrochemistry, and a Doctorat d'Etat in Physical Science from the University of Strasbourg, France, and a Master in Physical-Chemistry from the Research Center for Advanced Studies in Mexico City. Prior to joining the University of Poitiers in 1997, he worked as a Senior Scientist at the Hahn-Meitner-Institut-Berlin (now Helmholtz-Zentrum Berlin) in Germany, doing research in photoelectrochemical system for 13 years. He has developed materials in the nanodivided scale based on transition metal and chalcogenides in metallic and semiconducting form in Germany and France for over more than three decades. Using facile chemical routes, he has pioneered the use of non-platinum and non-precious metal catalysts center chalcogenides in order to understand the complex process of the oxygen reduction reaction process, an important technological reaction in low-temperature fuel cells. Other interests of such materials in combination with light harvesting are devoted to split water into hydrogen and oxygen, and photocatalysis. His other main research interests are (photo)electrochemistry and (photo)electrocatalysis of novel materials using various ex situ and in situ techniques, fuel generation, interfacial characterization, and surface analytical techniques. He is the author of over 200 publications, chapters, editor of a two-volume e-book on electrochemistry in Spanish and holds two US, two German, and two French patents. He has received the awards of the National Polytechnic Institute-Mexico as a R&D distinguished graduate, and of the Mexican Council of Technology SNI-III recognition as a Mexican researcher working outside Mexico, and has been awarded the NM Emanuel Medal from the Russian Academy of Science. He is also a Member of the International Society of Electrochemistry, the Electrochemical Society, the Bunsengesellschaft für Physikalische Chemie, and of the International Academy of Electrochemical Energy Science (IAOEES) and Member of the Editorial board of Fuel Cells Journal.

About the Authors

Chapter 1
Introduction

1.1 Basic Concepts of Electrochemical Kinetics

The thermodynamic principles of electrochemistry can be found in a vast literature, e.g., [1–4]. This section is mainly devoted to delineate the kinetics concepts applied so far to the multi-electron charge transfer at the electrocatalytic materials interface, which is further discussed in the next chapters. Therefore, some basic concepts are recalled in this section.

1.1.1 The Interfacial Charge Transfers Kinetic Model

The charge transfer at the electrode/electrolyte interface is usually associated with other phenomena, e.g., adsorption–desorption, mass transfer, surface diffusion, crystallization. Moreover, the driving force at the interface is the internal difference potential between the electrode and the electrolyte, necessarily linked to the double-layer capacitance [5–12], so that at the electrode surface, S, the surface reaction rate (mol cm^{-2} s^{-1}), which involves in the reaction a stoichiometric number of species ν_i, is defined as:

$$v = \frac{1}{\nu_i}\frac{1}{S}\left(\frac{\partial n_i}{\partial t}\right)_{Interf} \tag{1.1}$$

Thus, the reduction and oxidation rates at the interface *(0, t)* of a general electrochemical reaction are, namely (a) $\nu_{ox}Ox + ne^- \rightarrow \nu_{red}Red$; and (b) $\nu_{red}Red \rightarrow \nu_{ox}Ox + ne^-$ are:

$$v_{red} = k_{red}C_{Ox}^{\nu_{ox}}(0,t) \tag{1.2a}$$

N. Alonso-Vante, *Chalcogenide Materials for Energy Conversion*, Nanostructure Science and Technology, https://doi.org/10.1007/978-3-319-89612-0_1

$$v_{ox} = k_{Ox} C_{\mathrm{Red}}^{v_{red}}(0, t) \tag{1.2b}$$

If we confine the species *Ox* and *Red* on a metallic electrode with $|\nu_{red}| = |\nu_{ox}| = 1$ and $|\nu_e| = n \geq 1$, a series of elementary steps, following the kinetic theory, must be taken into account in the final equation of the current–potential characteristics. In the simplified picture, the rate constants, k_{red} and k_{ox}, will depend on the applied electrode potential difference between the electrode surface and the electrolyte.

$$k_{red} = k^0 e^{-(1-\alpha_c)nf\left(E-E^0\right)} \tag{1.3a}$$

$$k_{ox} = k^0 e^{+\alpha_a nf\left(E-E^0\right)} \tag{1.3b}$$

Hence, k^0 is the reduction, and oxidation rate constant at the standard potential, E^0, and $f = F/RT = (96{,}487\ C\ mol^{-1}/8.314\ J\ mol^{-1}K^{-1} \times 298\ K) = 38.9\ V^{-1}$. The charge transfer coefficient, α, has been identified, respectively, as α_a, and α_c. This coefficient is a descriptor of the applied electrode potential at the solid/electrolyte interface affecting the cathodic and anodic activation energies through the free Gibbs energy of reaction, $\Delta G_r^{\neq}$, and can be rationalized as follows, Eqs. (1.4 and 1.5):

$$\Delta G_r^{\neq} = -nFE = -nF(E - E^\circ) = -nF\Delta E \tag{1.4}$$

$$\frac{\partial \Delta G_c^{\neq}}{\partial \Delta G_r^{\neq}} = \frac{\partial \Delta G_c^{\neq}}{nF\Delta E} = \alpha_c \quad \text{and} \quad \frac{\partial \Delta G_a^{\neq}}{\partial \Delta G_r^{\neq}} = \frac{\partial \Delta G_a^{\neq}}{nF\Delta E} = \alpha_a \tag{1.5}$$

One can assume that α_a and α_c are constants with values between 0 and 1, so that at any electrode potential $\alpha_a + \alpha_c = 1$. However, if α is potential dependent, $\alpha_a + \alpha_c \neq 1$.

Combining the reaction rate, v, with the flow of electrons, ($I = \pm nFSv$), where $\pm$ stands for anodic and cathodic, respectively, the total current, I_{total}, is given by:

$$I_{total} = I_a + I_c = nFS[k_{ox} C_{\mathrm{Red}}(0, t) - k_{red} C_{Ox}(0, t)] \tag{1.6}$$

The substitution of Eqs. (1.3a, b) in (1.6) gives the general Butler–Volmer equation (1.7).

$$I = nFSk^0 \left[C_{\mathrm{Red}}(0, t) e^{(1-\alpha_a)nf(E-E^\circ)} - C_{Ox}(0, t) e^{-\alpha_c nf(E-E^\circ)}\right] \tag{1.7}$$

1.1.2 At Equilibrium

One first consequence of the kinetic current, displayed in Eq. (1.7), is to demonstrate that at equilibrium, i.e., $E = E^0 = E_{eq}$, the total current is equal to 0. This simple assumption makes Eq. (1.7) become the well-known Nernst equation (1.8).

$$\frac{C_{Ox}(0,t)}{C_{Red}(0,t)} = e^{nf(E_{eq}-E^{\circ})} \tag{1.8}$$

It is interesting to note that $C_{Ox}(0,t)$ and $C_{Red}(0,t)$ in Eq. (1.8) refers to the concentration at the electrode surface. However, at equilibrium, the species concentration in the electrolyte bulk C^*_{Ox} and C^*_{Red} can also be used in Eq. (1.8).

1.1.3 Far from Equilibrium

1.1.3.1 Far from Equilibrium and Charge Transfer

The electron charge transfer driving force is given by the so-called overpotential parameter: $\eta = E - E_{eq}$. Substituting $E = E_{eq} + \eta$ in (1.7), and rearranging, we obtain.

$$I = nFSk^0 C^*_{Ox} e^{-\alpha_a nf(E_{eq}-E^{\circ})} \left[e^{(1-\alpha_a)nf\eta} - e^{-\alpha_c nf\eta}\right] \tag{1.9a}$$

or

$$I = I_0 \left[e^{(1-\alpha_a)nf\eta} - e^{-\alpha_c nf\eta}\right] \tag{1.9b}$$

where I_O divided by the electrode surface area, S, represents the exchange current density, j_O. This parameter is again a descriptor used to compare, in a volcano curve, the electrocatalytic properties of various materials toward a defined reaction. Taking the following values for $\alpha_a = \alpha_c = 0.5$, T = 298.15 K, n = 1, and $j_O = 2$ mA cm^{-2}, and $\eta = \pm 0.15$ V, the plot of Eq. (1.9b) (dark curve) is given in Fig. 1.1a. The partial contributions for the anodic and cathodic currents are given in red, so that one obtains the well-known Tafel expression by rewriting the Butler–Volmer equation (1.9b) for the anodic and cathodic overpotential, when $\eta \to \pm\infty$.

$$\pm\eta(j) = a \pm b \ \ln|j| \tag{1.10}$$

where b is the Tafel slope ($\pm 2.303\ RT/\alpha F$ or ($\pm 2.303/\alpha\ f$). For n = 1, the slope corresponds to 0.118 V/decade. Equation (1.10) is plotted in Fig. 1.1b, with an exchange current density of 1 mA cm^{-2}. For simple kinetics, the Tafel slope is

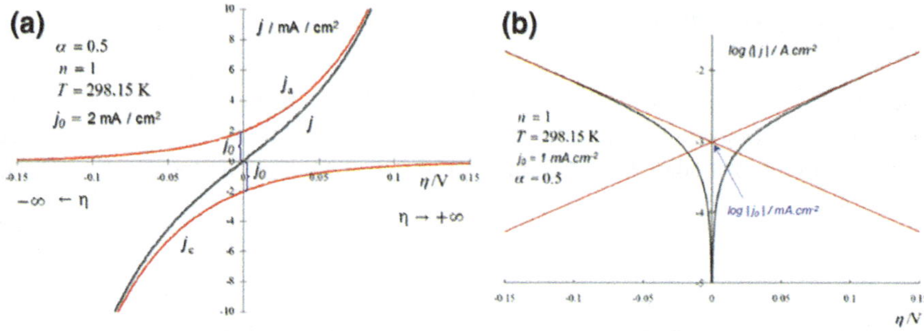

Fig. 1.1 **a** Plot of j versus the overpotential, η, calculated with Eq. (1.9) and **b** the Tafel representation of Eq. (1.10)

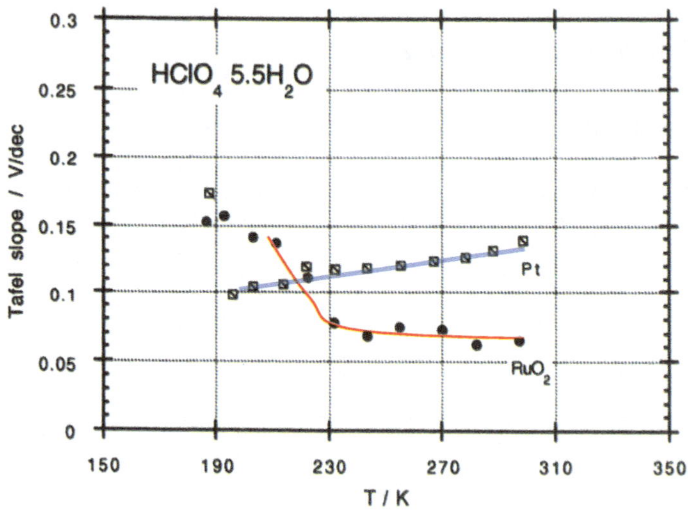

Fig. 1.2 Change of the Tafel slope with temperature for RuO_2 and Pt data adapted with permission from reference [13]. Copyright (1993) American Chemical Society

proportional to the temperature. This phenomenon is certainly true for outer-sphere charge transfer mechanisms, but in electrocatalysis of a multi-electron charge transfer process (e.g., water oxidation), the situation is more complex as demonstrated in Fig. 1.2 for platinum and ruthenium dioxide electrodes. The figure depicts the response obtained in a wide range of temperature at an applied electrode potential [13]. The latter one shows an anomalous behavior, whereas the former one follows the trend dictated by the temperature increase by b. This experimental fact, certainly, puts in evidence the role of the charge transfer coefficients as being descriptors of the surface complex chemistry during an electrocatalytic process.

1.1.3.2 Far from Equilibrium and Mass Overpotential

The application of an electrode potential at the interface to favor the charge transfer process means activating the reaction through an activation overpotential, η_{act}. This process, at the electrode/electrolyte interface, generates a concentration gradient that generates the concentration overpotential, η_{conc}. Departing from Eq. (1.7) in the equilibrium, we obtain for the exchange current Eq. (1.11):

$$I_0 = nFSk^0 C_{Ox}^{*(1-\alpha_a)} C_{\mathrm{Red}}^{*\alpha_c} \tag{1.11}$$

Rearranging Eqs. (1.7) over (1.11) it turns out that:

$$I = I_0 \left[\frac{C_{\mathrm{Red}}(0,t)}{C_{\mathrm{Red}}^*} e^{(1-\alpha_a)nf\eta_a} - \frac{C_{Ox}(0,t)}{C_{Ox}^*} e^{-\alpha_c nf\eta_c} \right] \tag{1.12}$$

Herein, the concept of mass transfer overpotential comes up clearly, using the definition of overpotential and the Nernst equation.

$$\eta_{conc} = E - E_{eq} = \frac{1}{nf} \ln \left(\frac{C_{Ox}(0,t)}{C_{Ox}^*} \Big/ \frac{C_{\mathrm{Red}}(0,t)}{C_{\mathrm{Red}}^*} \right) \tag{1.13}$$

With the definition of a limiting current, $I_l = nFSD\frac{C^*}{\delta}$, Eq. (1.13) can be represented as:

$$\eta_{conc} = \frac{1}{nf} \ln \left[-\left(\frac{D_{Ox}}{D_{\mathrm{Red}}} \right)^{2/3} \frac{C_{Ox}^*}{C_{\mathrm{Red}}^*} \cdot \frac{I - I_{l,c}}{I_{l,a} - I} \right] \tag{1.14}$$

One generally encounters in all steady-state experimental data (e.g., generated via the rotating disk electrode technique), the effect of the concentration gradient, and Eq. (1.12) is the general one to be taken into account to extract kinetic parameters.

1.2 Chalcogens and Metal Chalcogenides

The term chalcogen applies to the elements (O, S, Se, Te, and Po) placed in group 16 of the Periodic Table. The compounds formed by these elements are known as chalcogenides, and herein we make mainly reference to inorganic compounds based on or involving sulfur, selenium, and tellurium of the type (M_aX_b; where M is a transition metal, and X: S, Se, Te). A recent account on the history, and occurrence of chalcogens has been published [14]. Herein, we concentrate on the discussion regarding the electronic structure and the interfacial reactivity for electrocatalytic purposes of 2D (e.g., MoS_2, …), 3D (e.g., RuS_2, …) as well as surface coordinated transition metal centers (e.g., Ru_xSe_y, …).

1.2.1 The Bond Ionicity or Covalency

The degree of interaction of the transition metal valence *d*-, *p*-, and *s*-electrons with the chalcogen valence *p*-electrons determines the conduction and valence bands energy position of the chalcogenide materials in the solid state. Focusing on the binding energies data extracted from the literature [15] and schematized in Fig. 1.3 we can qualitatively observe that an increasing covalency is obtained when *d*-states of the metal with 2*p* states of the chalcogen proceed from oxygen to tellurium. Transition metal chalcogenide (TMC) compounds, in the groups 8–10 (or 8B) of the Periodic Table, can present species having oxidation states X^{2-} or $(X_2)^{2-}$. In other words, the cation is reduced and the anion is oxidized, going from layered, e.g., TiS_2 with Ti^{4+} and S^{2-}, to bulk FeS_2 with Fe^{2+} and $(S_2)^{2-}$. This phenomenon is attributed to the intra-atomic and interatomic interaction of electrons, and this latter describes the behavior of the majority of chalcogenide compounds, their band energy positions, and gap. Hence, the hybridization degree of chalcogen *p*-states with the transition metal *d*-states favor the metal–metal interaction and thus the magnitude of the band gap, leading to modulate the optoelectronic properties of TMC. This complex interplay, however, with respect to the filling of the metal *d* bands determine why semiconductors, as well as metallic conductors are obtained [16, 17].

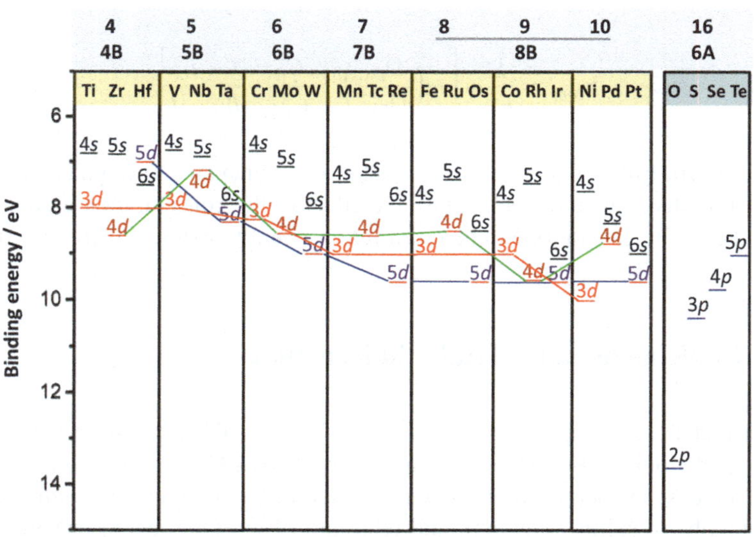

Fig. 1.3 Atomic ionization potential for the transition metals and chalcogens. Data extracted from reference [15]

1.2.2 The 2D, 3D Compounds

A portion of the Periodic Table, as that depicted in Fig. 1.4, gives an overview of the various TMC structures with two-dimensional (2D) and three-dimensional (3D) structures. The 2D-TMCs in MX_2 is actually composed of individual M atomic layers coordinated between two chalcogen (X) atomic layers, Fig. 1.5. This stacking configuration leads to two types of crystal structures, namely trigonal prismatic (Tp) (2H), and octahedral (Oh) (1T); see Fig. 1.6. As discussed in the literature, a similar TMC, e.g., 2H-MoS_2 can be converted to 1T- MoS_2 [18]. This phase transformation is of extreme importance in the materials physics and chemistry domain, since the conductivity of MoS_2 changes from semiconducting to conducting material. This means that a sole material can be used as a photocatalyst as well as an electrocatalyst for green fuel generation.

Based on a quantitative band structure calculation [19–21], a simplified electronic structure is shown in Fig. 1.7 for M: Mo, Pt; X: S, and Se [20]. A systematic variation in the valence band, via photoelectron spectroscopy, of such materials was summarized earlier [22]. The main feature for the Oh coordination symmetry is that the *d*-states are placed above a lower energy band made up of *p*-states from chalcogen, whereas on the other hand, in the Tp symmetry, a split of the t_{2g} energy band into a lower d_z^2 energy band and higher *d*-band derived from d_{xy}, and $d_x^2 - d_y^2$ states is separated by a hybridization gap.

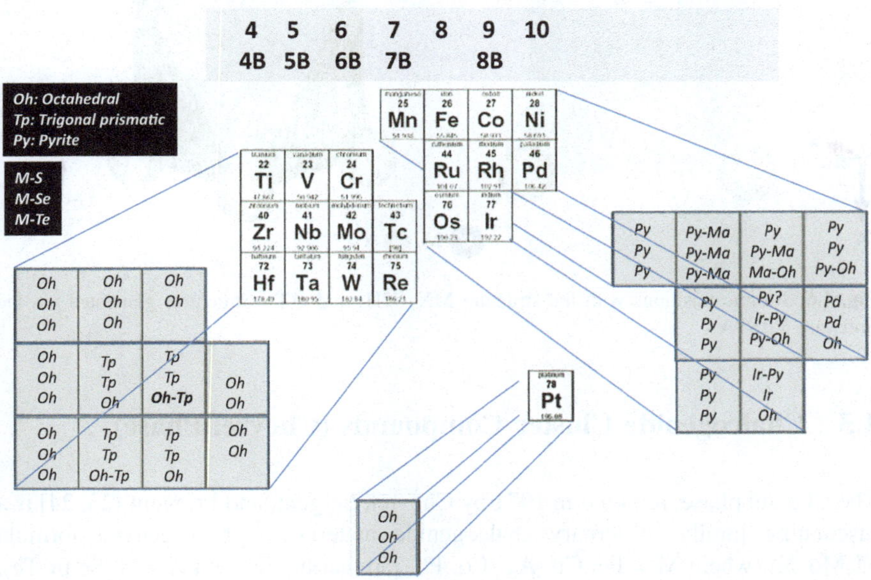

Fig. 1.4 Coordination symmetry (Oh, Tp, Py) of some transition metal dichalcogenides (MX_2). The right block shows essentially 3D structures. The block on the left and bottom forms 2D structures (layered compounds). Modified from reference [16]. Copyright (1992) Published by Elsevier B.V

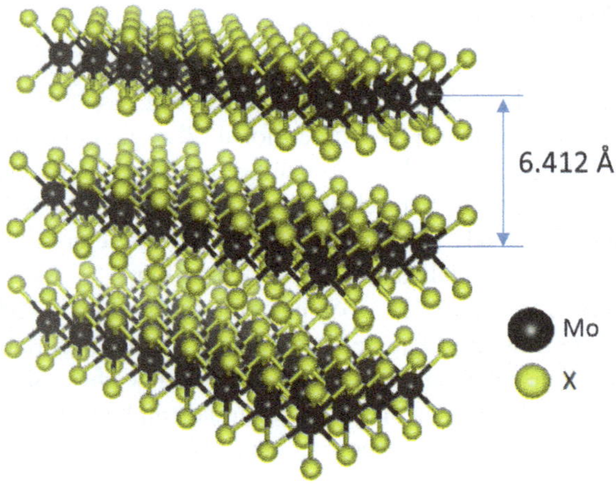

Fig. 1.5 A layered compound based on MoS_2. structure generated via the Software VESTA

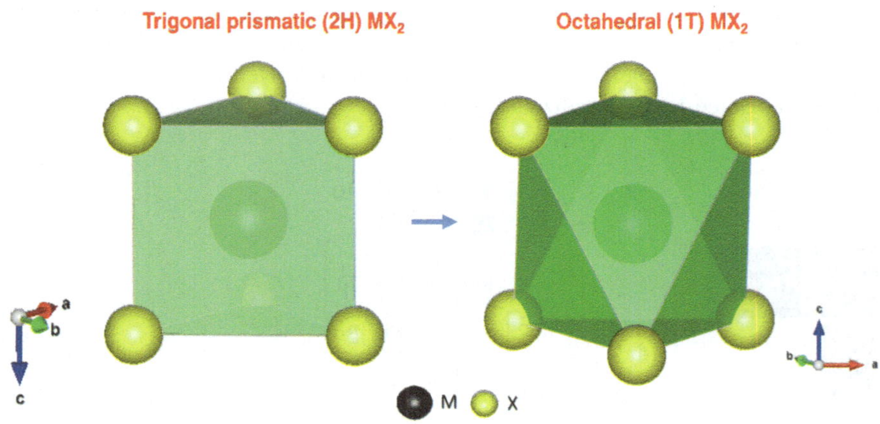

Fig. 1.6 Atomic positions with the structure MX_2 (2H) and (1T). Structures generated via the Software VESTA

1.3 Chalcogenide Cluster Compounds (Chevrel Phase)

The Chevrel phase, reported in 1971 by Chevrel, Sergent, and Pringent [23, 24] is a fascinating family of ternary chalcogenide materials with a general formula $M_xMo_6X_8$ (where M = Fe, Cu, Ag, Co, Pb, rare earth, etc., and X = S, Se or Te). Due to the variability of their dimension this phase possesses interesting physical properties, and has been investigated since the 1970s toward their superconducting properties with large critical magnetic fields [25]; and band structure [26–29].

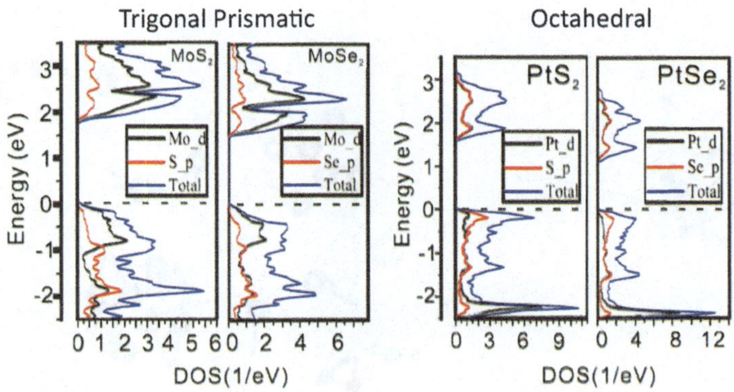

Fig. 1.7 Total density of states (DOS) of some MoX_2 and PtX_2 (X = S, Se). Adapted from reference [20]. Copyright (2016) under Creative Commons Attribution License

1.3.1 Structural Description

The structures are made up of blocks of slightly distorted cubes or clusters of Mo_6X_8 units, with eight X atoms at the cube corners and six Mo atoms slightly outside the middle of the cube faces (Fig. 1.8). At room temperature, most of them have rhombohedral structures and many distort to triclinic structures. As shown in Fig. 1.9 a network of interconnecting channels, parallel to the rhombohedral axes (C3), runs between the Mo_6X_8 clusters which contain interstitial sites or cavities, where M atoms reside. Such cavities share within the rhombohedral axis (C3), opposite pseudo squares. One can recognize two (Mo–Mo) distances describing the molybdenum cluster. The first distance, R1, belongs to two-molybdenum atoms found on the same plane perpendicular to the ternary axis (C3), while the second belongs to two adjoining planes, R2, in the cluster, namely: 0.266, 0.271 nm, respectively.

Fig. 1.8 [Mo_6X_8] cluster unit of the Chevrel phase. Structures generated via the Software VESTA

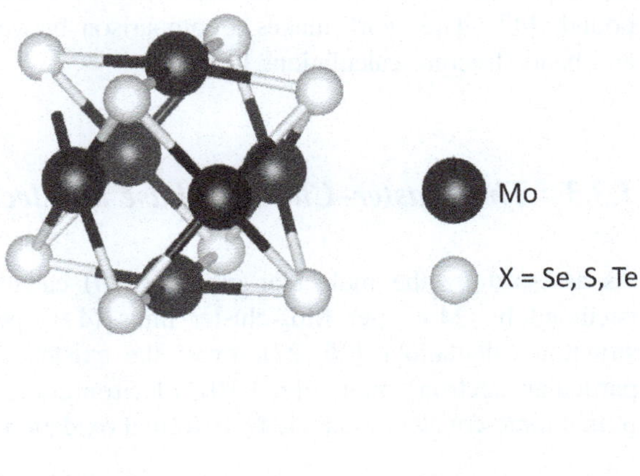

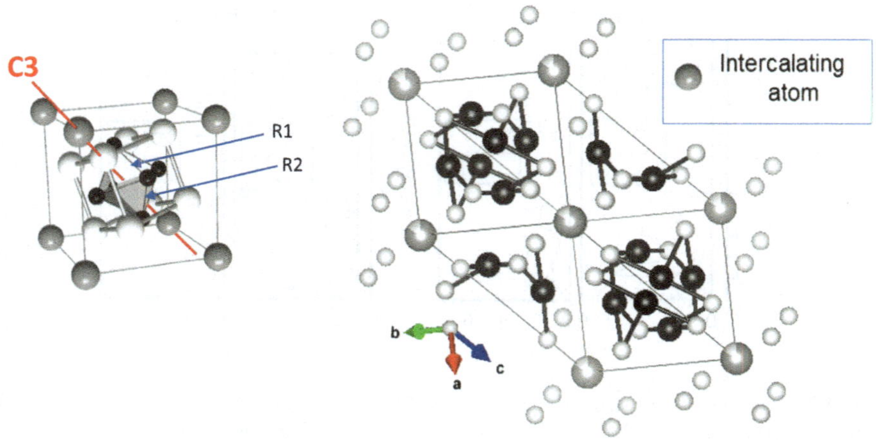

Fig. 1.9 [Mo_6X_8] cluster unit of the Chevrel phase indicating the ternary axis and the two Mo–Mo distances R1 and R2. The cluster network is highlighted at the right. Structures generated via the Software VESTA

1.3.2 Electronic Properties

The unit Mo_6X_8 (X = S, Se and Te), consisting of [Mo_6] octahedron clusters surrounded by eight chalcogens arranged in a distorted cube, cf. Figure 1.9, was theoretically investigated by ab initio methods, as certified by studies on electronic band structure [26, 27, 29], and crystal structure [30, 31]. Such studies revealed that the [Mo_6] cluster behaves as a redox center. Hence, their interest increased in the battery field community, as cathodes for Li batteries [32–36], additionally for processes such as, hydrodesulphurization catalysis, [37–39], and electrocatalysis [40–46]. The theoretical band structure calculations revealed a density of states (DOS) near the Fermi level (E_F) due to the Mo 4*d* character (Fig. 1.10). The splitting between the bonding and antibonding states of the Mo 4*d* band depends on the number of electrons in the cluster. Certainly, this characteristic led recently to the hope that thermoelectric materials can be found in the Chevrel phase compounds [47]. This work makes a comparison between the photoemission spectra and band structure calculations [47].

1.3.3 The Cluster–Chevrel Phase as Electron Reservoir

As assessed by the molecular orbital (MO) calculation, all bonding states are occupied by 24 e^- per Mo_6-cluster units (4 e^- per molybdenum atom). Band structure calculations [26, 27] reveal the existence of a band gap (E_g) in that particular electron count (Fig. 1.10). Electron counting rules are based on a simplified ionic-covalent model [31]. A formal oxidation state of −2 can be assigned to

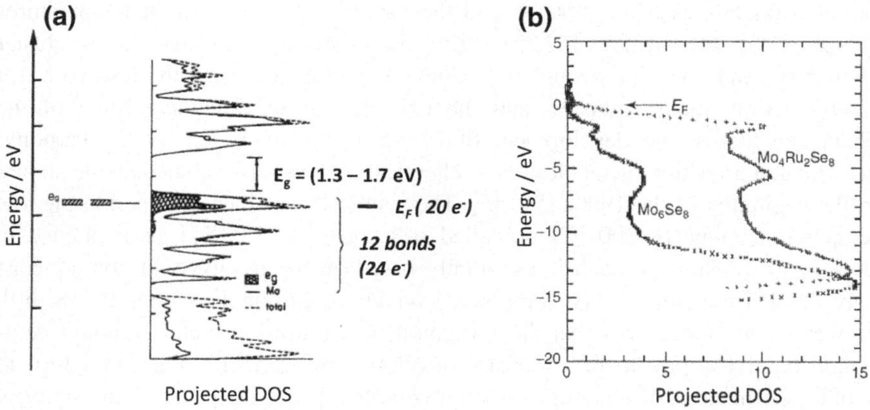

Fig. 1.10 **a** Projected density of states of Mo_6S_8 Chevrel-phase cluster. **b** Valence band spectra (showing the density of states (DOS) in the (Mo, Ru) 4*d* bands in the energy interval from 0 to −9 eV) of Mo_6Se_8 (20 e^-) and $Mo_4Ru_2Se_8$ (24 e^-). Figure (**a**) with permission from reference [27] Copyright (1983) American Chemical Society; figure (b) with permission from references [40, 41] Copyright (1987) American Chemical Society

the chalcogen X, i.e., 16 negative charges for the Mo_6-cluster unit which are balanced by the valence electrons of the metal atoms (e.g., Mo). Consequently, for Mo_6X_8 the remaining electrons in the intra-cluster bonding are: $(6 \times 6) - 16 = 20$ or 3.33 e^- per molybdenum atom. However, in the ruthenium containing pseudo-ternary cluster, (Mo_4Ru_2), as in $Mo_4Ru_2Se_8$, the electron counting is: $(4 \times 6) + (2 \times 8) - 16 = 24$, i.e., 4 e^- per cluster. This latter is a semiconductor (E_g = 1.3 eV) as reported some time ago [41]. This "magic" number of 4 e^- per cluster is also met in the ternary compound $Cu_4Mo_6S_8$, the semiconducting nature of which has not been identified yet. These electrons are essentially derived from the transition metal *d*-states [27] and available at the Fermi level, E_F (Fig. 1.10a). The increased density of states at E_F is attained with a valence electron counting (VEC) of 4 e^- per cluster, as demonstrated by valence band spectra measurements [40, 41] (Fig. 1.10b). The delocalization of electrons in the cluster offers a pool of charges that can be engaged in an electrocatalytic reaction with a minimum electronic relaxation. In this sense, Chevrel phases can be considered as models for the investigation of charge transfer electrocatalysis based on metal center clusters, and as cathodes for intercalation processes in batteries [48, 49], and other areas of physics and chemistry [50].

1.4 Soft Chemistry Synthesis of Chalcogen-Like Materials

In earlier days, the solid-state chemical route was preferentially used, to generate chalcogenide cluster materials [23, 51, 52]. For instance, the reaction to generate a Chevrel phase was performed in vacuum evacuated, sealed, silica tubes heated at initial temperatures of ca. 500 °C and then subjected to reaction at temperatures $\geq$900 °C for several days [52, 53]. With the knowledge acquired on the cluster chemistry and physics using the Chevrel phase as electron reservoir for multi-electron charge transfer, and the evidence of the chemical force of the chalcogen atoms, the development of a chemical path, based on the carbonyl coordinated transition metal clusters as chemical precursor, for chalcogenide started in the beginning of the 1990s [54, 55]. The synthesis was then confined to the low temperature range (< 200 °C), so-called soft chemistry [54–57], as explained in Sect. 1.4.1.2. This approach is essentially based on the pyrolysis in non-aqueous solvent of a metallic cluster compound containing carbonyl as ligands [58, 59]. However, one recognizes that the formation of a transition metal carbonyl compound has its origin in the chemistry developed by Longoni, and Chini [60] to synthesize dianions of platinum carbonyl complex: $[Pt_3(CO)_6]_{\sim n}^{2-}$. This chemical route is named herein, the carbonyl chemical route, and is devoted to the chalcogenide compounds.

1.4.1 The Carbonyl Chemical Route

The basis of this approach consists in pyrolyzing in mild conditions the chemical precursor, a transition metal complex containing as ligands carbonyl: $[M_x(CO)_y]$. The chemical reaction in an organic solvent during the pyrolysis of the chemical precursor can be affected by the solvent itself, since this latter can act as a ligand (L), and be coordinated to the metal center. This leads to the formation of a "critical" complex, $[M_x(CO)_{y-1}L_1]$, which nature is of importance to understand the final product $[M_xL_1]$, in form of nanoparticles, as schematically represented in Fig. 1.11. Various examples exist in the literature and they will be briefly described.

1.4.1.1 Ruthenium-Based Alloy Materials

We consider a key factor the use of transition metal complexes as chemical precursors, as a departure point for the synthesis of nanodivided materials (monometallic, bimetallic or chalcogenide species). A bimetallic cluster compound

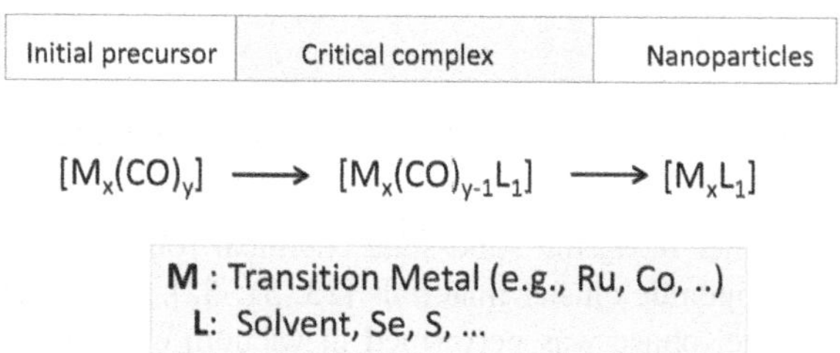

Fig. 1.11 Schematics of the pyrolyzing process either in a gas or in a non-aqueous solvent phase for the generation of chalcogenide cluster-like materials

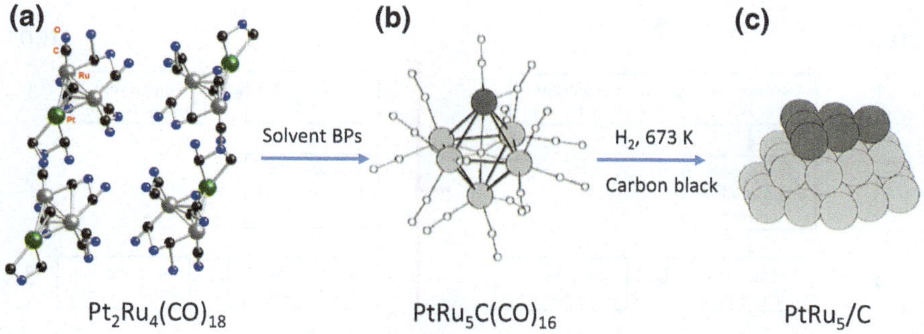

Fig. 1.12 Evolution of the transition metal cluster $Pt_2Ru_4(CO)_{18}$ **a** in a solvent (boiling point—BP) to form the carbide complex (**b**), and finally the pyrolysis in gas phase to the generation of the bimetallic carbon supported material (**c**). Figure (**a**) generated from the XRD data of reference [61]; figures (**b**) and (**c**) with permission from reference [63] Copyright (1997) American Chemical Society

was designed by Adams et al. [61] by reacting $Ru(CO)_5$ and $Pt(Cyclooctadiene)_2$, to obtain $Pt_2Ru_4(CO)_{18}$. Indeed, this cluster in a solvent can further evolve to form an intermediate complex, carbide $PtRu_5C(CO)_{16}$, prior to obtain the bimetallic catalyst, as shown in Fig. 1.12. The product PtRu nanoparticles can be supported, e.g., onto carbonaceous supports. The studies on this system revealed that this pyrolytic pathway can lead to the formation of highly dispersed nanomaterials with an increased crystalline disorder [62–64].

1.4.1.2 Ruthenium-Based Chalcogenide Materials

The genesis of the transition metal chalcogenide electrocatalysts, a process that precedes the above described one, takes its origin on the well-ordered Chevrel phase clusters (see Sect. 1.3.1). Indeed, simple pyrolysis in non-aqueous solvents under mild conditions (< 200 °C) of a transition metal cluster compound suffices to produce, at the boiling temperature of the solvent, a nanodivided product that contains coordinated chalcogen atoms (S, Se or Te). This process is shown schematically in Fig. 1.13. In the specific case of ruthenium, the chemical precursor was the tris-ruthenium dodecacarbonyl: $Ru_3(CO)_{12}$. Consequently, for the synthesis of Ru_xSe_y, one can inscribe, following the scheme in Fig. 1.11, the pyrolytic reaction in a determined non-aqueous solvent:

$$Ru_3(CO)_{12} + Se \rightarrow \left[Ru_xSe_y - CO_z\right] \rightarrow Ru_xSe_y + zCO \qquad (1.15)$$

This process can be monitored in xylene (Xyl) and/or 1,2 dichlorobenzene (DCB) solvents. It was later recognized that the reaction of ($Ru_3(CO)_{12}$) and elemental selenium dissolved in an organic solvent leads to a polynuclear chemical precursor in the first stages of the reaction. This is the so-called critical complex in

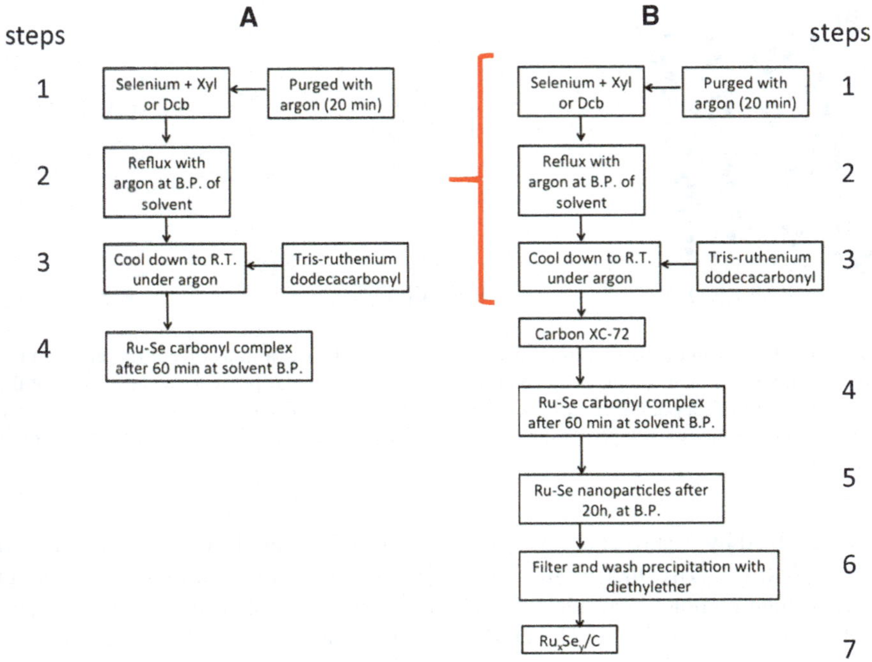

Fig. 1.13 Frame A shows the steps leading to the formation of a heteronuclear complex, and additionally frame B shows the completion (up to step 5) of the pyrolysis in the solvent, and in presence of a support, e.g., carbon. The supported material is obtained after completion in step 7

step 4, frame A, in Fig. 1.13. The presence of this "critical complex" was assessed by ^{13}C NMR, showing the chemical shift $\delta = 198.88$ ppm, characteristic of $Ru_4Se_2(CO)_{11}$, and FTIR analyses [65, 66]. The generation of this "critical complex" was obtained with a high yield in 1,2 dichlorobenzene (DCB). Chemically speaking, the simple pyrolysis technique, using a proper solvent's boiling point (BP), shows some advantages to produce this high nuclear complex as compared to other data reported in the literature [67–69]. However, by keeping the boiling temperature of the solvent in refluxing conditions under argon or under nitrogen during 20 h of reaction (step 5, frame B, Fig. 1.13) the carbonyl ligands are lost, and the resultant product is a cluster-like material: Ru_xSe_y. What is important here is to disperse the material, by adding a conducting support, e.g., carbonaceous, or oxide materials, as indicated, between steps 3 and 4, Fig. 1.13, to finally obtain the supported chalcogenide Ru_xSe_y/C (step 7 in frame B, Fig. 1.13). This process has been embraced by other research groups in the world [70–81] and further extended using aqueous medium [82]. In these works, however, the stoichiometry of the majority of the pyrolyzed products remained rather descriptive. Therefore, the rationalization of the "critical complex" or real chemical precursor, at least with respect to the Ru_xSe_y cluster-like, was assessed, in order to determine the surface chemistry of the material. With the knowledge of the nature of the "critical complex": $Ru_4Se_2(CO)_{11}$, its simple pyrolysis [58] led to x≈2 and y≈1. However, the chemical signature of such cluster-like materials was obtained combining various physical–chemical analytical techniques.

1.4.2 Toward Chalcogenide Cluster Surface Coordination

Taking example on the relatively well-studied ruthenium chalcogenide (Ru_xSe_y), the pyrolysis of $Ru_4Se_2(CO)_{11}$, in a solvent or in a gas phase, favors the elimination of the carbonyl ligands. The formation dynamics of Ru_xSe_y cluster-like compound was investigated in situ under temperature-controlled conditions, with regard to the initial fragmentation of cluster framework evolution followed by the preferential nucleation of selenium atoms at ruthenium centers [58] as indicated in Fig. 1.14. The thermal decomposition process can be subdivided into three regions, namely: (a) 20–30 min (150–240 °C) breakdown of the carbonyl and formation of ruthenium clusters; (b) 30–65 min (240–440 °C) slow growth of the preformed cluster by coalescence; and (c) 65–80 min (440–500 °C) formation of a phase separation. One observes at t = 30 min a sudden completion of the metallic cluster formation. This phenomenon (right scheme in the figure) leads to the formation of segregated core structures and finds precedent in other bimetallic systems as discussed in

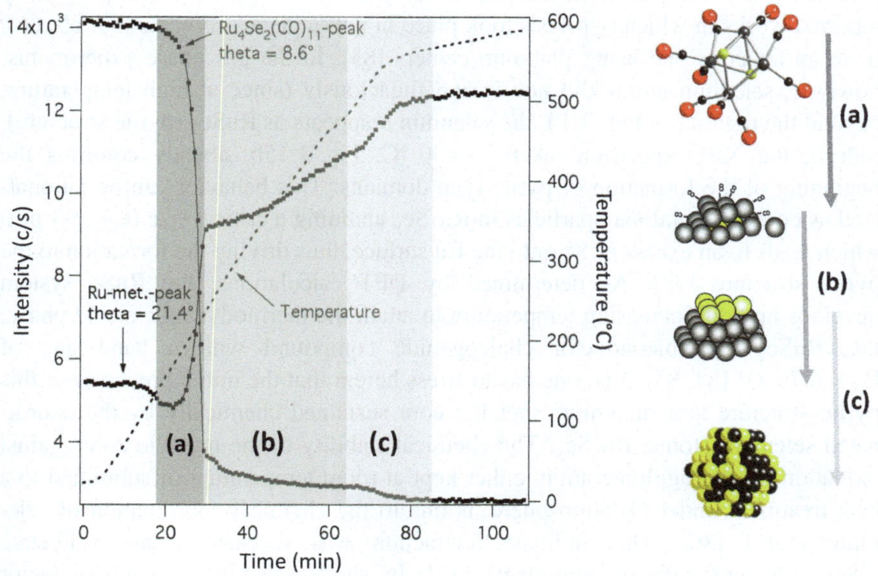

Fig. 1.14 Intensity variation of the diffraction peak 112 of carbonyl complex precursor $Ru_4Se_2(CO)_{11}$ at Theta = 8.6°; and around the strongest peak of *hcp* ruthenium at theta = 21.4°. The measurements were performed as a function of time and temperature under nitrogen. Figure with permission from reference [58] Copyright (2007) American Chemical Society

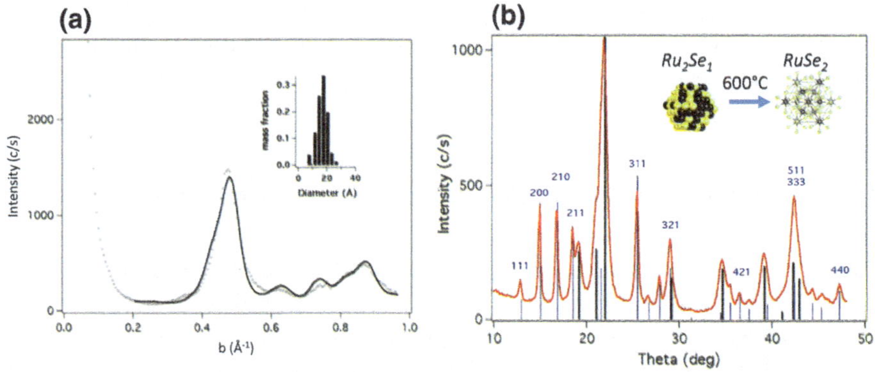

Fig. 1.15 **a** Diffraction pattern of Ru_xSe_y after pyrolysis at 220 °C of $Ru_4Se_2(CO)_{11}$. The inset shows the mass fractions of the *hcp* Ru model particles versus their diameter. From reference (58); **b** wide angle XRD pattern of Ru_xSe_y after 600 °C treatment. The thick vertical bars show the *hcp* structure of Ru_xSe_y. The presence of the $RuSe_2$ pyrite phase is present. Figures with permission from reference [58] Copyright (2007) American Chemical Society

Sect. 1.4.1.1. Since the end product leads, apparently, to a nanometallic core—*hcp* Ruthenium—Fig. 1.15a, it clearly indicates that chalcogen atoms must coordinate onto ruthenium surface atoms (the scattering power of selenium is ca. 60% that of ruthenium), with a narrow size distribution in the non-supported chalcogenide particles with a compositional distribution of 2:1 in full agreement with previously reported works in which pyrolysis took place in non-aqueous solvents [59, 65, 66], or in an aqueous one using platinum centers [83]. In the gas phase experiments, however, selenium atoms did not escape fugaciously (since at high temperature, beyond the region (c), Fig. 1.14, the selenium reappears as $RuSe_2$ (pyrite structure). Indeed, the XRD spectrum taken at 430 °C, Fig. 1.15b, already confirms the beginning of the formation of pyrite-type domains. This behavior can be rationalized as core Ru metal nanoparticles in Ru_xSe_y attaining a critical size (e.g., >3 nm) which leads to an excess of Se onto the Ru surface, thus driving the formation to the pyrite structure [58]. As determined by DFT calculations, the RuSe system develops up with increasing temperature to attain the thermodynamic stable phase, i.e., $RuSe_2$ a semiconductor chalcogenide compound with a band gap of $E_g = 0.76$ eV [84, 85]. Yet, one has to stress herein that the initial precursor of this pyrite structure is a ruthenium metallic core sustained chemically by the coordinated selenium atoms: Ru_xSe_y. The chemical stability of the metallic core against oxidation, and/or agglomeration, either kept at room temperature or submitted to a heat treatment under O_2-atmosphere, is due to the chemical coordination of selenium atoms [86]. This intimate interaction was revealed using solid-state ^{77}Se-NMR and XPS measurements [87]. In short, selenium, a semiconductor ($E_g = 1.99$ eV), becomes metallic because charge transfer takes place from the ruthenium metallic core. This fact explains the low liability of ruthenium nanoparticles to coordinate oxygen or be oxidized. The beneficial chalcogenide surface coordination was herein assessed and tested on other systems, such as Pt_xSe_y [83], Pt_xS_y [88, 89], Rh_xS_y [90, 91], Ir_xSe_y [92], Pd_xX_y (X: S, Se, Te) [93].

1.5 Other Chalcogenide Systems

The rich chemistry of chalcogens materials is exemplarily illustrated below for the sake of their fundamental understanding and evaluation toward future applications in energy storage and converting devices.

1.5.1 Amorphous–Glassy Chalcogenides

Excluding oxygen, the elements in group 6A (or16), Fig. 1.3, react with more electropositive elements, e.g., Ag, to form chalcogenides. Typical representatives of this kind of materials are AgInSbTe and GeSbTe. These materials can contain one or more chalcogens and are crystalline or amorphous (glassy), metallic or semiconducting, and conductors of ions and electrons. They are thus classified as covalent networks. The switch between the amorphous state (glassy) and the crystalline one can encode binary information useful to data storage because of the change in optical as well as electrical properties. Clearly, these relevant physical properties will depend on the electronic structure, and since they also show

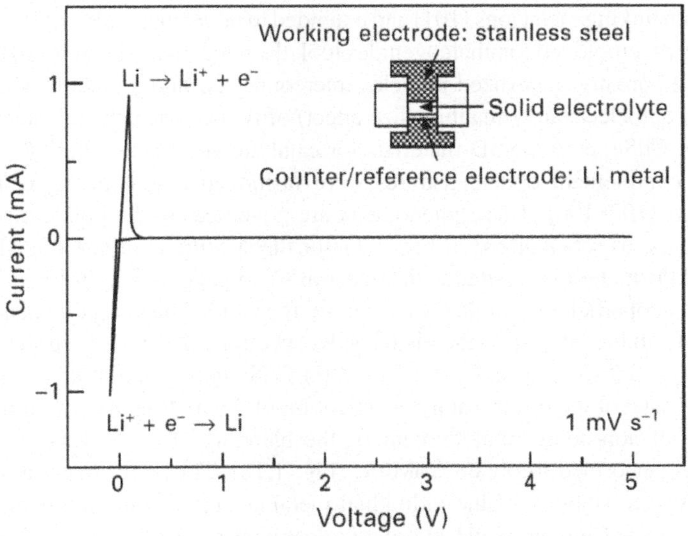

Fig. 1.16 Current–potential behavior of $70Li_2S{\cdot}30P_2S_5$ glass ceramic. With permission from reference [98] Copyright © 2014 Woodhead Publishing Limited

significant ionic conductivity, the switch between amorphous and crystalline states is also interesting for electrochemical redox switches. A comprehensive account of synthesis, properties, and applications was given recently, for example [94, 95]. Since these materials are connected to electronics, optics, optoelectronics, biomedical areas, physicochemical sensing, and data storage, the link between these materials with those mentioned in the above sections is their potential application as solid electrolytes in batteries. In this connection, silver-doped chalcogenide glasses (Ag_2S–GeS_2, Ag_2Se–$GeSe_2$) were discussed by Frumar and Wagner [96], Pradel and Ribes [97]; and (Li_2S–SiS_2–Li_4SiO_4, Li_2S–P_2S_5) by Tatsumisago and Hayashi [98]. For the sake of illustration, the current–potential characteristic first cycle of the system containing $70Li_2S$–$30P_2S_5$ is shown in Fig. 1.16, where a wide electrochemical window over 5 V was measured. This kind of bulk-type all-solid-state battery was reported to show an excellent performance and stability.

1.5.2 Amorphous–Gel Chalcogenide: Chalcogels

The generation of homogeneous nanonetworks or aerogels is well explored for oxides that build interconnecting structures consisting of micropore (< 2 nm), meso-pore (2–50 nm), and macro-pore (>50 nm). The current state of materials science for chalcogenides, in this context, has been extended to synthesize gel structures, since it is believed that the solgel process leading to aerogel formation provides a powerful and general approach for the creation of porous nanoarchitectures. Various strategies were envisaged to achieve the synthesis of chalcogenide gels, namely thiolysis [99], nanoparticle condensation [100], and metathesis reactions between soluble chalcogenide with linking metal ions [101] and extended to novel materials [102]. Therefore, the stratagem employed for chalcogenides took the name of chalcogels [101]. Hence, combining porosity (nanosized particles interconnected into a random 3D architecture) with optoelectronic (quantum size effect) of typical chalcogenide semiconductors (CdS, CdSe, etc.) the 3D materials for catalytic and photovoltaic devices with quantum confinement optical features were fashioned as compared to their bulk counterpart [103–106]. These phenomena are governed by the bulk density of the nanonetwork, as schematized in Fig. 1.17 for the specific case of CdSe [103]. The density of the material is related to the material's compactness; accordingly, the band gap from nanoparticles to single crystal crosses the various stages of dimensionality or spanning from the infrared to the visible solar spectrum. The CdSe aerogel has a gap energy, $E_g = 2.2$ eV, versus $E_g = 1.74$ eV for CdSe bulk material.

For the sake of the electrocatalytic activation of the hydrogen evolution reaction, by means of non-noble catalytic centers, the blending of chalcogels of CoS_x and MoS_x was performed to obtain $CoMoS_x$ (Fig. 1.18). This latter material combines the activity and stability of the individual metal centers of the former materials to attain a so-called universal pH hydrogen electrode, in alkaline, and acid environment [107], demonstrated by a similar overpotential, at a defined current density (Fig. 1.18).

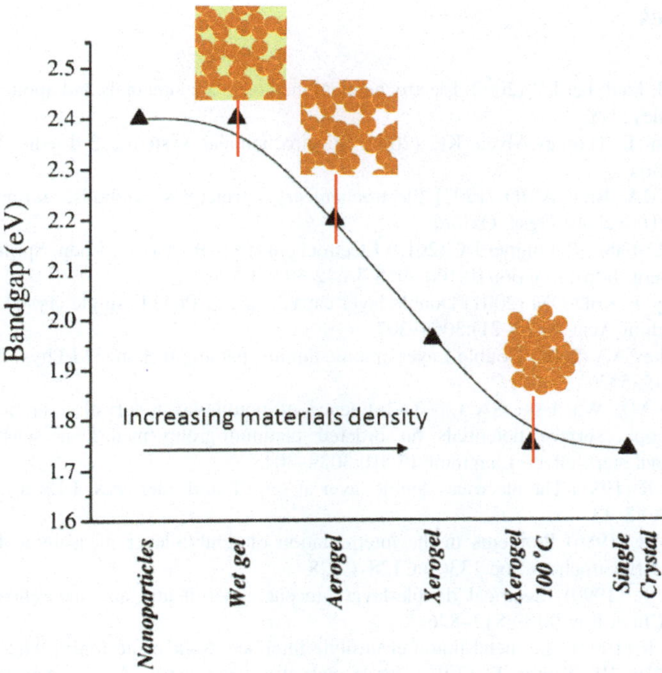

Fig. 1.17 Effect of the material's density and the optical band-gap energy on the various morphologies of CdSe. Adapted from reference [103]

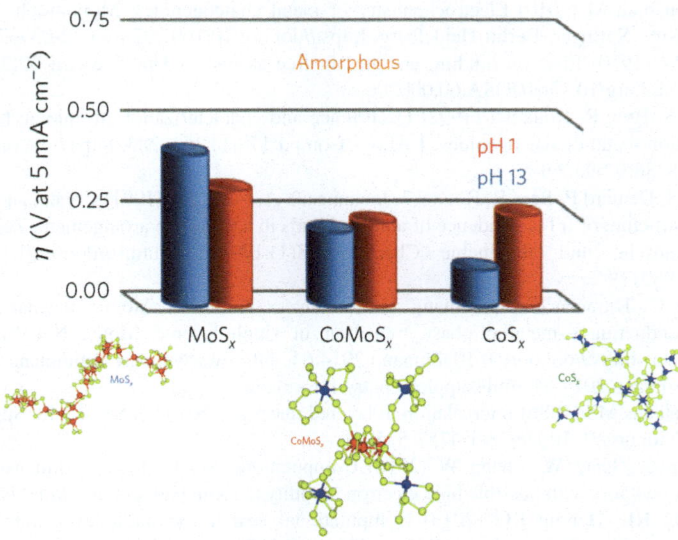

Fig. 1.18 Hydrogen evolution reaction activity on various chalcogels at a current density of 5 $mAcm^{-2}$ in 0.1M $HClO_4$ (pH 1) and 0.1M KOH (pH 13). With permission from Reference [107] Copyright © 2015, Rights Managed by Nature Publishing Group

References

1. Bard AJ, Faulkner LR (2002) Electrochemical methods: fundamentals and applications, 2nd edn. Wiley, NY
2. Newman J, Thomas-Alyea KE (2004) Electrochemical systems, 3rd edn. Wiley, Inc Publication
3. Brett CMA, Brett AMO (1993) Electrochemistry—principles, methods, and applications. Oxford University Press, Oxford
4. Lefrou C, Fabry P, Poignet J-C (2012) Electrochemistry—the basics, 1 edn. Springer, Berlin Heidelberg. https://doi.org/10.1007/978-3-642-30250-3
5. Pajkossy T, Kolb DM (2001) Double layer capacitance of Pt(111) single crystal electrodes. Electrochim Acta 46(20–21):3063–3071
6. Kornyshev AA (2007) Double-Layer in ionic liquids: paradigm change? J Phys Chem B 111 (20):5545–5557
7. Weaver MJ, Wasileski SA (2001) Influence of double-layer solvation on local versus macroscopic surface potentials on ordered platinum-group metals as sensed by the vibrational stark effect. Langmuir 17(10):3039–3043
8. Parsons R (1981) The electrical double-layer at solid-liquid interfaces. J Electroanal Chem 118 (Feb):3–18
9. Parsons R (1986) Problems in the interpretation of double-layer measurements on solid metals. J Electrochem Soc 133(3):C128–C128
10. Parsons R (1990) Electrical double-layer—recent experimental and theoretical developments. Chem Rev 90(5):813–826
11. Parsons R (1997) The metal-liquid electrolyte interface. Solid State Ionics 94(1–4):91–98
12. Schmickler W, Santos E (2010) Interfacial electrochemistry, 2 edn. Springer, Berlin Heidelberg. https://doi.org/10.1007/978-3-642-04937-8
13. Alonso-Vante N, Colell H, Stimming U, Tributsch H (1993) Anomalous low-temperature kinetic effects for oxygen evolution on RuO_2 and Pt electrodes. J Phys Chem 97(29): 7381–7384
14. Bouroushian M (2010) Electrochemistry of metal chalcogenides. Monographs in electrochemistry. Springer, Berlin Heidelberg. https://doi.org/10.1007/978-3-642-03967-6
15. Lotz W (1970) Electron binding energies in free atoms*. J Opt Soc Am 60(2):206–210. https://doi.org/10.1364/JOSA.60.000206
16. Jobic S, Brec R, Rouxel J (1992) Occurrence and characterization of anionic bondings in transition metal dichalcogenides. J Alloys Compd 178(1):253–283. https://doi.org/10.1016/0925-8388(92)90269-F
17. Jobic S, Deniard P, Brec R, Rouxel J, Jouanneaux A, Fitch AN (1991) Crystal and electronic band structure of $IrTe_2$: evidence of anionic bonds in a CdI_2-like arrangement. Zeitschrift für anorganische und allgemeine Chemie 598(1):199–215. https://doi.org/10.1002/zaac.19915980119
18. Lin Y-C, Dumcenco DO, Huang Y-S, Suenaga K (2014) Atomic mechanism of the semiconducting-to-metallic phase transition in single-layered MoS_2. Nat Nano 9 (5): 391–396. https://doi.org/10.1038/nnano.2014.64. http://www.nature.com/nnano/journal/v9/n5/abs/nnano.2014.64.html#supplementary-information
19. Dresselhaus MS (1986) Intercalation in layered materials. Nato Sci Ser B 148. Springer, US. https://doi.org/10.1007/978-1-4757-5556-5
20. Huang Z, Zhang W, Zhang W (2016) Computational search for two-dimensional MX2 semiconductors with possible high electron mobility at room temperature. Materials 9(9):716
21. Zhuang HL, Hennig RG (2013) Computational search for single-layer transition-metal dichalcogenide photocatalysts. J Phys Chem C 117(40):20440–20445. https://doi.org/10.1021/jp405808a

22. Ley L, Cardona M, Pollak RA (1979) Photoemission in semiconductors. In: Ley L, Cardona M (eds) Photoemission in solids II: case studies. Springer, Berlin, Heidelberg, pp 11–172. https://doi.org/10.1007/3-540-09202-1_2
23. Chevrel R, Sergent M, Prigent J (1971) Sur de nouvelles phases sulfurées ternaires du molybdène. J Solid State Chem 3(4):515–519. https://doi.org/10.1016/0022-4596(71)90095-8
24. Chevrel R, Sergent M (1982) Chemistry and structure of ternary molybdenum chalcogenides. In: Fischer Ø, Maple MB (eds) Superconductivity in ternary compounds I, vol 32. Springer, Berlin Heidelberg, pp 25–86
25. Fischer Ø, Maple MB (eds) (1982) Superconductivity in ternary compounds I, vol 32. Structural, electronic, and lattice properties. Springer, Berlin Heidelberg. https://doi.org/10.1007/978-3-642-81868-4
26. Nohl H, Klose W, Andersen OK (1982) Band Structures of $M_xMo_6X_a$ - and $M_2Mo_6X_6$-Cluster compounds. In: Fischer Ø, Maple MB (eds) Superconductivity in ternary compounds I, vol 32. Springer, Berlin Heidelberg, pp 165–222
27. Hughbanks T, Hoffmann R (1983) Molybdenum chalcogenides: clusters, chains, and extended solids. The approach to bonding in three dimensions. J Am Chem Soc 105(5): 1150–1162. https://doi.org/10.1021/ja00343a014
28. Kganyago KR (2004) A theoretical study of alkali metal intercalated layered metal dichalcogenides and chevrel phase molybdenum chalcogenides. University of the North, Turfloop
29. Cai S-H, Liu C-W (1995) Theoretical investigation on the band structures of several Chevrel-phase compounds. J Chem Soc, Faraday Trans 91(3):479–483. https://doi.org/10.1039/FT9959100479
30. Le Berre F, Peña O, Hamard C, Horyń R, Wojakowski A (1997) Superconducting and structural properties of rare-earth-based Chevrel-phase selenides $REMo_6Se_8$: first single crystal studies. J Alloys Compd 262–263:406–409. https://doi.org/10.1016/S0925-8388(97)00344-7
31. Yvon K (1982) Structure and bonding of ternary superconductors. In: Fischer Ø, Maple MB (eds) Superconductivity in ternary compounds I, vol 32. Springer, Berlin Heidelberg, pp 87–112
32. Nagao M, Kitaura H, Hayashi A, Tatsumisago M (2009) Characterization of all-solid-state lithium secondary batteries using $Cu_xMo_6S_{8-y}$ electrode and $Li_2S–P_2S_5$ solid electrolyte. J Power Sour 189(1):672–675. https://doi.org/10.1016/j.jpowsour.2008.09.025
33. Wakihara M, Ikuta H, Uchida T (1993) Nickel Chevrel-phase sulfides $NiyMo_6S_{8-z}$ and $Ni_yMo_6S_{8-z}O_n$ as the cathode of lithium secondary batteries. J Power Sour 44(1–3): 651–655. https://doi.org/10.1016/0378-7753(93)80215-B
34. Guohua L, Ikuta H, Uchida T, Wakihara M (1995) Re-examination of copper Chevrel-phase sulfides as cathode in lithium secondary batteries. J Power Sour 54(2):519–521. https://doi.org/10.1016/0378-7753(94)02139-T
35. Wakihara M, Uchida T, Suzuki K, Taniguchi M (1989) A rechargeable lithium battery employing iron Chevrel phase compound ($Fe_{1.25}Mo_6S_{7.8}$) as the cathode. Electrochim Acta 34(6):867–869. https://doi.org/10.1016/0013-4686(89)87121-X
36. Suzuki K, Iijima T, Wakihara M (1998) Chromium Chevrel phase sulfide ($Cr_xMo_6S_{8-y}$) as the cathode with long cycle life in lithium rechargeable batteries. Solid State Ionics 109(3–4):311–320. https://doi.org/10.1016/S0167-2738(98)00074-5
37. Kochubey DI, Rogov VA, Babenko VP (2009) Low-temperature synthesis of supported hydrodesulfurization catalysts based on chevrel phases. Kinet Catal 50(2):270–274. https://doi.org/10.1134/s0023158409020189
38. Benson JW, Schrader GL, Angelici RJ (1995) Studies of the mechanism of thiophene hydrodesulfurization: 2H NMR and mass spectral analysis of 1,3-butadiene produced in the deuterodesulfurization (DDS) of thiophene over $PbMo_6S_8$ catalyst. J Mol Catal A: Chem 96(3):283–299. https://doi.org/10.1016/1381-1169(94)00025-5

39. Harel-Michaud V, Pesnel-Leroux G, Burel L, Chevrel R, Geantet C, Cattenot M, Vrinat M (2001) Nickel Chevrel phase supported on porous alumina. J Alloys Compd 317–318: 195–200. https://doi.org/10.1016/S0925-8388(00)01324-4
40. Jaegermann W, Pettenkofer C, Alonso Vante N, Schwarzlose T, Tributsch H (1990) Chevrel phase type compounds: electronic, chemical and structural factors in oxygen reduction electrocatalysis. Ber Bunsenges Phys Chem 94(4):513–520. https://doi.org/10.1002/bbpc.19900940416
41. Alonso-Vante N, Jaegermann W, Tributsch H, Hönle W, Yvon K (1987) Electrocatalysis of oxygen reduction by chalcogenides containing mixed transition metal clusters. J Am Chem Soc 109(11):3251–3257
42. Tributsch H (1997) Challenges for (photo)electrocatalysis research. Catal Today 39(3): 177–186. https://doi.org/10.1016/s0920-5861(97)00099-0
43. Alonso-Vante N (1996) Electrocatalyse par l'intermédiaire des centres métalliques de composés de métaux de transition. Réduction de l'oxygène moléculaire. J Chim Phys Phys-Chim Biol 93(4):702–710
44. Alonso-Vante N, Tributsch H (1986) Energy conversion catalysis using semiconducting transition metal cluster compounds. Nature 323(6087):431–432
45. Alonso-Vante N, Schubert B, Tributsch H, Perrin A (1988) Influence of d-state density and chemistry of transition metal cluster selenides on electrocatalysis. J Catal 112(2):384–391
46. Alonso-Vante N, Schubert B, Tributsch H (1989) Transition metal cluster materials for multi-electron transfer catalysis. Mater Chem Phys 22(3–4):281–307
47. Kobayashi K, Fujimori A, Ohtani T, Dasgupta I, Jepsen O, Andersen OK (2001) Electronic structure of the Chevrel-phase compounds $Sn_xMo_6Se_{7.5}$: photoemission spectroscopy and band-structure calculations. Phys Rev B 63(19):195109
48. Thole F, Wan LF, Prendergast D (2015) Re-examining the Chevrel phase Mo_6S_8 cathode for Mg intercalation from an electronic structure perspective. Phys Chem Chem Phys 17(35):22548–22551. https://doi.org/10.1039/C5CP03046C
49. Wan LF, Wright J, Perdue BR, Fister TT, Kim S, Apblett CA, Prendergast D (2016) Revealing electronic structure changes in Chevrel phase cathodes upon Mg insertion using X-ray absorption spectroscopy. Phys Chem Chem Phys 18(26):17326–17329. https://doi.org/10.1039/C6CP02412B
50. Peña O (2015) Chevrel phases: past, present and future. Physica C (Amsterdam, Neth) 514:95–112. https://doi.org/10.1016/j.physc.2015.02.019
51. Chevrel R, Sergent M, Prigent J (1974) Un nouveau sulfure de molybdene: Mo_3S_4 preparation, proprietes et structure cristalline. Mater Res Bull 9(11):1487–1498. https://doi.org/10.1016/0025-5408(74)90095-6
52. Sergent M, Chevrel R (1973) Sur de nouvelles phases séléniées ternaires du molybdène. J Solid State Chem 6(3):433–437. https://doi.org/10.1016/0022-4596(73)90235-1
53. Fischer C, Gocke E, Stege U, Schöllhorn R (1993) Molybdenum cluster chalcogenides: in situ X-Ray studies on the formation of $Cu_xMo_6S_8$ via electron/ion transfer. J Solid State Chem 102(1):54–68. doi:http://dx.doi.org/10.1006/jssc.1993.1007
54. Alonso-Vante N (1998) Inert for selective oxygen reduction of oxygen and method for the production thereof. Germany Patent WO1997DE02453 19971016; DE19961044628 19961017
55. Alonso-Vante N, Giersig M, Tributsch H (1991) Thin layer semiconducting cluster electrocatalysts for oxygen reduction. J Electrochem Soc 138(2):639–640
56. Alonso-Vante N, Tributsch H (1994) Electrode materials and strategies for (Photo)electrochemistry. In: Lipkowski J, Ross P (eds) Electrochemistry of novel materials, vol III. VCH, New York, pp 1–63
57. Alonso-Vante N (2006) Carbonyl tailored electrocatalysts. Fuel Cells 6(3–4):182–189
58. Vogel W, Kaghazchi P, Jacob T, Alonso-Vante N (2007) Genesis of Ru_xSe_y nanoparticles by pyrolysis of $Ru_4Se_2(CO)_{11}$: a combined X-ray in Situ and DFT study. J Phys Chem C 111:3908–3913

59. Vogel W, Le Rhun V, Garnier E, Alonso-Vante N (2001) Ru clusters synthesized chemically from dissolved carbonyl. in situ study of a novel electrocatalyst in the gas phase and in electrochemical environment. J Phys Chem B 105(22):5238–5243
60. Longoni G, Chini P (1976) Synthesis and chemical characterization of platinum carbonyl dianions $[Pt_3(CO)_6]n^{2-}$ (n = - 10, 6, 5, 4, 3, 2, 1). a new series of inorganic oligomers. J Am Chem Soc 98(23):7225–7231
61. Adams RD, Chen G, Wu W (1993) The synthesis and structural analysis of $Pt_2Ru_4(CO)_{18}$ and the products obtained from its reactions with 1, 2-bis(diphenylphosphino)ethane. J Cluster Sci 4(2):119–132
62. Nashner MS, Somerville DM, Lane PD, Adler DL, Shapley JR, Nuzzo RG (1996) Bimetallic catalyst particle nanostructure. evolution from molecular cluster precursors. J Am Chem Soc 118(51):12964–12974
63. Nashner MS, Frenkel AI, Adler DL, Shapley JR, Nuzzo RG (1997) Structural characterization of carbon-supported platinum–ruthenium nanoparticles from the molecular cluster precursor $PtRu_5C(CO)_{16}$. J Am Chem Soc 119(33):7760–7771
64. Hills CW, Nashner MS, Frenkel AI, Shapley JR, Nuzzo RG (1999) Carbon support effects on bimetallic Pt–Ru nanoparticles formed from molecular precursors. Langmuir 15(3): 690–700
65. Le Rhun V (2001) Réduction électrocatalytique de l'oxygène moléculaire sur des électrodes à amas métaalliques en présence de méthanol. University of Poiters, Poitiers, PhD
66. Le Rhun V, Alonso-Vante N (2000) Tailoring of nanodivided electrocatalyst materials based on transition metal. J New Mater Electrochem Syst 3(4):331–336
67. Layer TM, Lewis J, Martin A, Raithby PR, Wong WT (1992) The chemistry of the selenium-containing cluster $[Ru_4\text{-}(\mu_4\text{-}Se)_2(CO)_8(\mu\text{-}CO)_3]$: crystal structures of $[Ru_3(\mu_3\text{-}Se)_2(CO)_7(Ph_2PC_2PPh_2)]$, $[Ru_4(\mu_4\text{-}Se)_2(CO)_{10}(SbPh_3)]$ and $[Ru_3(\mu_3\text{-}Se)_2(CO)_9]$. J Chem Soc, Dalton Trans 24:3411–3417. https://doi.org/10.1039/DT9920003411
68. Johnson BFG, Layer TM, Lewis J, Martin A, Raithby PR (1992) Synthesis and characterisation of novel selenium-containing clusters; crystal structures of $[Ru_4(\mu_4\text{-}Se)_2(CO)_8(\mu\text{-}CO)_3]$ and $[Ru_3(\mu_3\text{-}Se)_2(CO)_7(Ph_2P(CH_2)_3PPh_2)]$. J Organomet Chem 429(3): C41–C45
69. Layer TM, Lewis J, Martin A, Raithby PR, Wong WT (1993) Synthesis of novel selenium-containing osmium carbonyl clusters: X-ray structures of $[Os_5(\mu\text{-}H)\ (CO)_{15}(\mu_3\text{-}SePh)]$ and $[Os_6(\mu\text{-}H)(CO)_{18}(\mu\text{-}SeH)]$. J Organomet Chem 444(1–2)
70. Ramírez-Raya SD, Solorza-Feria O, Ordoñez-Regil E, Benaissa M, Valverde SMF (1998) Synthesis and characterization of $W_{0.12}Ru_{2.1}Se$ and $W_{0.013}Ru_{1.27}Se$ electrocatalysts. Nanostruct Mater 10(8):1337–1346
71. Solorza-Feria O, Ramírez-Raya S, Rivera-Noriega R, Ordoñez-Regil E, Fernández-Valverde SM (1997) Kinetic studies of molecular oxygen reduction on $W_{0.013}Ru_{1.27}Se$ thin films chemically synthesized. Thin Solid Films 311(1–2):164–170. doi:http://dx.doi.org/10.1016/S0040-6090(97)00685-8
72. Suarez-Alcantara K, Rodriguez-Castellanos A, Duron-Torres S, Solorza-Feria O (2007) $Ru_xCr_ySe_z$ electrocatalyst loading and stability effects on the electrochemical performance in a PEMFC. J Power Sour 171(2):381–387
73. Rodriguez FJ, Sebastian PJ, Solorza O, Perez R (1998) Mo-Ru-W chalcogenide electrodes prepared by chemical synthesis and screen printing for fuel cell applications. Int J Hydrogen Energy 23(11):1031–1035
74. Castellanos RH, Campero A, Solorza-Feria O (1998) Synthesis of W-Se-Os carbonyl electrocatalyst for oxygen reduction in 0.5 M H_2SO_4. Int J Hydrogen Energy 23(11): 1037–1040
75. Kolary-Zurowska A, Zieleniak A, Miecznikowski K, Baranowska B, Lewera A, Fiechter S, Bogdanoff P, Dorbandt I, Marassi R, Kulesza PJ (2007) Activation of methanol-tolerant carbon-supported $RuSe_x$ electrocatalytic nanoparticles towards more efficient oxygen reduction. J Solid State Electrochem 11(7):915–921

76. Kulesza PJ, Miecznikowski K, Baranowska B, Skunik M, Kolary-Zurowska A, Lewera A, Karnicka K, Chojak M, Rutkowska I, Fiechter S, Bogdanoff P, Dorbandt I, Zehl G, Hiesgen R, Dirk E, Nagabhushana KS, Boennemann H (2007) Electroreduction of oxygen at tungsten oxide modified carbon-supported $RuSe_x$ nanoparticles. J Appl Electrochem 37(12):1439–1446
77. Bron M, Bogdanoff P, Fiechter S, Dorbandt I, Hilgendorff M, Schulenburg H, Tributsch H (2001) Influence of selenium on the catalytic properties of ruthenium-based cluster catalysts for oxygen reduction. J Electroanal Chem 500(1–2):510–517
78. Bron M, Bogdanoff P, Fiechter S, Hilgendorff M, Radnik J, Dorbandt I, Schulenburg H, Tributsch H (2001) Carbon supported catalysts for oxygen reduction in acidic media prepared by thermolysis of $Ru_3(CO)_{12}$. J Electroanal Chem 517(1–2):85–94
79. Fiechter S, Dorbandt I, Bogdanoff P, Zehl G, Schulenburg H, Tributsch H, Bron M, Radnik J, Fieber-Erdmann M (2007) Surface modified ruthenium nanoparticles: structural investigation and surface analysis of a novel catalyst for oxygen reduction. J Phys Chem C 111(1):477–487
80. Zehl G, Schmithals G, Hoell A, Haas S, Hartnig C, Dorbandt I, Bogdanoff P, Fiechter S (2007) On the structure of carbon-supported selenium-modified ruthenium nanoparticles as electrocatalysts for oxygen reduction in fuel cells. Angew Chem Int Ed 46(38):7311–7314
81. Zehl G, Bogdanoff P, Dorbandt I, Fiechter S, Wippermann K, Hartnig C (2007) Carbon supported Ru-Se as methanol tolerant catalysts for DMFC cathodes. part I: preparation and characterization of catalysts. J Appl Electrochem 37(12):1475–1484
82. Zaikovskii VI, Nagabhushana KS, Kriventsov VV, Loponov KN, Cherepanova SV, Kvon RI, Bonnemann H, Kochubey DI, Savinova ER (2006) Synthesis and structural characterization of Se-modified carbon-supported Ru nanoparticles for the oxygen reduction reaction. J Phys Chem B 110(13):6881–6890
83. Ma J, Canaff C, Alonso-Vante N (2013) The effect of tuning and origin of tolerance to organics of platinum catalytic centers modified by selenium. Physica Status Solidi (a) 211(9):2030–2034. https://doi.org/10.1002/pssa.201330148
84. Fiechter S, Kühne HM (1987) Crystal growth of RuX_2 (X = S, Se, Te) by chemical vapour transport and high temperature solution growth. J Cryst Growth 83(4):517–522. https://doi.org/10.1016/0022-0248(87)90246-6
85. Sheu J-S, Shih Y-S, Lin S-S, Huang Y-S (1991) Growth and characterization of $RuSe_2$ single crystals. Mater Res Bull 26(1):11–17. https://doi.org/10.1016/0025-5408(91)90033-I
86. Dassenoy F, Vogel W, Alonso-Vante N (2002) Structural studies and stability of cluster-like Ru_xSe_y electrocatalysts. J Phys Chem B 106(47):12152–12157
87. Babu PK, Lewera A, Jong HC, Hunger R, Jaegermann W, Alonso-Vante N, Wieckowski A, Oldfield E (2007) Selenium becomes metallic in Ru-Se fuel cell catalysts: An EC-NMR and XPS investigation. J Am Chem Soc 129(49):15140–15141
88. Gochi-Ponce Y, Alonso-Nunez G, Alonso-Vante N (2006) Synthesis and electrochemical characterization of a novel platinum chalcogenide electrocatalyst with an enhanced tolerance to methanol in the oxygen reduction reaction. Electrochem Commun 8(9):1487–1491
89. Gago AS, Gochi-Ponce Y, Feng Y-J, Esquivel JP, Sabaté N, Santander J, Alonso-Vante N (2012) Tolerant chalcogenide cathodes of membraneless micro fuel cells. Chemsuschem 5(8):1488–1494. https://doi.org/10.1002/cssc.201200009
90. Ziegelbauer JM, Gatewood D, Gullá AF, Guinel MJF, Ernst F, Ramaker DE, Mukerjee S (2009) Fundamental investigation of oxygen reduction reaction on rhodium sulfide-based chalcogenides. J Phys Chem C 113(17):6955–6968
91. Papageorgopoulos DC, Liu F, Conrad O (2007) Reprint of "a study of Rh_xS_y/C and $RuSe_x$/C as methanol-tolerant oxygen reduction catalysts for mixed-reactant fuel cell applications". Electrochim Acta 53(2):1037–1041
92. Lee K, Zhang L, Zhang JJ (2007) A novel methanol-tolerant Ir-Se chalcogenide electrocatalyst for oxygen reduction. J Power Sour 165(1):108–113

93. Ohtani T, Ikeda K, Hayashi Y, Fukui Y (2007) Mechanochemical preparation of palladium chalcogenides. Mater Res Bull 42(11):1930–1934. https://doi.org/10.1016/j.materresbull.2006.12.012
94. Adam JL, Zhang XH (eds) (2014) Chalcogenide glasses. Preparation, properties and applications, 1st edn. Woodhead Publishing. doi:http://dx.doi.org/10.1016/B978-0-85709-345-5.50023-4
95. Frumar M, Frumarova B, Wagner T (2011) 4.07—amorphous and glassy semiconducting chalcogenides A2—Bhattacharya, Pallab. In: Fornari R, Kamimura H (eds) Comprehensive semiconductor science and technology. Elsevier, Amsterdam, pp 206–261. doi:http://dx.doi.org/10.1016/B978-0-44-453153-7.00122-X
96. Frumar M, Wagner T (2003) Ag doped chalcogenide glasses and their applications. Curr Opin Solid State Mater Sci 7(2):117–126. https://doi.org/10.1016/S1359-0286(03)00044-5
97. Pradel A, Ribes M (2014) 7—Ionic conductivity of chalcogenide glasses. In: Chalcogenide glasses. Woodhead Publishing, pp 169–208. doi:http://dx.doi.org/10.1533/9780857093561.1.169
98. Tatsumisago M, Hayashi A (2014) 19 - Chalcogenide glasses as electrolytes for batteries. In: Chalcogenide glasses. Woodhead Publishing, pp 632–654. doi:http://dx.doi.org/10.1533/9780857093561.2.632
99. Stanić V, Pierre AC, Etsell TH, Mikula RJ (2000) Influence of reaction parameters on the microstructure of the germanium disulfide gel. J Am Ceram Soc 83(7):1790–1796. https://doi.org/10.1111/j.1151-2916.2000.tb01465.x
100. Gacoin T, Lahlil K, Larregaray P, Boilot JP (2001) Transformation of CdS colloids: sols, gels, and precipitates. J Phys Chem B 105(42):10228–10235. https://doi.org/10.1021/jp0117381
101. Bag S, Trikalitis PN, Chupas PJ, Armatas GS, Kanatzidis MG (2007) Porous semiconducting gels and aerogels from chalcogenide clusters. Science 317(5837):490–493. https://doi.org/10.1126/science.1142535
102. Polychronopoulou K, Malliakas CD, He J, Kanatzidis MG (2012) Selective surfaces: quaternary Co(Ni)MoS-based chalcogels with divalent (Pb^{2+}, Cd^{2+}, Pd^{2+}) and trivalent (Cr^{3+}, Bi^{3+}) metals for gas separation. Chem Mater 24(17):3380–3392. https://doi.org/10.1021/cm301444p
103. Brock SL, Arachchige IU, Kalebaila KK (2006) Metal chalcogenide gels, xerogels and aerogels. Comments Inorg Chem 27(5–6):103–126. https://doi.org/10.1080/02603590601084434
104. Mohanan JL, Arachchige IU, Brock SL (2005) Porous semiconductor chalcogenide aerogels. Science 307(5708):397
105. Arachchige IU, Brock SL (2007) Sol-Gel methods for the assembly of metal chalcogenide quantum dots. Acc Chem Res 40(9):801–809. https://doi.org/10.1021/ar600028s
106. Bag S, Arachchige IU, Kanatzidis MG (2008) Aerogels from metal chalcogenides and their emerging unique properties. J Mater Chem 18(31):3628–3632. https://doi.org/10.1039/B804011G
107. Staszak-Jirkovsky J, Malliakas CD, Lopes PP, Danilovic N, Kota SS, Chang K-C, Genorio B, Strmcnik D, Stamenkovic VR, Kanatzidis MG, Markovic NM (2016) Design of active and stable Co-Mo-S_x chalcogels as pH-universal catalysts for the hydrogen evolution reaction. Nat Mater 15(2):197–203. https://doi.org/10.1038/nmat4481 http://www.nature.com/nmat/journal/v15/n2/abs/nmat4481.html#supplementary-information

Chapter 2
Fuel Cell Electrocatalysis

2.1 Generalities

In a fuel cell (FC) system, the combustion process of a fuel at low temperatures, e.g., hydrogen, and the typical oxidant, oxygen from air, is not governed by the thermodynamic cycles occurring in combustion engines [1]. The energy available in such systems can be well described by the Gibbs free energy. As it will be described below, one important drawback is the high activation barrier, especially at the cathode side. Consequently, the need of abundant, cheap, and stable materials with electrocatalytic properties is of paramount importance. The consequence is that the majority of the electrochemical reactions of technological interest are multi-electron charge transfer processes. This is the case of production of electricity and water from their constituents: hydrogen and oxygen. The reverse reaction, however, needs energy to generate hydrogen (fuel) and oxygen. Thus, this cycle, named water cycle, is the achievement of a dream summarized in Grove's quote: "*decomposition of water by means of its composition,*" schematized in the well-spread gaseous voltaic battery, Fig. 2.1, published in 1842 [2], and its working principle was discovered by the German-Swiss scientist Christian Friedrich Schönbein in 1838 [3]. These two systems, i.e., fuel cell (FC) or discharging mode—and electrolyzer (EC) or charging mode have been investigated separately in the last decades, in acid [using proton conducting membranes (PEM)] and alkaline media [anion conducting membranes (AEM)] (see Fig. 2.2a, b). Moreover, the interest in the development of a H_2/O_2 energy storage system led to unify both systems in one, so-called unitized regenerative cell (URC), Fig. 2.2c as reported by Mitlitsk et al. [4, 5]. With the advent of cheap, active, and stable electrode materials development, this system is attracting a lot of interest, as recently reviewed [6].

As schematized in Fig. 2.2a, the anode and the cathode in a fuel cell are separated by a membrane serving as electrolyte. At the anode, different fuels can be used, such as small organics (e.g., methanol, formic acid) and hydrogen. These species are oxidized delivering the electrons that travel at the external circuit to the

N. Alonso-Vante, *Chalcogenide Materials for Energy Conversion*, Nanostructure Science and Technology, https://doi.org/10.1007/978-3-319-89612-0_2

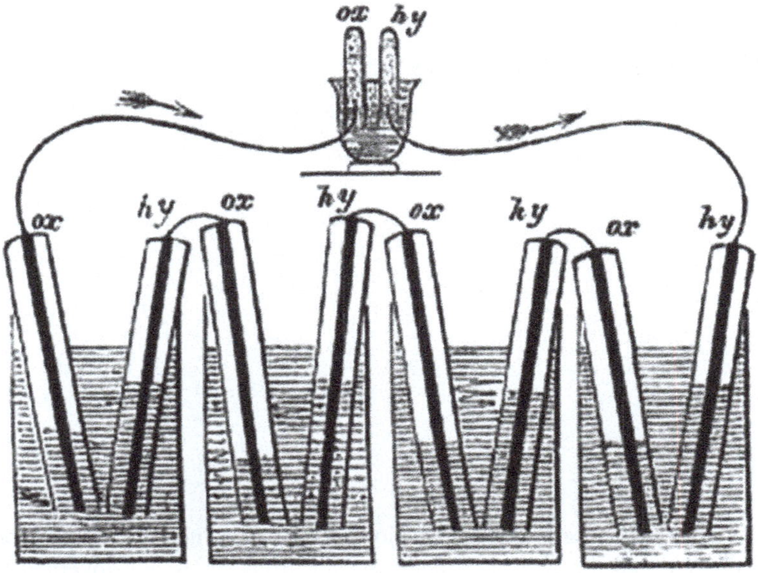

Fig. 2.1 Arrangement of 26-cells was the minimum number required to electrolyze water, Ref. [2]

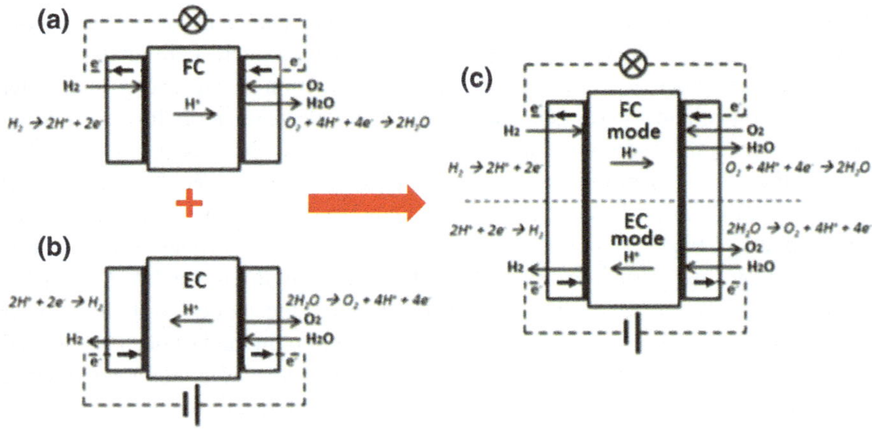

Fig. 2.2 Inward the polymer electrolyte membrane (PEM) system: **a** fuel cell (FC); **b** electrolyzer Cell (EC); **c** unitized regenerative cell (URC). Discharging mode: FC; recharging mode: EC

cathode. These electrons are received by one oxygen molecule to be reduced at the cathode. The product of the combustion of hydrogen and oxygen in a fuel cell is water if the number of electrons exchanged fulfills the reaction requirement (see below). Summing up, the processes being performed in a FC are the hydrogen oxidation reaction (HOR) and the oxygen reduction reaction (ORR). The former process means fuel consumption; therefore, its production using the water electrolysis process, see Fig. 2.2b, is necessary, provided cheap, stable, and active electrocatalysts are available to cost competitive with the presently used hydrogen generated via fossil fuel sources [7, 8].

2.2 Thermodynamics and Kinetics

The water cycle, shortly described in Sect. 2.1, is based on the elementary hydrogen and oxygen reactions. Hydrogen is oxidized at the anode (HOR), whereas oxygen is reduced at the cathode (ORR). These reactions are associated with the production of electricity and water in a fuel cell [9–11], Figs. 2.2a and 2.3, curves b, whereas the stored chemical energy is released in an electrolyzer cell, Figs. 2.2b and 2.3, curves a. Figure 2.3 clearly depicts the potential region where HER/HOR as well as ORR/OER can take place in a fuel cell (FC) system, and an electrolyzer cell (EC) system. The thermodynamic "gap" of water, i.e., 1.229 V can sensitively be affected on both ends (HOR/ORR) and (HER/OER) in low-temperature FCs and ECs, respectively, due to reaction kinetics hampering, which increases the reaction overpotential at the anode and cathode (cf. Sect. 1.1.3), depending on the working mode conditions of the system. This phenomenon is intimately related to the electrocatalytic properties of the electrode materials, vide supra.

The free energy change of one molecule of water, Eq. (2.1), under standard conditions is $\Delta G = +\,237.2$ kJ/mol.

$$H_2O \rightarrow \frac{1}{2}\,O_2 + H_2 \qquad (2.1)$$

Using the relation $\Delta G = -nF\Delta E°$, $\Delta E° = 1.229$ V per electron transferred.

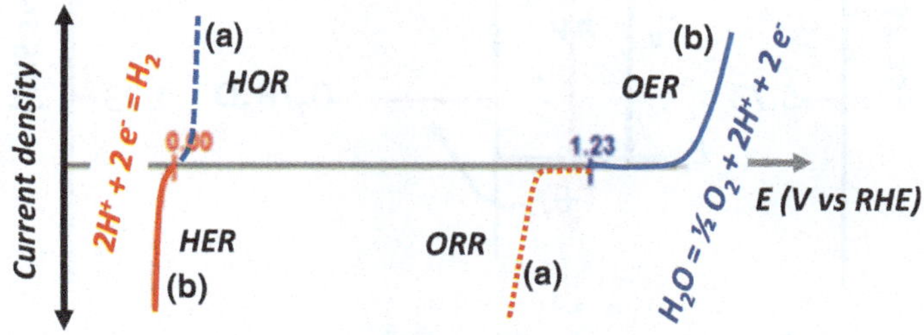

Fig. 2.3 Schematics of polarization curves **a** fuel cell and **b** electrolyzer modes

Equation (2.1) is the sum of Eqs. (2.2) and (2.3).

$$H_2O \rightarrow \frac{1}{2} O_2 + 2H^+ + 2e^- \quad (OER) \tag{2.2}$$

$$2H^+ + 2e^- \rightarrow H_2 \quad (HER) \tag{2.3}$$

The reactions (2.2) and (2.3) can be driven by light (see Chap. 3) via a semiconductor, which must absorb light with photon energies of >1.23 eV, otherwise, converting energy into H_2 and O_2.

$$H_2O + 2h^+ \rightarrow \frac{1}{2} O_2 + 2H^+ \quad (OER) \tag{2.4}$$

$$2H^+ + 2e^- \rightarrow H_2 \quad (HER) \tag{2.5}$$

The photoprocess must generate two electron–hole (e^--h^+) pairs per molecule of hydrogen, i.e., an energy of 2 × 1.229 eV = 2.46 eV; or four electron–hole pairs per molecule of oxygen, i.e., an energy of 4 × 1.229 eV = 4.92 eV. This means that a single semiconductor material, must have a gap energy large enough to split water, and energy band positions (conduction band—E_c, valence band—E_v) straddling the electrochemical redox potential of reactions (2.4) and (2.5), see Fig. 2.4, thus depicting a nice principle of the artificial photosynthesis or artificial leaf [12] where solar energy is used for producing a storable fuel. In addition, Fig. 2.4 illustrates the important relation between two scales, i.e., the one with increasing electron energy (eV)—vacuum scale and the other with increasing electrode potential (V/SHE)—SHE scale. This concept follows the definition of the

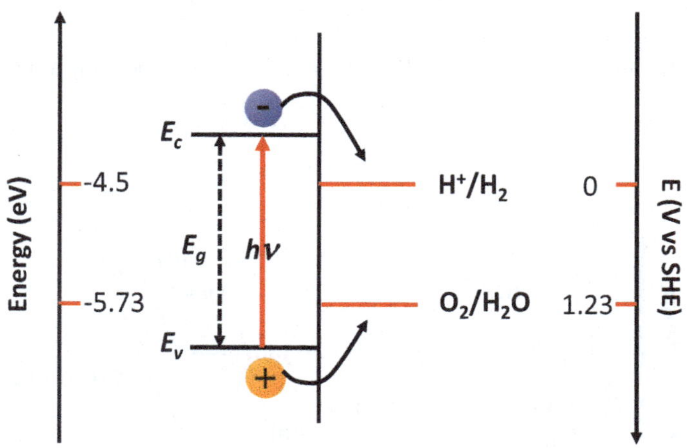

Fig. 2.4 Water splitting process in acid medium (pH 0) under illumination. The right vertical axis shows the redox electrode potential versus the standard hydrogen electrode and the left vertical axis shows the vacuum scale of electron energy. The accepted energy reference of −4.5 eV corresponds to the hydrogen redox potential 0 V

absolute electrode potential, and its value for the standard hydrogen electrode (SHE) is given by Eq. (2.6) [13, 14].

$$E_{Ox/\mathrm{Red}} = -eE_{Ox/\mathrm{Red}}(abs) = -eE_{Ox/\mathrm{Red}}(SHE) - 4.5 \tag{2.6}$$

2.2.1 *Hydrogen Evolution/Oxidation Reaction (HER/HOR)*

The HER and HOR (in acid medium) are treated as a combination of three steps: (i) Volmer step, where proton (H^+) or hydronium (H_3O^+) ions discharge on the electrode surface, M, building an adsorbed intermediate, M-H_{ad}:

$$\mathrm{H_3O^+ + M + e^- \rightleftarrows M - H_{ad} + H_2O} \tag{2.7}$$

This step is followed by, either the Heyrovsky step:

$$\mathrm{H_3O^+ + M - H_{ad} + e^- \rightleftarrows M + H_2 + H_2O} \tag{2.8}$$

Or the Tafel recombination step:

$$\mathrm{2M - H_{ad} \rightleftarrows 2M + H_2 + OH^-} \tag{2.9}$$

In alkaline medium, these reactions are analogous, except that the species $\mathrm{M - H_{ad}}$ in the Volmer reaction is formed by the discharge of water molecules:

$$\mathrm{H_2O + M + e^- \rightleftarrows M - H_{ad} + OH^-} \tag{2.10}$$

The reaction steps described in Eqs. (2.7)–(2.10) depend on the electronic properties of the electrode surface, so that the rate controlling steps are usually evaluated through the Tafel slope (see Sect. 1.1.3). The chemical adsorption and desorption of H atoms are competitive processes, and a good electrocatalyst, according to the Sabatier principle [15], should form a sufficiently strong bond (M-H_{ad}) for facilitating the electron transfer process, but it should be also weak enough to assure bond breaking obtaining the product at a minimum thermodynamic overpotential, η_T.

Following the described thermodynamic constraints [16], the equilibrium potential for Volmer and Heyrovsky steps in (2.7) and (2.8) is:

$$\frac{E^0_{H_3O^+/H_{ads}} + E^0_{ads,H_3O^+/H_2}}{2} = E^0_{H_3O^+/H_2} = 0 \tag{2.11}$$

The free energy change of (H_3O^+ + e^-) in Eq. (2.7) in the electrochemical scale [cf. (2.6)] is related to an energy in the vacuum scale (-$eE_{Ox/Red}$ (abs)), so that the standard equilibrium potential for the Volmer reaction can be written as:

$$E^0_{H_3O^+/H_{ads}} = -\frac{\Delta G^0_{H_{ads}}}{e} \quad (2.12)$$

The adsorption energy of H_{ads} is $\Delta G^0_{H_{ads}}$, thus if $\Delta G^0_{H_{ads}} < 0$, then $E^0_{H_3O^+/H_{ads}} > 0$, shaping the well-known hydrogen underpotential deposition region, H_{upd}, a phenomenon generally observed on transition metal electrodes, e.g., Pt [17], Pd, Ir, Ru, e.g., Pt in Fig. 2.5. It is to be noticed that the standard equilibrium, considering the Volmer step, as thermodynamically favorable at E = 0, the less thermodynamically favorable step is the Heyrovsky one, so that a concept of a thermodynamic overpotential, η_T, can be defined as:

$$\eta_T = E^0_{H_3O^+, H_{ads}/H_2} - E^0_{H_3O^+/H_{ads}} \quad (2.13)$$

In this sense, an electrocatalytic material can be considered as excellent if $\eta_T = \Delta G^0_{Hads}/e = 0$. However, any electrocatalyst with $\Delta G^0_{Hads}/e \neq 0$ will show a $\eta_T > 0$ indicating that Volmer or Heyrovsky step may show a standard potential at equilibrium < or > than 0, as schematized in Fig. 2.6.

Illustrating the classical Sabatier principle [18], assessed for electrocatalysis [15, 19], a plot of η_T versus the adsorption energy, $\Delta G^0_{H_{ads}}$, shows the tradeoff between the strong- versus weak-binding of H_{ads} species in a volcano relationship (see Fig. 2.7). In the classical electrochemical literature [20–23] such a volcano curve is

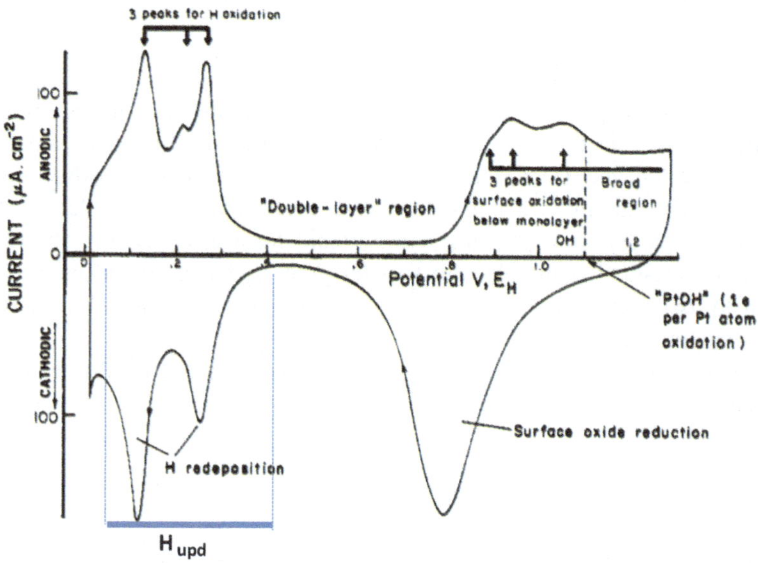

Fig. 2.5 Surface electrochemistry for Pt electrode in pure aqueous 0.5 M H_2SO_4 solution at 25 °C; The cyclic voltammetry sweep rate = 50 mV s^{-1} (With permission from Ref. [17] Copyright © 1973, American Chemical Society)

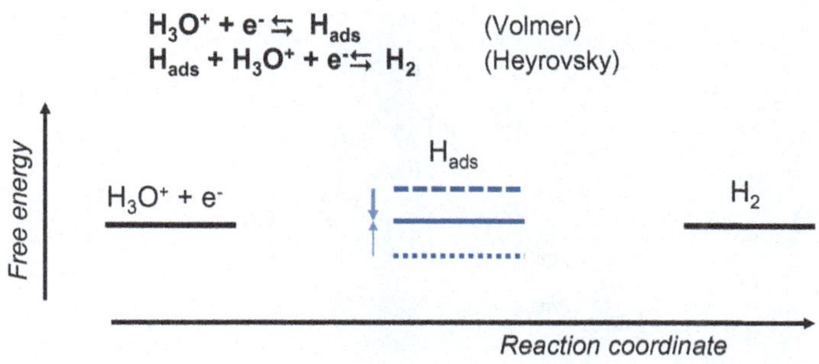

Fig. 2.6 A schematized variation (dotted, dashed lines) of the free energy of Volmer, and Heyrovsky steps, relative to each other. The optimized condition is fulfilled when the energy level tends to reach the same free energy, as to have a thermodynamic overpotential, $\eta_T = 0$

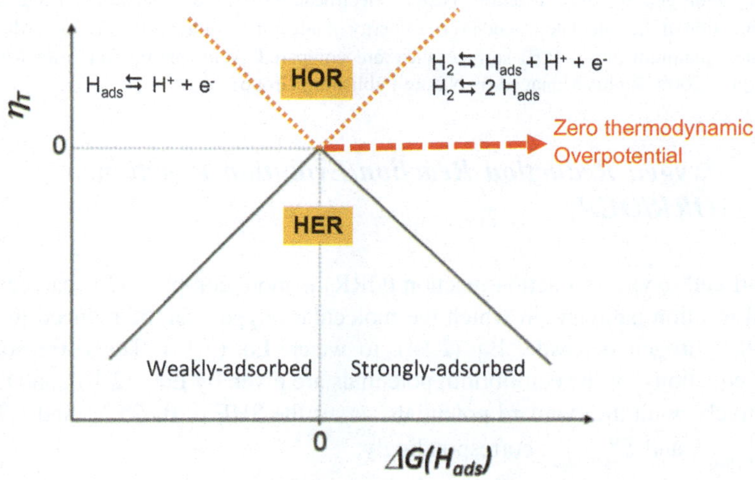

Fig. 2.7 Thermodynamic volcano plot, adapted from Refs. [15, 22–24]

represented by a kinetic parameter, e.g., exchange current density, j_O (cf. Sect. 1.1.3.2) versus the strength of the metal–hydrogen bond. Since a similar analysis of Eqs. (2.7) and (2.8) can be applied for the HOR [16], Fig. 2.7 also shows the volcano “inverted” plot for this process. The adsorption energy, $\Delta G^0_{H_{ads}}$, known as the descriptor was calculated by the density functional theory (DFT) [24, 25]. Such calculation assesses again that there is not any good alternative to the platinum electrode material, as shown in Fig. 2.8. Indeed, plotting the exchange current density versus $\Delta G^0_{H_{ads}}$ obtained by DFT shows a volcano with an optimum at zero free energy of adsorption. One can recall that the potential determining step is different on each side of the volcano maximum; otherwise, an indication of the magnitude of overpotential of the reaction caused by, e.g., $\Delta G^0_{H_{ads}} > 0$ (Volmer reaction); $\Delta G^0_{H_{ads}} < 0$ (Heyrovsky or Tafel reaction).

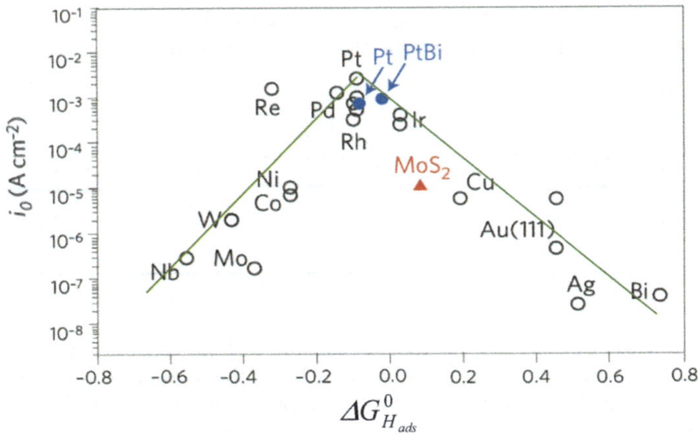

Fig. 2.8 Hydrogen evolution reaction (HER)—electrocatalysts—experimental exchange current, j_0, as a function of the calculated standard free energy of adsorption of hydrogen, $\Delta G^0_{H_{ads}}$. Marked in blue color, platinum and a PtBi surface alloy are compared (With permission from Ref. [25] Copyright © 2009, Rights Managed by Nature Publishing Group)

2.2.2 *Oxygen Reduction Reaction/Evolution Reaction (ORR/OER)*

The half-cell oxygen reduction reaction (ORR) is more complex. One can consider a serial reaction pathway, in which the molecular oxygen can be reduced to water through hydrogen peroxide, Eq. (2.14), to water, Eq. (2.15). The corresponding Nernst equations for the equilibrium potentials are given by Eqs. (2.16), and (2.17), respectively, with the standard potentials versus the SHE of 0.695 V, and 1.763 V for, $E^0_{O_2/H_2O_2}$, and $E^0_{H_2O_2/H_2O}$, correspondingly.

$$O_2 + 2H^+ + 2e^- \rightleftarrows 2H_2O_2 \tag{2.14}$$

$$H_2O_2 + 2H^+ + 2e^- \rightleftarrows 2H_2O \tag{2.15}$$

$$E_{O_2/H_2O_2} = E^0_{O_2/H_2O_2} - \frac{RT}{2F}\ln\frac{a^2_{H_2O}}{a_{H_2O_2}a^2_{H^+}} \tag{2.16}$$

$$E_{H_2O_2/H_2O} = E^0_{H_2O_2/H_2O} - \frac{RT}{2F}\ln\frac{a^2_{H_2O}}{a_{H_2O_2}a^2_{H^+}} \tag{2.17}$$

Herein, a_i is the activity of the species i in water. So that, for the global reaction at equilibrium, Eq. (2.18), the standard potential is $E^0_{O_2/H_2O}$ = 1.229 V.

$$O_2 + 4H^+ + 4e^- \rightleftarrows 2H_2O \tag{2.18}$$

$$E_{O_2/H_2O} = E^0_{O_2/H_2O} - \frac{RT}{4F}\ln\frac{a^2_{H_2O}}{a_{O_2}a^4_{H^+}} \tag{2.19}$$

For each pH-unit increase of the electrolyte solution, the equilibrium potential becomes more negative by 59.2 mV versus the SHE. On the other hand, if a_{O_2} is associated with the partial pressure of oxygen, p_{O_2}, the equilibrium potential will be shifted 14.8 mV by a tenfold decrease of p_{O_2}.

In the standard conditions (all $a = 1$), the equilibrium potential is equal to the standard potential, and this latter parameter (using the energy scale in eV) can be used to define the free energy of the oxygen molecule, as follows:

$$eE^0_{O_2/H_2O} = \frac{[\Delta G(O_2) - 2\Delta G(H_2O)]}{4} = 1.229\,\text{eV} \tag{2.20}$$

Taking $\Delta G(H_2O) = 0\,\text{eV}$, this leads to $\Delta G(O_2) = 4 \times 1.23 = 4.92\,\text{eV}$. This energy is to be overcome in order to produce water in the ORR process. However, multi-electron processes generating multi-intermediate species are in the interplay of this complex mechanism.

To understand this, let us start with the reaction represented in Eq. (2.21).

$$O_2 + e^- \rightleftarrows O_2^- \tag{2.21}$$

This simple charge transfer forms the superoxide ion [O_2^-], typically encountered in non-aqueous electrochemistry, e.g., acetonitrile on platinum and glassy carbon electrodes [26], or on the cluster chalcogenide: Chevrel phase [27]. Hydrogen superoxide is generated with the addition, into the non-aqueous solvent, of dry protons from 96% H_2SO_4, so that under this condition, we consider as a first step the following reaction:

$$O_2 + H^+ + e^- \rightleftarrows HO_{2(ads)} \tag{2.22a}$$

Considering that such reactions take place at a catalytic site (*) on the electrode surface, we can rewrite Eqs. (2.22a) as (2.22b):

$$(*) - O_{2(ads)} + (H^+ + e^-) \rightleftarrows (*) - O_2H_{(ads)} \tag{2.22b}$$

which subsequently with an excess of protons forms the hydrogen peroxide gives

$$(*) - O_2H_{(ads)} + (H^+ + e^-) \rightleftarrows (*) - O_{(ads)} + H_2O \text{ or } ((*) - H_2O_2) \tag{2.23}$$

Up to here, this mechanism involves the production of hydrogen peroxide, or the so-called series pathway, as suggested many years ago [28]. This is the most popular intermediate measured electrochemically on most ORR electrocatalysts via the rotating disk electrode technique [29, 30]. Most catalytic centers (precious or non-precious ones), reported so far, produce hydrogen peroxide. An analysis of such centers was summarized by the present author [31]. There is strong evidence that the rate determining step is via the hydrogenation of adsorbed oxygen in step (2.22b).

Further hydrogenation to water as the final product can be represented by Eqs. (2.24) and (2.25).

$$(*) - O_{(ads)} + (H^+ + e^-) \rightleftarrows (*) - OH_{(ads)} \quad (2.24)$$

$$(*) - OH_{(ads)} + (H^+ + e^-) \rightleftarrows (*) + H_2O \quad (2.25)$$

Water in alkaline medium serves as proton-donor reactant for the ORR process (cf. Eq. (2.18) in acid medium),

$$O_2 + 2H_2O + 4e^- \rightleftarrows 4OH^- \quad (2.26)$$

Applying a similar reasoning regarding the equilibrium potentials, discussed above (cf. Eq. (2.11)), the thermodynamic redox potential of O_2/H_2O can be associated with the reactions steps, Eqs. (2.22b)–(2.25) with Eq. (2.27):

$$\frac{E^0_{O_2/O_2H_{ads}} + E^0_{O_2H_{ads}/OH_{ads}} + 2E^0_{OH_{ads}/H_2O}}{4} = E^0_{O_2/H_2O} = 1.229\,\text{V} \quad (2.27)$$

This simple relation teaches us that the thermodynamic overpotential is linked to the determining steps among reactions (2.22b)–(2.25), meaning that the conditions of an ideal electrocatalyst [16, 32] must have a thermodynamic overpotential, η_T, of zero. This can be achieved if $\Delta G(O_2) = 4.92\,\text{eV}$, and $\Delta G(H_2O) = 0\,\text{eV}$, so that $\Delta G(O_2H_{ads}) = 3 \times 1.229 = 3.69\,\text{eV}$, $\Delta G(O_{ads}) = 2 \times 1.229 = 2.46\,\text{eV}$, and $\Delta G(OH_{ads}) = 1 \times 1.229 = 1.23\,\text{eV}$. It was established by DFT calculations that the binding energies of species O_{ads}, OH_{ads}, O_2H_{ads} are not independent [33], but they can be interrelated into a "simple" linear equation, so-called scaling relationship, as follows:

$$G(O_2H_{ads}) = 0.5 \times \Delta G(O_{ads}) + K_{O_2H} \quad (2.28)$$

$$\Delta G(OH_{ads}) = 0.5 \times \Delta G(O_{ads}) + K_{OH} \quad (2.29)$$

Since these equations are the result of chemical bonding principles, the factor 0.5 makes reference to the surface double bonding of two-unpaired valence electrons of oxygen, whereas one-unpaired valence electron is to be found in OH_{ads} and O_2H_{ads} entities. Furthermore, the difference $\Delta K^{O_2H}_{OH}$ in energy between the adsorbed species K_{O_2H} and K_{OH} is independent of the nature of the material electrocatalysts (e.g., metal, oxides, chalcogenides, etc. [34]).

$$\Delta K_{OH}^{O_2H} = K_{O_2H} - K_{OH} \approx 3.2 \pm 0.2\,\mathrm{eV} \tag{2.30}$$

The energetic balance applying Eqs. (2.28) and (2.20) to Eqs. (2.22b)–(2.25) identifies the thermodynamic overpotential, η_T, of zero can be obtained if $\Delta K_{OH}^{O_2H} =$ 2.46 eV. However, one clearly sees that the thermodynamic overpotential is:

$$\eta_T = \frac{1}{2e}\left(\Delta K_{OH}^{O_2H} - 2.46\right) \tag{2.31}$$

The thermodynamic overpotential, η_T, as a function of the descriptor, $\Delta G(O_{ads})$, for reactions in Eqs. (2.22b) and (2.25), shows the two crossing points (see Fig. 2.9). Apparently, the difference $\eta_T = 0.37$ V is the one that leads to the unavoidable overpotential observed experimentally and assessed theoretically [35], on the precious and non-precious metal centers known so far, Fig. 2.10a and the way up to decrease such overpotential by improving the ORR activity of Pt, the best ORR electrocatalytic electrode material, through its alloys Pt:M (M: Co, Ni, Fe, etc.) [36], as shown by the theoretical prediction—dashed line—in Fig. 2.10b. It is to be noted that the experimental data, shown in this figure, correspond to measured kinetic currents on a range of catalysts with platinum "skins" (Pt monolayers supported on single-crystal electrodes). At the top of the volcano plot, it is the best ever man-made electrocatalyst: Pt_3Ni (111).

With respect to the oxygen evolution reaction (OER), this process occurs at a potential higher than 1.23 V/SHE [cf. Eq. (2.1)] on most oxide-covered metals [37], and oxide electrodes [38–40]. Moreover, particular attention has been paid to put into a perspective the electronic structure of the materials that govern the binding strength of the various intermediate species generated in the multi-electron charge transfer process. The understanding of this aspect can lead to find the optimal electrocatalytic material [41, 42].

The reaction steps invoked in the literature are in fact similar to those considered above for the ORR in the reverse direction, starting with the water dissociation [cf. Eq. (2.25)].

$$(*) + H_2O \rightleftarrows (*) - OH_{(ads)} + (H^+ + e^-) \tag{2.32}$$

Followed by a further oxidation of OH_{ads} species:

$$(*) - OH_{(ads)} \rightleftarrows (*) - O_{(ads)} + (H^+ + e^-) \tag{2.33}$$

To obtain the superoxide intermediate,

$$(*) - O_{(ads)} + H_2O \rightleftarrows (*) - OOH_{(ads)} + (H^+ + e^-) \tag{2.34}$$

that acts as the molecular oxygen evolution precursor.

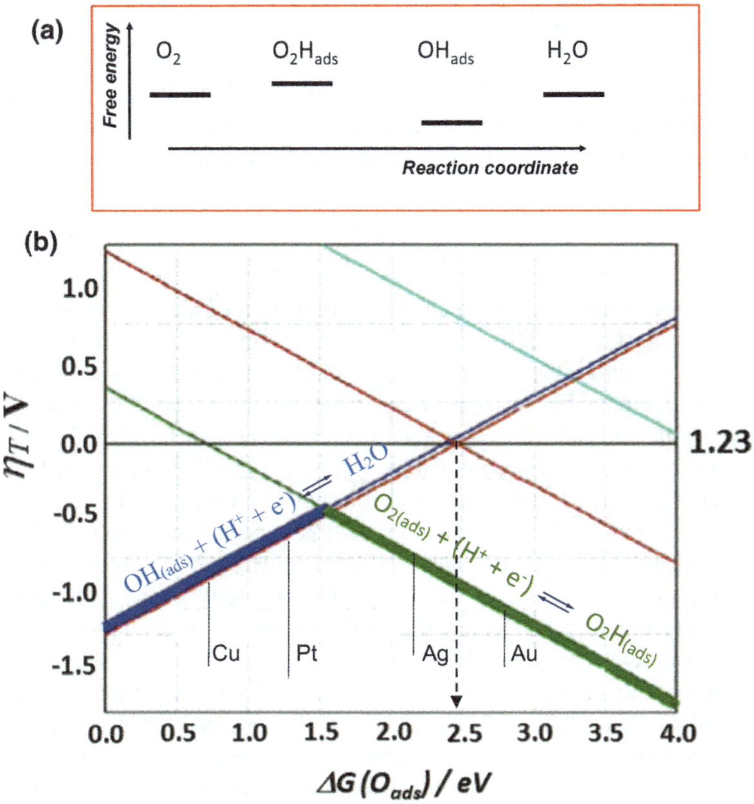

Fig. 2.9 **a** Lining up energy levels show that the thermodynamic overpotential now depends on the ability of the catalyst to bind oxygen. **b** The thermodynamic overpotential as a function of the binding energy of the descriptor. Cu, Pt, Ag, and Au are schematized on the volcano curve (Figure adapted with permission of Ref. [16] Copyright © 2010 Elsevier B.V.)

$$(*) - \mathrm{OOH_{(ads)}} \rightleftarrows \mathrm{O_2} + (\mathrm{H^+} + \mathrm{e^-}) \tag{2.35}$$

Here again, similar considerations as those presented for the ORR in search of the optimal OER electrocatalyst were taken into account. Applying the scaling relationship, with free energies with reference to surface oxides [43], we have,

$$\Delta G(O_2H_{ads}) \approx 0.64 \times \Delta G(O_{ads}) + 2.40\,\mathrm{eV} \tag{2.36}$$

$$\Delta G(OH_{ads}) \approx 0.61 \times \Delta G(O_{ads}) - 0.58\,\mathrm{eV} \tag{2.37}$$

The zero thermodynamic overpotential must fulfill the following data: $\Delta G(OH_{ads}) = 1 \times 1.229 = 1.23\,\mathrm{eV}$, $\Delta G(O_{ads}) = 2 \times 1.229 = 2.46\,\mathrm{eV}$, and $\Delta G(O_2H_{ads}) = 3 \times 1.229 = 3.69\,\mathrm{eV}$.

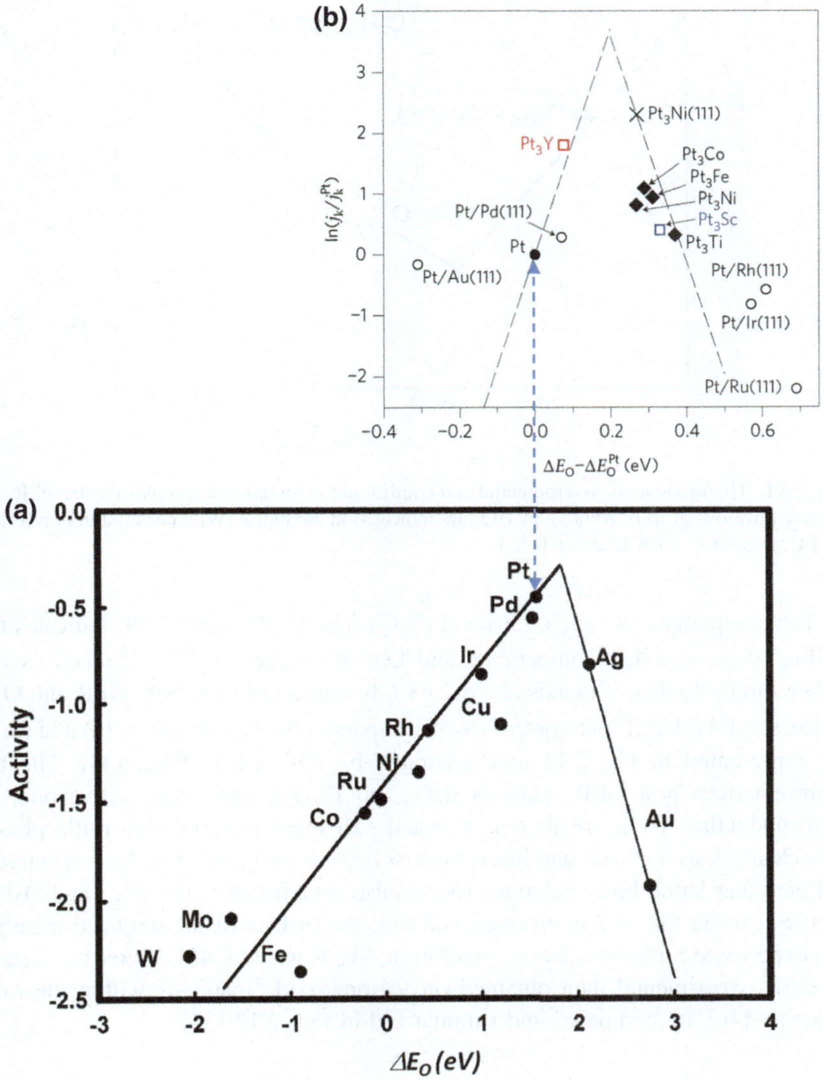

Fig. 2.10 **a** Oxygen reduction activity trends as a function of the oxygen binding energy, ΔE_O. With permission from Ref. [35] Copyright © 2004, American Chemical Society; **b** relative to Pt, the experimental kinetic current density (circles: monolayers supported on single-crystal metal electrodes; diamonds: polycrystalline alloys annealed in ultra-high vacuum before immersion in the electrochemical cell; crosses: bulk $Pt_3Ni(111)$ alloys annealed in ultra-high vacuum before immersion) for a range of alloy electrocatalysts with Pt "skins," in addition to Pt_3Y and Pt_3Sc, are plotted as a function of the calculated oxygen adsorption energy, ΔE_O. The dashed lines are theoretical predictions [35] with permission of Ref. [36] Copyright © 2009, Rights Managed by Nature Publishing Group

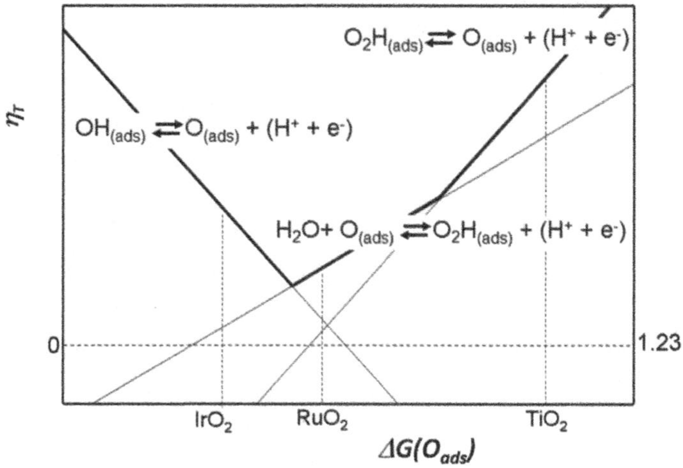

Fig. 2.11 Thermodynamic overpotential and equilibrium potential volcano plot for the OER. The relative positions of IrO_2; RuO_2; and TiO_2 are indicated in the figure (With permission of reference [43] Copyright © 2006 Elsevier B.V.)

The comparison of Eqs. (2.36) and (2.37) with (2.28), and (2.29) indicates that scaling slope is rather comparable, and that the oxide $\Delta K_{OH}^{O_2H} = 2.98\,\text{eV}$ is also rather similar to that of metals (ca. 3.2 eV). It appears that for both OER and ORR processes, this $\Delta K_{OH}^{O_2H}$ parameter is a critical parameter. Equations (2.36) and (2.37) are represented in Fig. 2.11 and generated by DFT calculations [43]. The two representatives best OER catalysts (IrO_2 and RuO_2), and a less active one, the semiconducting TiO_2, are therein depicted. All these materials are rutile phases. One clearly sees that IrO_2 and RuO_2 bind oxygen strongly and weakly, respectively. On the other hand, linear relations can be also established with OH_{ads} and O_2H_{ads} species. At the top of this inversed volcano, the surface of the oxidized materials can experience further oxidation/dissolution, like RuO_2 [44, 45]. Moreover, a series of OER experimental data obtained on various oxide materials within the rutile structure [46] are compared and summarized in Fig. 2.12 [34].

2.3 H_2/O_2 Fuel Cell Electrochemistry

A fuel cell (FC) is a device designed to produce electricity from a chemical reaction, and it belongs to a variety of electrochemical power sources. Unlike galvanic cells working as batteries (primary), and accumulator or storage batteries (secondary), the FC uses a continuous flow of gaseous or liquid (fuel) reactants with a continuous elimination of the products. Its operation is, in principle, sustained as far as the reactants are provided. The FC systems are indeed significant because they offer a series of advantages, namely (i) a high efficiency of fuel utilization, (ii) a

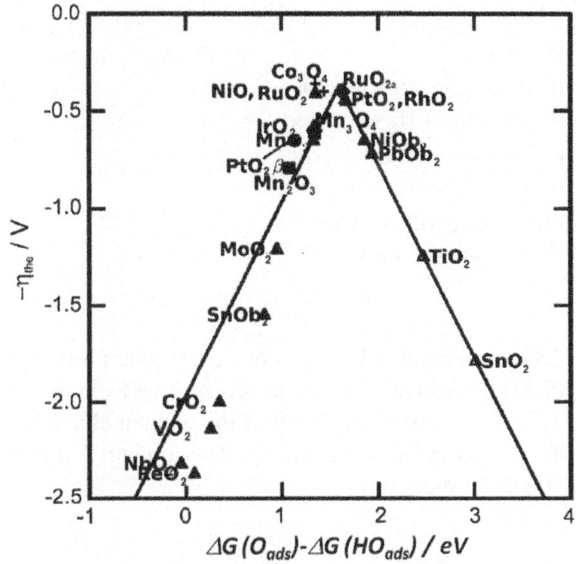

Fig. 2.12 Activity trends toward the oxygen evolution reaction, for rutile, anatase, Co_3O_4, Mn_xO_y, and other oxides as a function of ΔG (O_{ads})-$\Delta G(HO_{ads})$ (With permission of Ref. [34] Copyright © 2011 WILEY-VCH Verlag GmbH & Co. KGaA, Weinheim)

large operation power ranging from microwatts to megawatts, and (iii) to further under development to find out the best catalyst materials among the most abundant elements in nature. In this book, we will limit our discussion to hydrogen–oxygen electrochemistry on low-temperature fuel cell principles. Taking advantage of the principles outlined in Sect. 2.2, we can sum up that the overall reaction at the anode (negative) is the hydrogen oxidation reaction—HOR,

$$H_2 \rightarrow 2H^+ + 2e^- \quad (2.38)$$

and at the cathode (positive) the oxygen reduction reaction (ORR).

$$O_2 + 4H^+ + 4e \rightarrow 2H_2O \quad (2.39)$$

In this manner, a closed electrical circuit is generated (cf. Fig. 2.2a). Otherwise, positive electrical current flows from the anode to the cathode, so that the chemical reaction, producing the current, is:

$$2H_2 + O_2 \rightarrow 2H_2O \quad (2.40)$$

The reaction of 2 mol of hydrogen and 1 mol of oxygen, at normal conditions of pressure (1 Bar) and temperature (298 K), produces 2 mol of water (36 g). The thermal energy (or reaction enthalpy, $-\Delta H$) set free in Eq. (2.40) when it takes place as a direct chemical reaction totals 285.8 kJ/mol, Table 2.1.

The free energy $-\Delta G$ amounts 237.1 kJ/mol, value corresponding to a maximum electrical energy W_e^{max}. In other words, the maximum attainable thermodynamic efficiency, $\eta_{Therm} = 83\%$.

Table 2.1 Some thermodynamic data of H_2/O_2 fuel cell

Reaction	*T (K)*	*ΔG (kJ/mol)*	*ΔH (kJ/mol)*	*ΔS (J/K)*	n	*E° (V)*	η_{therm}
$H_2 + ½\ O_2 \longrightarrow H_2O(l)$	298	−237.1	−285.8	−162	2	1.23	0.83[a]
$H_2 + ½\ O_2 \longrightarrow H_2O(g)$	298	−229	−242	−44	2	1.18	0.95[b]
$H_2 + ½\ O_2 \longrightarrow H_2O(g)$	378						
$H_2 + ½\ O_2 \longrightarrow H_2O(g)$	1273						

[a]LHV—Lower Heating Value
[b]HHV—Higher Heating Value

One example of such devices is the proton exchange membrane fuel cell (PEMFC), which is a current prospective technology in the automotive application [47]. The electrode potential of the oxygen electrode, E_c, is more positive than that of the hydrogen electrode, E_a. This potential difference delivers the cell voltage (U) of the fuel cell.

$$U = E_c - E_a \tag{2.41}$$

The FC operating in discharging (or consumption) mode means that reactions (2.38) and (2.39) take place continuously. In thermodynamics, the potential difference of the cell is called electromotive force (EMF = $E_c^0 - E_a^0$) of the cell. If we consider the hydrogen–oxygen system, this EMF corresponds to free energy of the current producing reaction in Eq. (2.40), i.e., 1.229 V. However, for practical reasons, the voltage of an out of operation cell is named open-circuit voltage (OCV), U_0. So that $U_0 <$ EMF. Figure 2.13a clearly illustrates this fact, where the current sets on at a value less than E_c^0 owing to the lack of equilibrium of the oxygen electrode. This is the reason why the operating working voltage of a FC, U_i is even lower due to various factors that can be easily identified in some regions indicated in Fig. 2.13a, namely (1) the activation polarization a dominated region—where the main losses are due to the overpotentials required by the reactions at the anode and cathode, (2) the ohmic polarization dominated region—where the losses are mainly due to the internal resistance of the cell (*jR* drop), (3) concentration polarization dominated region—where the losses can be mainly attributed to mass transport effects. Most important from the materials' perspective is the large drop from the thermodynamic limit observed in the activation polarization region, predominantly attributed to the high overpotential for the oxygen reduction reaction at the cathode, as discussed in Sect. 2.2.2.

The voltage of the cell, U_i, will be lower the higher the current density, j, as observed in the current–voltage characteristic of a fuel cell shown in Fig. 2.13a, curve (1). This characteristic can be simplified by the linear equation: $U_i = U_0 - jR$ where R, in a first approximation, is associated with the apparent internal resistance of the cell. Moreover, Fig. 2.13a shows indeed that the shape of the current–voltage is S-shaped. One recognizes, however, that the OCV or U_0 and the operating voltage, U_i, are considered important parameters in a fuel cell. On the other hand,

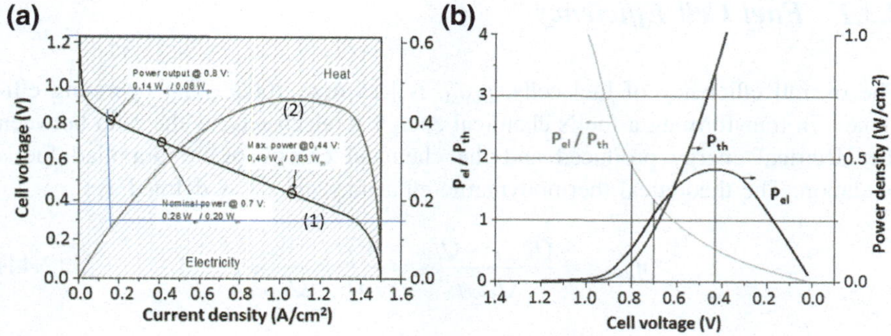

Fig. 2.13 **a** Electricity and **b** heat generation in fuel cells, see text

the difference between E^0 and U_i will depend on the nature of the reaction kinetics, and on the nature of the electrode. In the PEMFC, a voltage of 0.6–0.7 V is commonly used, resulting in significant energy losses due to heat generation (Fig. 2.13b). At a working voltage of $U_i = 0.65$ V about the same amount of electrical energy and heat is generated. The flowing current at this voltage at the external load with a resistance R_{ext} is determined by Ohm's law.

$$I = \frac{U_i}{R_{ext}} = \frac{U_0}{R + R_{ext}} \quad (2.42)$$

During the fuel consumption in the FC, the power $P = U_i \times I$ (Power = Voltage × Current) or (Watt (= Joule/s) = Volt × Coulomb/s) is delivered. Using Eq. (2.42), the power can be written as,

$$P = \frac{U_i^2 R}{(R + R_{ext})^2} \quad (2.43)$$

This equation shows that decreasing the external resistance, the voltage decreases; hence, the power–current relation goes through a maximum; Fig. 2.13a, curve (2). The current, I, and the power, P, are both determined by the load (R_{ext}) selected by the user. However, the maximum current I_{max} and the associated P_{max} establish important features of the cells. Indeed, this performance sets a critical lower bound of the cell voltage to a critical value, as to avoid overheating. Attempts to go to even lower cell voltages in order to increase the electrical power density by increasing the current density (to a maximum power point) have been made. Therefore, in an operating fuel cell, the sustainability of the thermal conditions can be established by fixing a lower limiting current.

For hydrogen-based fuel cells (PEMFCs) the generation of heat is a technical problem that is usually solved by cooling arrangements.

2.3.1 Fuel Cell Efficiency

The overall efficiency of fuel cells, η_{total}, is less than 100%. The operating efficiency, in transforming a fuel's chemical energy to electricity, is the ratio between the electrical energy produced and the chemical energy of the supplied fuel's oxidation. The theoretical thermodynamic efficiency, η_{therm}, is defined as:

$$\eta_{therm} = \frac{Q_{react} - Q_{lat}}{Q_{react}} = \frac{\Delta G}{\Delta H} = \frac{W_{el}}{Q_{react}} \tag{2.44}$$

where Q_{react} the total thermal energy produced during a chemical reaction—the reaction enthalpy ($\Delta H = \Delta G + T\Delta S$) is set free as electrical energy, W_{el} (the current flows at the external circuit), whereas the remaining part of the reaction evolved in form of latent heat of reaction, is the reaction entropy, Q_{lat} ($-T\Delta S$). This latent heat is similar to the Carnot heat in heat engines. Equation (2.44) shows that, in an electrochemical system, a big portion of the chemical energy is converted directly into electricity, since Q_{lat} is small compared to Q_{react} (cf. Table 2.1). For most FC systems, this last parameter is negative, meaning that U_i will drop with a temperature increase. The η_{therm} will never be larger than 90%, because of the negative temperature coefficient. Since a FC is not operating reversibly, we must also define a voltage efficiency, η_v, as:

$$\eta_v = \frac{U_i}{E^0} = \frac{U_i}{EMF} \tag{2.45}$$

So that the real fuel cell efficiency η should be $\eta = \eta_{\text{therm}} \times \eta_{\text{v}}$. For H_2/O_2 FC, the value of EMF = E^0 is 1.229 V. In a real system, it is worth to mention that the overall efficiency η_{total} of a fuel cell depends, besides η_{therm}, and η_{v}, on two other additional factors, e.g., the efficiency due to the reactant utilization, or coulombic efficiency, η_{coul}; and the efficiency of the design, η_{design} ($\eta_{\text{total}} = \eta \times \eta_{\text{coul}} \times \eta_{\text{design}}$). It is thus clear that efforts to improve a FC system must be concentrated to increase η_{v} (via electrocatalysis); η_{coul} (via selectivity or tolerance of materials), and η_{design} (via system and stack design).

2.3.2 Fuel Cell Parameters

In order to associate fuel cell power generators, as those summarized in Table 2.2, it is convenient to normalize the relative rates in terms of current producing reactions to a surface (A).

These usual parameters to compare the relative efficiencies of FCs are current density, $j = I/A$ (mA/cm^2), and power density, $p = P/A$ (mW/cm^2). This later can be expressed as power density referred to unit mass, $p_m = P/M$ (mW/kg), or to unit volume, $p_v = P/V$ (mW/L).

Table 2.2 Summary of most common fuel cells (FCs) technologies

	PEMFC	DMFC	AFC	PAFC	MCFC	SOFC
Fuel	H_2	CH_3OH	H_2	H_2	H_2, CO, CH_4, HC[(a)]	H_2, CO, CH_4, HC[(a)]
Electrolyte	(*Nafion*)	(*Nafion*)	KOH	H_3PO_4	Li–, K_2CO_3	ZrO_2–Y_2O_3
Charge carried in electrolyte	H^+	H^+	OH^-	H^+	CO_3^{2-}	O^{2-}
Temp. (°C)	50–100	50–90	60–120	175–200	650	1000
η (%)	35–60	<50	35–55	35–45	45–55	50–60
Unit Size (KW)	0.1–500	≪1	<5	5–2000	800–2000	>2.5
Applications	Portable, stationary and automotive	Portable, mobile and stationary use	Space, military, submarines, transport	Medium to large power generation	Large power generation	Medium to large power generation

[a]Hydrocarbons

2.3.3 *Electrode Reactions*

The electrochemical reactions taking place in a fuel cell determine the performance of the system. Focusing on the H_2/O_2 FCs, Table 2.3 shows the reactions that occur in acid fuel cells: polymer electrolyte fuel cells (PEMFC), phosphoric acid fuel cells (PAFC), and alkaline fuel cells (AFC). It is worth to note, that all these systems require noble metal electrocatalysts for optimum reaction rates at the anode and at the cathode.

2.4 H_2/O_2 Anode and Cathode Electrochemistry

Most research systems are based on the recording of current–potential characteristics, under well-defined conditions (e.g., temperature, mass transport, etc.). This section will be devoted to highlight some important tasks to perform in measuring

Table 2.3 Electrochemical reactions in acid and alkaline FCs

Fuel cell type	Anode reaction	Cathode reaction
Acid (PEMFC, PAFC)	$H_2 \longrightarrow 2H^+ + 2e^-$	$\frac{1}{2}O_2 + 2H^+ + 2e^- \longrightarrow H_2O$
Alkaline (AFC)	$H_2 + 2(OH^-) \longrightarrow H_2O + 2e^-$	$\frac{1}{2}O_2 + H_2O + 2e^- \longrightarrow 2(OH^-)$

electrocatalytic materials (applicable to precious or non-precious catalytic centers) in form of nanodivided powders. The main drawbacks are to mimic the electrocatalytic nanoparticles (NPs) behavior, embedded on a gas diffusion layer of a membrane electrode assembly (MEA) of a PEMFC, using the thin-film rotating disk electrode (RDE), and rotating ring-disk electrode (RRDE) technique, to evaluate the corresponding electrocatalytic activity of supported NPs in half-cell arrangement. The supports can be of different nature, e.g., carbon, graphene, oxides, and composites (carbon-oxide; graphene oxide). This means that the electrochemical study is based on the preparation of inks. The replication across the laboratories requires an experimental protocol which requires the ink formulation, the quality of the film, and the procedure to perform the electrochemical tests [48].

The learning curve in electrocatalysis is to identify the surface reactivity of the electrode materials. The action of the electrocatalyst can be due to structural or chemical modification of the surface of the electrode in addition to the influence of the ions present in the electrolyte. In other words, structural effects can be associated with the changes of the electronic state at the surface (d-band occupation, d-band center shift) or by the variation of geometric nature at the surface that occurs in crystal-defined planes [49–52], alloy [53–55], strain effects [56–58]. More details are given in Sects. 2.4.2 and 2.4.3.

2.4.1 Turn Over Frequency

Having taken the necessary precautions to prepare inks and electrodes, the electrocatalytic activity of a given material must be described using the concept of turn over frequency (TOF) number. To appreciate this concept, one can imagine working with thin films of variable thickness, which can be obtained by successive pipetting an ink onto the disk (e.g., glassy carbon, as substrate) of the electrode, as shown in Fig. 2.14. This leads to a deposited mass loading variation, effect of which is to increase the number of NPs in the thin layer. Such examples were tested with carbon Vulcan-supported platinum NPs with a defined thickness, x_0, and varying mass loading [31], or with cobalt chalcogenide ($CoSe_2$) [59] varying thickness and a defined mass loading. The increase (with similar size) of NPs represents an increase of the number of sites (catalytic ones). Since these NPs are confined in a defined thin-film space with a determined thickness, we are dealing with a quasi-volume electrode, so that the electrocatalytic activity can be established as the current per unit volume (A/cm^3). This parameter can be related to the site density (site/cm^3) with Eq. (2.46).

$$Activity = (Site\,density)\,(TOF)\,q \tag{2.46}$$

where q is the elemental charge (1.6×10^{-19} C/e$^-$), TOF in (e$^-$/Site/s) is related to the kinetic current density. Taking the electrochemical reaction being carried out on a catalytic "Site":

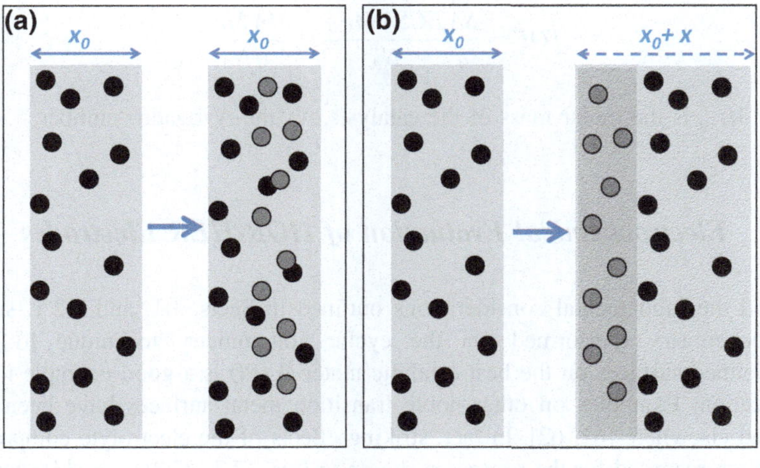

Fig. 2.14 Schematics of the catalytic site density variation in carbon matrix support. **a** loading increase through direct synthesis; **b** loading increase through successive layer depositions (With permission of Ref. [31] Copyright © 2010 WILEY-VCH Verlag GmbH & Co. KGaA, Weinheim)

$$\upsilon_O Ox + ne^- + Site \rightarrow \upsilon_R \text{Red} \tag{2.47}$$

The rate of electrons at the "Site" can be written as:

$$v_{e^-} = \frac{1}{Site} \frac{1}{n} \frac{dn_e}{dt} \tag{2.48}$$

This equation measures the flow of (e^-/Site/s), otherwise, the so-called TOF. The active "Site" can be expressed per square centimeter of electrocatalyst surface: $Site\ cm^{-2}$, or site per cubic centimeter of electrocatalyst volume: $Site\ cm^{-3}$. Clearly, n_e is the concentration of electrons ($e^-\ cm^{-2}$ or $e^-\ cm^{-3}$), and t, the reaction time (s).

The activity of an electrocatalytic material can be expressed as a current density, j, $\left(A.cm_{geo}^{-2}\right)$. This parameter can be, indeed, further reported as mass activity (MA = $A.mg_{Cat}^{-1}$), and specific activity (SA = $\mu A.cm_{Cat}^{-2}$), so that the electrochemical active surface area, ECSA, combines MA and SA, as follows:

$$ECSA = \frac{MA}{SA} = \frac{10^9 cm_{Cat}^2}{gr_{Cat}} \tag{2.49}$$

Making use of the current densities, easily determined from current–potential experiments, one can calculate TOF as:

$$TOF = \frac{SA}{q}\frac{ECSA\, M_{Cat}}{N_A} = \frac{MA\, M_{Cat}}{q\, N_A} \tag{2.50}$$

where M_{Cat} is the molar mass of the catalysts, N_A the Avogadro number.

2.4.2 Electrochemical Evaluation of HOR-HER Electrodes

Beyond the fundamental considerations outlined in Sects. 1.1, and 2.2.1, surface electrochemistry (performed via the cyclic voltammetry technique [60]) on well-defined surfaces on the best catalytic material (Pt) is a good example to start this section. Examples on other noble transition metal surfaces have been summarized elsewhere [61, 62]. In fact, striking effects of the electrolyte composition have been recorded for the processes shown in Eqs. (2.7)–(2.10) on (*hkl*) surfaces of e.g., platinum, since these surfaces present a long-range atomic order and are sensitive to the anion adsorption. Figure 2.15 summarizes the fingerprints of low-*hkl* index of platinum submitted to a scan rate of 50 mV/s in acid containing sulfate (SO_4^{2-} or HSO_4^-) and perchlorate (ClO_4^-) anions. The responses of (100), (111), and (110) correspond to the H_{upd} region (cf. Fig. 2.5) for the polycrystalline material. In this latter, two main couples of reversible adsorption processes appear between 0.4 V/SHE and the potential where the hydrogen evolution reaction sets on (0 V/SHE). The peak centered at 0.27 V/SHE is due to the strongly adsorbed hydrogen atoms, whereas the peak centered at 0.12 V/SHE corresponds to weakly adsorbed hydrogen. On well-cleaned polycrystalline surfaces a third peak (at ca. 0.22 V/SHE) appears on the ongoing positive scan. At this point, it is interesting to recall that the charge under the H_{upd} region of 210 μC cm^{-2} [63] corresponding to a full H monolayer coverage is used to determine the real surface area necessary to obtain the TOF factor (see Sect. 2.4.1). Now, comparing the shape of the H_{upd} regions in Fig. 2.5 (recorded in sulfuric acid) with the shapes presented in Fig. 2.15, one learns that the H-coverage takes place on surfaces with different symmetries. Each *hkl*, indeed, shows a different adsorption energy, since Pt atoms are arranged in different unit cells, leading to a difference in the overlapping between the electronic local density of a "Site" with the adsorbed hydrogen. Hence, as a result, we perceive that the strong adsorbed hydrogen peak corresponds to the electrochemical response at the (100) surface, whereas the weakly adsorbed peak to that generated on (110) and (111) surfaces. Moreover, as shown in Table 2.4, the charge density is compatible in all surfaces. The average charge density is 197 μC cm^{-2}, approaching the value of the polycrystalline surface.

The H monolayer coverage on these surfaces is affected by the composition of the electrolyte solution, pointing out the effect of the specific adsorption of the anions. It is worth to recall that an important amount of work combining ultra-high vacuum (UHV) and electrochemistry experiments [64] led to unravel the interesting so-called butterfly feature on the voltammogram recorded on the Pt (111) surface in

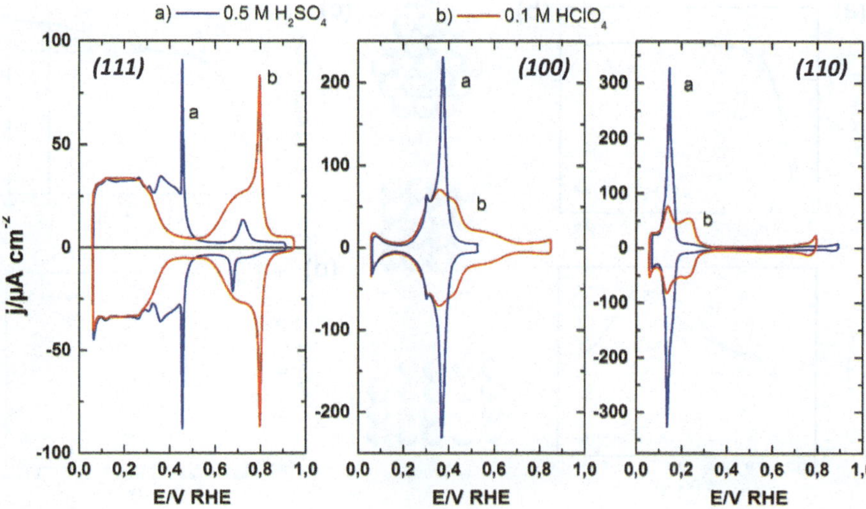

Fig. 2.15 Cyclic voltammograms of (hkl) Pt surfaces recorded in acid at 50 mV/s. The specific anions adsorption affects the shape of the H_{upd} region (Adapted from Ref. [65] Copyright © 2011, Springer)

Table 2.4 Charge density on *hkl* surface of Pt (fcc)

Surface (*hkl*)	n (atom/cm^2)	Surf. Conc. (mol/cm^2)	Charge density (μC/cm^2)
(100)	1.28×10^{15}	2.13×10^{-9}	205
(110)	0.92×10^{15}	1.53×10^{-9}	147
(111)	1.50×10^{15}	2.49×10^{-9}	240

sulfuric acid. Such spikes reveal the transition of the SO_4^{2-} (or HSO_4^-) adlayer between disordered and ordered states. A historical account of platinum surface electrochemistry was provided in 2011 [65, 66].

Around 0 V/RHE, the difference of the HER/HOR on Pt (*hkl*) is clearly obvious in acid and in alkaline solutions, Fig. 2.16, assessing the role of the surface geometry on the kinetic rates [67]. It is also clearly discernible, although less sensitive that in acid medium such an effect is less pronounced as compared to the effect in alkaline medium. Summarizing the effects, the degree of electroactivity decreases in the order (110) > (100) > (111) in both media. Considering the mechanism of HOR and HER, discussed in Sect. 2.2.1, one can distinguish the potential intervals between H_{upd} and H_{opd} regions (cf. Fig. 2.15) at the (*khl*) surfaces. Since the more open surface (110) is the HOR/HER most active one, the top sites, as indicated in Fig. 2.15b must be available for the hydrogen interaction (adsorption/bond making and breaking) favoring the H_{opd} process, thus leaving on (110) a dual active site. This is not the case for (100) and (111), since they

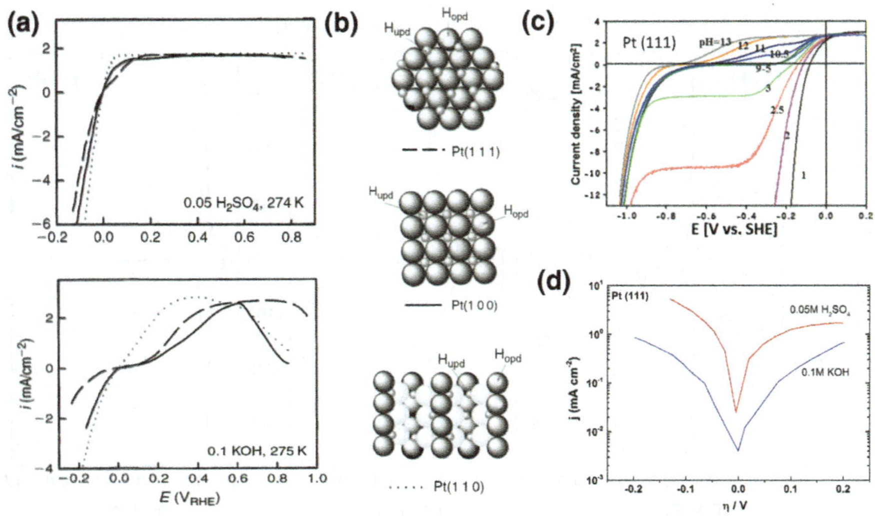

Fig. 2.16 **a** Current–potential curves for the HER and HOR on Pt (hkl) in acid (top), and alkaline (bottom) media, and **b** surface ideal representations of Pt (hkl) showing H_{upd}, and H_{opd} sites, with permission of Ref. [67] Copyright © 2002 Elsevier Science B.V. **c** The hydrogen oxidation reaction rate by promotion of hydroxyl adsorption, with permission from Ref. [70] Copyright © 2013, Rights Managed by Nature Publishing Group. **d** Tafel plot of data extracted from curve for Pt (111) of this figure (**a**)

apparently only have one type active site, where Heyrovsky–Volmer mechanism must be operative. Another interesting feature of the HOR/HER on Pt (hkl) surfaces is the fact that the kinetics in alkaline medium are lower than in acid medium. The effect can be relied to OH_{ads} species blocking the active sites for the adsorption of hydrogen [68]. This path makes it difficult to find non-precious metal centers to activate the HOR. However, tailoring efficient dissociative adsorption sites (metalhydroxyl oxide) for water (hydrogen is discharged by water in alkaline medium) is required to augment HOR/HER kinetics [69, 70]. Particularly interesting to consider is the fact that the variation of pH and, as such, the HOR/HER current–potential simulations were performed to cover the whole pH range [70].

Considering total reactions (2.51) and (2.52), and a partial pressure of hydrogen of 1 Bar:

$$2H_3O^+ + 2e^- \underset{K_1}{\overset{K_2}{\rightleftarrows}} H_2 + 2H_2O \tag{2.51}$$

$$2H_2O + 2e^- \underset{K_1}{\overset{K_2}{\rightleftarrows}} H_2 + 2HO^- \tag{2.52}$$

The corresponding anodic, and cathodic kinetic current densities in acid, Eq. (2.53) and in alkaline, Eq. (2.54), can be expressed as:

$$j_a = 2FK_1[H_2]_{x=0}\exp\left[\alpha_a\left(\frac{F}{RT}(E - E_2^0) - \ln[H_2]^{1/2}\right)\right] \tag{2.53a}$$

$$j_c = -2FK_2[H_3O^+]_{x=0}\exp\left[-\alpha_c\left(\frac{F}{RT}(E - E_2^0) - \ln[H_2]^{1/2}\right)\right] \tag{2.53b}$$

$$j_a = 2FK_1[H_2]_{x=0}[OH^-]_{x=0}\exp\left[\alpha_a\left(\frac{F}{RT}(E - E_1^0) - \ln[H_2]^{1/2}\right)\right] \tag{2.54a}$$

$$j_c = -2FK_2[H_2O]_{x=0}\exp\left[-\alpha_c\left(\frac{F}{RT}(E - E_1^0) - \ln[H_2]^{1/2}\right)\right] \tag{2.54b}$$

All the symbols have their respective meaning. The constants K_1 and K_2 refer to the coverage of the H_{ads}, or OH_{ads} species. The effect of mass transfer is added via the Levich equation: $j = 0.62nFD^{2/3}\nu^{-1/6}[C_{ads} - C_{bulk}]_{x=0}\omega^{1/2}$ to the simulation curves (see Fig. 2.16c).

For this, HOR/HER apparently simple reaction, the surface electrochemistry on well-defined surfaces teaches us that the surface must be tailored for dual functional catalytic sites to activate a proton donor to bridge the pH gap. In this sense, the material in nanodivided form is essential to develop devices for the real-world application and explorations in acid and alkaline medium using carbon-supported platinum, iridium, and palladium nanoparticles [71]. Figure 2.17 contrasts the electrochemical response obtained via the RDE technique of carbon-supported Pt, Ir, and Pd nanoparticles in acid and alkaline medium. Here again, a striking difference in the HOR/HER electroactivity is observed in acid *versus* alkaline medium for all carbon-supported samples. The Tafel plot [cf. Sect. 1.1.3.1, Eq. (1.9b)], in Fig. 2.17c, d clearly displays that at an overpotential η = 0.05 V the anodic current densities for Pt are ca. 200 times, for Ir ca. 160 times, and for Pd ca. 150 times more active in acid than in alkaline media. Similar differences can be read at the cathodic side, i.e., η = −0.05 V. At this point, it is also worth to add that on Pt(111), see Fig. 2.16d, the HER is ca. a factor of 20 times (η = −0.10 V) higher in acid than in alkaline environment. Nanoparticulated materials, indeed, increase the number of active sites. Summing up, the order of magnitude is similar for these three metals, implying a similar binding energy of the descriptor ($\Delta G^0_{H_{ads}}$) discussed in Sect. 2.2.1. From a fundamental point of view, and under the light of the results [71] depicted in Fig. 2.17, it is shown that the higher oxophilicity of iridium does not enhance the HOR/HER activity; therefore, the HOR/HER mechanism to be favored is the Volmer step (cf. Eq. (2.7) in Sect. 2.2.1), thus pointing toward a similar mechanism in acid as well as in alkaline medium.

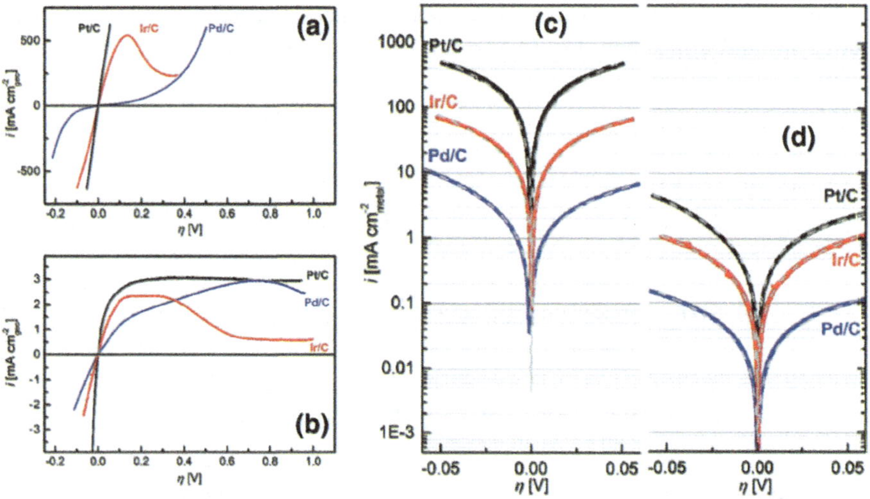

Fig. 2.17 **a** Inward a PEMFCs arrangement the polarization curves (2 mV/s) in a H_2-saturated atmosphere at ambient pressure and 313 K on various electrocatalysts, namely 5 wt% Pt/C (ECSA = 120 × 10^4 cm^2/g), 20 wt% Ir/C (ECSA = 59 × 10^4 cm^2/g) and 10 wt% Pd/C (ECSA = 105 × 10^4 cm^2/g). **b** 0.1 M NaOH at 10 mV/s at 1600 rpm, **c** and **d** are the corresponding Tafel plots of **a** and **b**, respectively (With permission of Ref. [71] CC Attribution 3.0)

2.4.3 Electrochemical Evaluation of ORR Electrodes

This section will be focused on the typical ORR electrocatalytic response on materials in the nanodivided form. Indeed, the evaluation of the electrochemical performance of electrode materials for the ORR process is typically performed by the use of the rotating disk electrode (RDE) and the rotating ring-disk (RRDE) electrode techniques [30]. Generally, typical ORR characteristics under mass transport conditions are obtained, as shown in Fig. 2.18a. One can recognize three regions in which the ORR process is controlled, namely (i) kinetic controlled (activation region), where the ORR rate is quite slow; (ii) the mixed kinetic diffusion where the reaction accelerates (increase of the current) as the overpotential increases; (iii) the diffusion-controlled region, the current density is determined by the diffusion rate of reactants and reaches a plateau at a determined rotation rate of the RDE. On the same figure, one can read the half-wave potential, $E_{1/2}$. This parameter is used to verify and compare in a qualitative way the activity of the electrocatalysts. The more active materials should show under the same measurement conditions a more positive half-wave potential value. Figure 2.19a shows that $E_{1/2}$ shifts to a positive potential with the mass loading or site density [cf. Fig. 2.14 and Eq. (2.46)]. It is worth to mention that a lower $E_{1/2}$ is equivalent to a strong indication which favors the ORR 2-electron pathway, or formation of hydrogen peroxide as assessed by RRDE experiments (see Fig. 2.19b). This phenomenon,

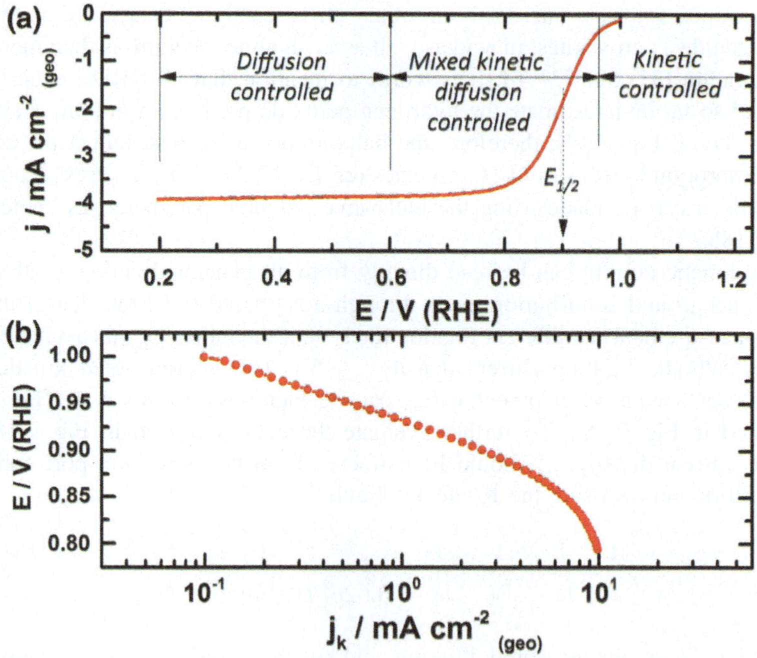

Fig. 2.18 **a** ORR polarization curve obtained on 20 wt%/C in 0.1 M $HClO_4$ acid medium at 900 rpm at a scan rate of 5 mV/s. **b** Corresponding Tafel plot of the kinetic current density, after mass correction

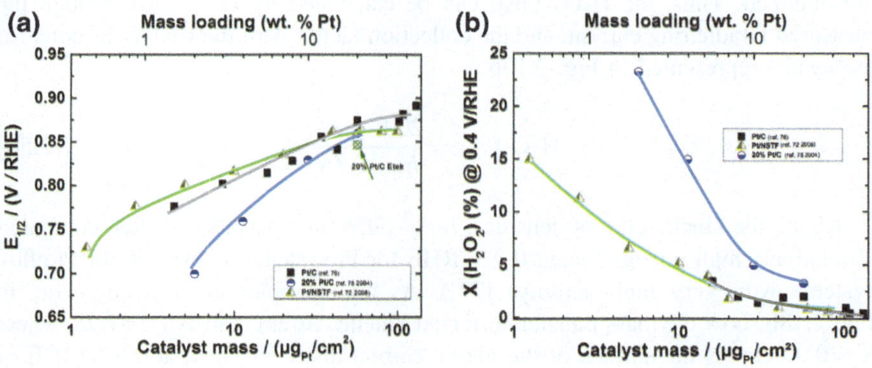

Fig. 2.19 **a** Half-wave potential of RDE current potential characteristics of Pt-based electrodes as a function of the mass loading. **b** Hydrogen peroxide detected via the ring electrode of the RRDE technique (Data extracted from Refs. [72, 75]. Modified figure with permission of Ref. [76] Copyright © 2009 Elsevier Ltd.)

which appears to be universal on other ORR catalytic centers (e.g., metals, chalcogenides, perovskites, in acid as well as in alkaline medium) is documented in the literature [31, 59, 72–76]. It is worth to mention that all RRDE experiments reported so far underestimate the hydrogen peroxide production in porous systems (quasi-3D, cf. Fig. 2.14); therefore, the halfway potential must reflect the adsorption–desorption kinetics of O_2H_{ads} species (cf. Eq. (2.23)). Thus, a cross-laboratory comparison can be made using the half-wave potential parameter, as depicted in Fig. 2.18b.

The kinetic current can be read directly from all generated curves (with eliminated background contribution made through a N_2-based test from that of the O_2-based measurement) at different rotation rates, ω, if the ratio: $j/j_{l,c}$ (measured current density/cathodic limiting current density) $\leq$ 5%, (see region noted kinetic controlled) so that an onset potential E_{on} can be identified when $j = 0$. This is not indicated in Fig. 2.18a. To further evaluate the data generated in Fig. 2.18a the kinetic current density, j_k, should be extracted from the mass transport corrected polarization curves using the Koutecky–Levich Eq. (2.55) [60].

$$\frac{1}{j} = \frac{1}{j_K} + \frac{1}{j_{l,c}} = \frac{1}{j_K} + \frac{1}{0.62nFD_0^{2/3}\omega^{1/2}\nu^{-1/6}C_0^*} \tag{2.55}$$

where j, $j_{l,c}$, j_k are the measured, limiting, and kinetic current density, respectively, n is the electron transfer number, F is the Faraday constant (96,485 C/mol), D is the oxygen diffusion coefficient in the electrolyte, ν is the kinematic viscosity of the electrolyte, and C_O^* is the bulk concentration of oxygen. To evaluate quantitatively the hydrogen peroxide production, the collection of the current at the ring electrode, I_R, biased at a potential of 1.2 V/RHE of the RRDE system, and the disk current, I_D, are obtained. Thus, the H_2O_2 yield can be calculated by Eq. (2.56) through the measured faradic ring current, and the collection factor, N of the RRDE, as done for the results represented in Fig. 2.19b.

$$H_2O_2(\%) = \frac{200 \cdot I_R/N}{I_D + I_R/N} \tag{2.56}$$

Again, the kinetic current densities, j_K, of different catalysts are often compared at relatively high potential, e.g., 0.9 V/RHE for Pt systems (or 0.95 V (or Pt alloy systems with very high activity) [77]). As an example, the measurements in Fig. 2.18b show the mass transfer corrected kinetic current density ($j = I_D/S$, where S = 0.07 cm^2 is the surface of the glassy carbon disk). We read at 0.9 V/RHE of 2.68 mA/cm^2 (geo), or a kinetic current I_K = 0.187 mA. Using the H_{upd} region (cf. Fig. 2.5) to determine the real surface of platinum NPs deposited onto the glassy carbon disk, we can determine the specific activity (SA = 0.13 mA/cm^2_{Pt}). Having deposited 20 μg of 20 wt% catalysts ink onto the glassy carbon disk, the mass activity is MA = 46.7 mA/mg_{Pt}. Otherwise, using Eq. (2.49) the electrochemical active surface area of this system corresponds to ECSA = 33.6 $(m^2/g)_{Pt}$. The resulting turn over frequency, (i.e., electrons/site/s), is TOF = 0.09 s^{-1}. This latter data from RDE at 25 °C (298 K) compared to the literature value (TOF = 25 s^{-1} [77]), obtained from a PEM fuel cell system operating at 80 °C (353 K), seem too small. Therefore, a fair comparison is the parameter-specific activity (SA). Indeed, from Table 5 [77], for Pt/C(TKK) AS = 195 $\mu A/cm^2_{Pt}$ at 80 °C. The data deduced in Fig. 2.18b correspond to AS = 130 $\mu A/cm^2_{Pt}$ at 25 °C. Adjusting this last kinetic current to 80 °C using: $j_k \propto \exp(-E_{act}/k_B T)$, where a typical value of activation energy for the ORR is 0.3 eV [78], and then the ratio between the two kinetic currents can be formulated as:

$$\frac{\ln j_{k1}}{\ln j_{k2}} \approx \frac{E_{act}(T_1 - T_2)}{k_B T_1 T_2} \tag{2.57}$$

The calculated ratio is ca. 2.06. Thus, the temperature adjusted for the data in Fig. 2.18b at 0.9 V/RHE is $(AS)_{80°C}$ = 130 $\mu A/cm^2_{Pt}$ × 2 = 260 $\mu A/cm^2_{Pt}$. This is an interesting indicator that puts in evidence that the utilization of Pt NPs at the working RDE electrode (measured in half-cell) is far from being 100%, thus the low TOF value. However, in practical fuel cell applications, the mass activity is more important than the specific activity. Ways to develop benchmarking methods to test HOR/HER, ORR/OER electrocatalyst performance are still underway (see, e.g., Ref. [79]) to assure comparability with practical systems. That is the reason why defined targets (e.g., MA = 440 mA/mg_{Pt}) for the year 2020 for catalysts in automotive applications have been projected by the USA department of energy (DoE) [80].

For the cathodic process the determination of TOF for the hydrogen peroxide reduction (HPR) and the ORR processes, as a function of the catalyst mass loading was reported, and the results are summarized in Fig. 2.20 [81]. The data evaluation for each system was performed using the considerations mentioned above through Eqs. (2.46)–(2.50), and (2.55), (2.56). The TOF for the ORR process decreases

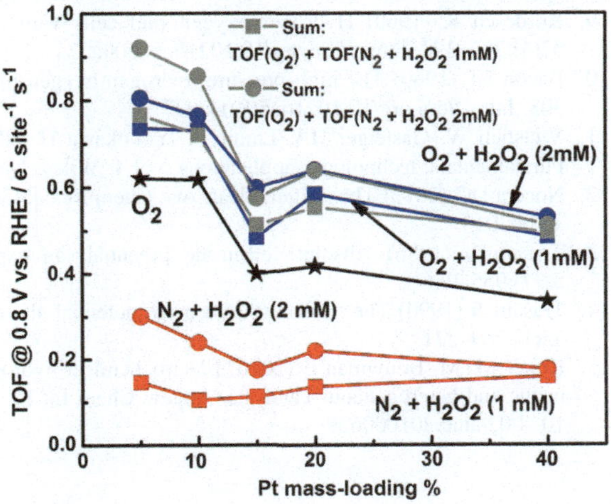

Fig. 2.20 TOF values taken at 0.8 V versus RHE for hydrogen peroxide reduction (HPR) and the oxygen reduction reaction (ORR) in 0.5 M H_2SO_4 on various mass loadings of Pt/C (With permission of Ref. [81]. Copyright © 2016 Wiley-VCH Verlag GmbH & Co. KGaA, Weinheim)

with the mass loading, whereas the HPR is less sensitive than ORR with the catalyst mass loading. One can assess at least two important facts: (i) 1 mM of hydrogen peroxide (HP) finds sufficient active sites, whereas 2 mM HP reveals less accessible active sites with the mass loading; (ii) the HOR and ORR apparently reflect the sum of the individual process. Therefore, this phenomenon clearly reveals that there is no interference for actives sites for ORR and HPR. However, it is clear that ORR is more perturbed by the lack of active sites with the mass loading, reflecting as phenomenon the distance among particles is different, because the NPs size and morphology are similar.

References

1. Kunze J, Stimming U (2009) Electrochemical versus heat-engine energy technology: a tribute to Wilhelm Ostwald's visionary statements. Angew Chem Int Ed 48(49):9230–9237. https://doi.org/10.1002/anie.200903603
2. Grove WR (1842) On a gaseous voltaic battery. Philos Mag J Sci 21(140):417–420. https://doi.org/10.1080/14786444208621600
3. Meldola R (1900) Christian Friedrich Schönbein, 1799–1868. Ein Blatt zur Geschichte des 19. Jahrhunderts. Nature 62:97–99. https://doi.org/10.1038/062097a0
4. Mitlitsky F, Myers B, Weisberg AH (1998) Regenerative fuel cell systems. Energy Fuels 12(1):56–71. https://doi.org/10.1021/ef970151w
5. Mitlitsky F, Myers B, Weisberg AH, Molter TM, Smith WF (1999) Reversible (unitised) PEM fuel cell devices. Fuel Cells Bull 2(11):6–11. https://doi.org/10.1016/S1464-2859(00)80110-8
6. Wang Y, Leung DYC, Xuan J, Wang H (2016) A review on unitized regenerative fuel cell technologies, part-A: Unitized regenerative proton exchange membrane fuel cells. Renew Sustain Energy Rev 65:961–977. https://doi.org/10.1016/j.rser.2016.07.046
7. Newman J, Hoertz PG, Bonino CA, Trainham JA (2012) Review: an economic perspective on liquid solar fuels. J Electrochem Soc 159(10):A1722–A1729. https://doi.org/10.1149/2.046210jes
8. Holladay JD, Hu J, King DL, Wang Y (2009) An overview of hydrogen production technologies. Catal Today 139(4):244–260. https://doi.org/10.1016/j.cattod.2008.08.039
9. Kordesch K (1960) Hydrogen-oxygen fuel cells with carbon electrodes. Ind Eng Chem 52(4):296–298. https://doi.org/10.1021/ie50604a024
10. Bacon FT (1960) The high pressure hydrogen-oxygen fuel cell. Ind Eng Chem 52(4):301–303. https://doi.org/10.1021/ie50604a027
11. Vielstich W, Gasteiger HA, Lamm A, Yokokawa H (eds) (2010) Handbook of fuel cells. Fundamentals, technology, applications, vol 4. Wiley, New York
12. Nocera DG (2012) The artificial leaf. Acc Chem Res 45(5):767–776. https://doi.org/10.1021/ar2003013
13. Trasatti S (1986) Absolute electrode potential: an explanatory note. Pure Appl Chem 58(7):955–966
14. Trasatti S (1990) The "absolute" electrode potential–the end of the story. Electrochim Acta 35(1):269–271
15. Koper MTM, Bouwman E (2010) Electrochemical hydrogen production: bridging homogeneous and heterogeneous catalysis. Angew Chem Int Ed 49(22):3723–3725. https://doi.org/10.1002/anie.201000629

16. Koper MTM (2011) Thermodynamic theory of multi-electron transfer reactions: implications for electrocatalysis. J Electroanal Chem 660(2):254–260. https://doi.org/10.1016/j.jelechem.2010.10.004
17. Conway BE, Angerstein-Kozlowska H, Sharp WBA, Criddle EE (1973) Ultrapurification of water for electrochemical and surface chemical work by catalytic pyrodistillation. Anal Chem 45(8):1331–1336. https://doi.org/10.1021/ac60330a025
18. Sabatier P (1911) Hydrogénations et déshydrogénations par catalyse. Ber Dtsch Chem Ges 44(3):1984–2001. https://doi.org/10.1002/cber.19110440303
19. Su H-Y, Gorlin Y, Man IC, Calle-Vallejo F, Norskov JK, Jaramillo TF, Rossmeisl J (2012) Identifying active surface phases for metal oxide electrocatalysts: a study of manganese oxide bi-functional catalysts for oxygen reduction and water oxidation catalysis. Phys Chem Chem Phys 14(40):14010–14022. https://doi.org/10.1039/C2CP40841D
20. Trasatti S (1972) Work function, electronegativity, and electrochemical behaviour of metals: III. Electrolytic hydrogen evolution in acid solutions. J Electroanal Chem 39(1):163–184. https://doi.org/10.1016/S0022-0728(72)80485-6
21. Trasatti S (2009) Electrochemical theory|Hydrogen Evolution A2 - Garche, Jürgen. In: Encyclopedia of electrochemical power sources. Elsevier, Amsterdam, pp 41–48. http://dx.doi.org/10.1016/B978-044452745-5.00022-8
22. Parsons R (1958) The rate of electrolytic hydrogen evolution and the heat of adsorption of hydrogen. Trans Faraday Soc 54:1053–1063. https://doi.org/10.1039/TF9585401053
23. Gerischer H (1958) Mechanismus der Elektrolytischen Wasserstoffabscheidung und Adsorptionsenergie von Atomarem Wasserstoff. Bull Soc Chim Belg 67(7–8):506–527. https://doi.org/10.1002/bscb.19580670714
24. Nørskov JK, Bligaard T, Logadottir A, Kitchin JR, Chen JG, Pandelov S, Stimming U (2005) Trends in the exchange current for hydrogen evolution. J Electrochem Soc 152(3):J23–J26
25. Nørskov JK, Bligaard T, Rossmeisl J, Christensen CH (2009) Towards the computational design of solid catalysts. Nat Chem 1(1):37–46
26. Cofre P, Sawyer DT (1986) Electrochemical reduction of dioxygen to perhydroxyl (HO_2.) in aprotic solvents that contain Broensted acids. Anal Chem 58(6):1057–1062. https://doi.org/10.1021/ac00297a017
27. Bungs M, Alonso-Vante N, Tributsch H (1990) Non-aqueous oxygen reduction electrochemistry of a transition metal chalcogenide cluster compound. Ber Bunsen-Ges Phys Chem 94:521–528
28. Wroblowa HS, Yen Chi P, Razumney G (1976) Electroreduction of oxygen: a new mechanistic criterion. J Electroanal Chem 69(2):195–201
29. Du C, Tan Q, Yin G, Zhang J (2014) 5—Rotating disk electrode method. In: Rotating electrode methods and oxygen reduction electrocatalysts. Elsevier, Amsterdam, pp 171–198. http://dx.doi.org/10.1016/B978-0-444-63278-4.00005-7
30. Jia Z, Yin G, Zhang J (2014) 6—Rotating ring-disk electrode method. In: Rotating electrode methods and oxygen reduction electrocatalysts. Elsevier, Amsterdam, pp 199–229. http://dx.doi.org/10.1016/B978-0-444-63278-4.00006-9
31. Alonso-Vante N (2010) Platinum and non-platinum nanomaterials for the molecular oxygen reduction reaction. ChemPhysChem 11(13):2732–2744
32. Anderson AB (2012) Insights into electrocatalysis. Phys Chem Chem Phys 14(4):1330–1338. https://doi.org/10.1039/C2CP23616H
33. Abild-Pedersen F, Greeley J, Studt F, Rossmeisl J, Munter TR, Moses PG, Skúlason E, Bligaard T, Nørskov JK (2007) Scaling properties of adsorption energies for hydrogen-containing molecules on transition-metal surfaces. Phys Rev Lett 99(1):016105
34. Man IC, Su H-Y, Calle-Vallejo F, Hansen HA, Martínez JI, Inoglu NG, Kitchin J, Jaramillo TF, Nørskov JK, Rossmeisl J (2011) Universality in oxygen evolution electrocatalysis on oxide surfaces. ChemCatChem 3(7):1159–1165. https://doi.org/10.1002/cctc.201000397

35. Nørskov JK, Rossmeisl J, Logadottir A, Lindqvist L, Kitchin JR, Bligaard T, Jonsson H (2004) Origin of the overpotential for oxygen reduction at a fuel-cell cathode. J Phys Chem B 108(46):17886–17892
36. Greeley J, Stephens IEL, Bondarenko AS, Johansson TP, Hansen HA, Jaramillo TF, Rossmeisl J, Chorkendorff I, Nørskov JK (2009) Alloys of platinum and early transition metals as oxygen reduction electrocatalysts. Nat Chem 1(7):552–556. http://www.nature.com/nchem/journal/v1/n7/suppinfo/nchem.367_S1.html
37. Trasatti S (1980) Electrocatalysis by oxides—attempt at a unifying approach. J Electroanal Chem 111(1):125–131. https://doi.org/10.1016/S0022-0728(80)80084-2
38. Ponce J, Rehspringer JL, Poillerat G, Gautier JL (2001) Electrochemical study of nickel–aluminium–manganese spinel $Ni_xAl_{1-x}Mn_2O_4$. Electrocatalytical properties for the oxygen evolution reaction and oxygen reduction reaction in alkaline media. Electrochim Acta 46 (22):3373–3380. https://doi.org/10.1016/s0013-4686(01)00530-8
39. Gorlin Y, Lassalle-Kaiser B, Benck JD, Gul S, Webb SM, Yachandra VK, Yano J, Jaramillo TF (2013) In situ x-ray absorption spectroscopy investigation of a bifunctional manganese oxide catalyst with high activity for electrochemical water oxidation and oxygen reduction. J Am Chem Soc 135(23):8525–8534. https://doi.org/10.1021/ja3104632
40. Suntivich J, Perry EE, Gasteiger HA, Shao-Horn Y (2013) The influence of the cation on the oxygen reduction and evolution activities of oxide surfaces in alkaline electrolyte. Electrocatalysis 4(1):49–55. https://doi.org/10.1007/s12678-012-0118-x
41. Hong WT, Risch M, Stoerzinger KA, Grimaud A, Suntivich J, Shao-Horn Y (2015) Toward the rational design of non-precious transition metal oxides for oxygen electrocatalysis. Energy Environ Sci 8(5):1404–1427. https://doi.org/10.1039/C4EE03869J
42. Busch M, Halck NB, Kramm UI, Siahrostami S, Krtil P, Rossmeisl J (2016) Beyond the top of the volcano?—A unified approach to electrocatalytic oxygen reduction and oxygen evolution. Nano Energy 29:126–135. https://doi.org/10.1016/j.nanoen.2016.04.011
43. Rossmeisl J, Qu ZW, Zhu H, Kroes GJ, Nørskov JK (2007) Electrolysis of water on oxide surfaces. J Electroanal Chem 607(1–2):83–89
44. Yeo RS, Orehotsky J, Visscher W, Srinivasan S (1981) Ruthenium-based mixed oxides as electrocatalysts for oxygen evolution in acid electrolytes. J Electrochem Soc 128(9): 1900–1904. https://doi.org/10.1149/1.2127761
45. Kötz R, Stucki S, Scherson D, Kolb DM (1984) In-situ identification of RuO_4 as the corrosion product during oxygen evolution on ruthenium in acid media. J Electroanal Chem 172(1):211–219. https://doi.org/10.1016/0022-0728(84)80187-4
46. Matsumoto Y, Sato E (1986) Electrocatalytic properties of transition metal oxides for oxygen evolution reaction. Mater Chem Phys 14(5):397–426. https://doi.org/10.1016/0254-0584(86)90045-3
47. Yoshida T, Kojima K (2015) Toyota MIRAI fuel cell vehicle and progress toward a future hydrogen society. Electrochem Soc Interf 24(2):45–49. https://doi.org/10.1149/2.F03152if
48. Garsany Y, Ge J, St-Pierre J, Rocheleau R, Swider-Lyons KE (2014) Analytical procedure for accurate comparison of rotating disk electrode results for the oxygen reduction activity of Pt/C. J Electrochem Soc 161(5):F628–F640. https://doi.org/10.1149/2.036405jes
49. Motoo S, Furuya N (1985) Electrochemistry of platinum single crystal surfaces: Part II. Structural effect on formic acid oxidation and poison formation on Pt (111), (100) and (110). J Electroanal Chem 184(2):303–316
50. Herrero E, Franaszczuk K, Wieckowski A (1994) Electrochemistry of methanol at low-index crystal planes of platinum—an integrated voltammetric and chronoamperometric study. J Phys Chem 98(19):5074–5083
51. Lamy C, Leger JM, Clavilier J, Parsons R (1983) Structural effects in electrocatalysis—a comparative-study of the oxidation of CO, HCOOH and CH_3OH on single-crystal Pt electrodes. J Electroanal Chem 150(1–2):71–77
52. Clavilier J, Faure R, Guinet G, Durand R (1980) Preparation of monocrystalline Pt microelectrodes and electrochemical study of the plane surfaces cut in the direction of the 111

and 110 planes. J Electroanal Chem 107(1):205–209. https://doi.org/10.1016/S0022-0728(79)80022-4

53. Huang S-Y, Chang C-M, Yeh C-T (2006) Promotion of platinum-ruthenium catalyst for electro-oxidation of methanol by ceria. J Catal 241(2):400–406
54. Alonso-Vante N (2008) Tailoring of metal cluster-like materials for the molecular oxygen reduction reaction. Pure Appl Chem 80(10):2103–2114
55. Wanjala BN, Fang B, Loukrakpam R, Chen Y, Engelhard M, Luo J, Yin J, Yang L, Shan S, Zhong C-J (2012) Role of metal coordination structures in enhancement of electrocatalytic activity of ternary nanoalloys for oxygen reduction reaction. ACS Catal 2(5):795–806. https://doi.org/10.1021/cs300080k
56. Kim J-H, Chang S, Kim Y-T (2014) Compressive strain as the main origin of enhanced oxygen reduction reaction activity for Pt electrocatalysts on chromium-doped titania support. Appl Catal B: Environ 158–159:112–118. https://doi.org/10.1016/j.apcatb.2014.04.003
57. Vogel W, Timperman L, Alonso-Vante N (2010) Probing metal substrate interaction of Pt nanoparticles: Structural XRD analysis and oxygen reduction reaction. Appl Catal A-Gen 377:167–173
58. Pedersen AF, Ulrikkeholm ET, Escudero-Escribano M, Johansson TP, Malacrida P, Pedersen CM, Hansen MH, Jensen KD, Rossmeisl J, Friebel D, Nilsson A, Chorkendorff I, Stephens IEL (2016) Probing the nanoscale structure of the catalytically active overlayer on Pt alloys with rare earths. Nano Energy 29:249–260. https://doi.org/10.1016/j.nanoen.2016.05.026
59. Feng Y, Alonso-Vante N (2012) Carbon-supported cubic $CoSe_2$ catalysts for oxygen reduction reaction in alkaline medium. Electrochim Acta 72:129–133. https://doi.org/10.1016/j.electacta.2012.04.003
60. Bard AJ, Faulkner LR (2002) Electrochemical methods: fundamentals and applications, 2nd edn. Wiley, New York
61. Kibler LA (2003) Preparation and characterization of noble metal single crystal electrode surfaces
62. Wolfschmidt H, Baier C, Gsell S, Fischer M, Schreck M, Stimming U (2010) STM, SECPM, AFM and electrochemistry on single crystalline surfaces. Materials 3(8):4196
63. Biegler T, Rand DAJ, Woods R (1971) Limiting oxygen coverage on platinized platinum; relevance to determination of real platinum area by hydrogen adsorption. J Electroanal Chem 29(2):269–277. https://doi.org/10.1016/S0022-0728(71)80089-X
64. Wagner FT, Ross PN Jr (1983) Leed analysis of electrode surfaces: structural effects of potentiodynamic cycling on Pt single crystals. J Electroanal Chem 150(1–2):141–164. https://doi.org/10.1016/S0022-0728(83)80198-3
65. Climent V, Feliu J (2011) Thirty years of platinum single crystal electrochemistry. J Solid State Electrochem 15(7):1297–1315. https://doi.org/10.1007/s10008-011-1372-1
66. Feliu JM, Herrero E (2011) Surface electrochemistry and reactivity. Contrib Sci 6:161–172
67. Marković NM, Ross PN Jr (2002) Surface science studies of model fuel cell electrocatalysts. Surf Sci Rep 45(4–6):117–229. https://doi.org/10.1016/S0167-5729(01)00022-X
68. Schmidt TJ, Ross PN, Markovic NM (2001) Temperature-dependent surface electrochemistry on Pt single crystals in alkaline electrolyte: part 1: CO oxidation. J Phys Chem B 105(48):12082–12086. https://doi.org/10.1021/jp0124037
69. Strmcnik D, Lopes PP, Genorio B, Stamenkovic VR, Markovic NM (2016) Design principles for hydrogen evolution reaction catalyst materials. Nano Energy 29:29–36. https://doi.org/10.1016/j.nanoen.2016.04.017
70. Strmcnik D, Uchimura M, Wang C, Subbaraman R, Danilovic N, van der V, Paulikas AP, Stamenkovic VR, Markovic NM (2013) Improving the hydrogen oxidation reaction rate by promotion of hydroxyl adsorption. Nat Chem 5(4):300–306. http://www.nature.com/nchem/journal/v5/n4/abs/nchem.1574.html#supplementary-information
71. Durst J, Siebel A, Simon C, Hasche F, Herranz J, Gasteiger HA (2014) New insights into the electrochemical hydrogen oxidation and evolution reaction mechanism. Energy Environ Sci 7(7):2255–2260. https://doi.org/10.1039/C4EE00440J

72. Bonakdarpour A, Dahn TR, Atanasoski RT, Debe MK, Dahn JR (2008) H_2O_2 release during oxygen reduction reaction on Pt nanoparticles. Electrochem Solid-State Lett 11(11): B208–B211
73. Bonakdarpour A, Lefevre M, Yang R, Jaouen F, Dahn T, Dodelet J-P, Dahn JR (2008) Impact of loading in RRDE experiments on Fe–N–C catalysts: two- or four-electron oxygen reduction? Electrochem Solid-State Lett 11(6):B105–B108
74. Bonakdarpour A, Delacote C, Yang R, Wieckowski A, Dahn JR (2008) Loading of Se/Ru/C electrocatalyst on a rotating ring-disk electrode and the loading impact on a H_2O_2 release during oxygen reduction reaction. Electrochem Commun 10(4):611–615
75. Inaba M, Yamada H, Tokunaga J, Tasaka A (2004) Effect of agglomeration of Pt/C catalyst on hydrogen peroxide formation. Electrochem Solid-State Lett 7(12):A474–A476
76. Timperman L, Feng YJ, Vogel W, Alonso-Vante N (2010) Substrate effect on oxygen reduction electrocatalysis. Electrochim Acta 55(26):7558–7563
77. Gasteiger HA, Kocha SS, Sompalli B, Wagner FT (2005) Activity benchmarks and requirements for Pt, Pt-alloy, and non-Pt oxygen reduction catalysts for PEMFCs. Appl Catal B 56(1–2):9–35
78. Alonso-Vante N, Colell H, Stimming U, Tributsch H (1993) Anomalous low-temperature kinetic effects for oxygen evolution on RuO_2 and Pt electrodes. J Phys Chem 97(29): 7381–7384
79. Pedersen CM, Escudero-Escribano M, Velázquez-Palenzuela A, Christensen LH, Chorkendorff I, Stephens IEL (2015) Benchmarking Pt-based electrocatalysts for low temperature fuel cell reactions with the rotating disk electrode: oxygen reduction and hydrogen oxidation in the presence of CO (review article). Electrochim Acta 179:647–657. https://doi.org/10.1016/j.electacta.2015.03.176
80. Debe MK (2012) Electrocatalyst approaches and challenges for automotive fuel cells. Nature 486(7401):43–51
81. Timperman L, Luo Y, Alonso-Vante N (2016) On the availability of active sites for the hydrogen peroxide and oxygen reduction reactions on highly dispersed platinum nanoparticles. ChemElectroChem 3(10):1705–1712. https://doi.org/10.1002/celc.201600144

Chapter 3
Environmental Catalysis

3.1 Introduction

Pollution abatement and pollution prevention enters into the field of environmental catalysis. Over the years, a series of technologies, e.g., catalytic sensors, fuel cell systems, photocatalytic reactors, aiming at monitoring and controlling these processes, have been developed and succinctly summarized in the literature [1, 2]. Within the "green chemistry hub," the use of light as a reagent applied to heterogeneous photocatalytic systems, for practical applications such as water decontamination, water splitting, carbon dioxide reduction, represents a paramount challenge to pursue [3–5] due to the environmental benefits it entails.

Sunlight is a unique natural, non-expensive, abundant, and endlessly renewable source of energy. For example, solar photons can be converted straightforward to heat, as shown in Fig. 3.1. This heat can be further converted, by indirect means, e.g., mechanical, or by direct means via photovoltaic devices to electricity, and/or directly to fuel generation (better known as solar fuels) by means of, e.g., photoelectrochemical systems (PECs). For the electricity and fuel production (e.g., hydrogen, CO_2 reduction products), one needs semiconducting materials able to capture efficiently the visible fraction of the solar spectrum (400–800 nm). Besides, a precondition of such semiconducting materials is their reasonable abundance on the earth's crust, with a low environmental impact through their life cycle, and stability with time under light irradiation, in the operating medium. Herein, we concentrate on some basic principles of artificial photosynthesis [6–9], but definitely, the challenge nowadays still remains to split water into hydrogen and oxygen using solar visible photons. Aiming at mimicking the natural photosynthesis, UV-sensitive oxide semiconductors were used for the photoelectrolysis of water. Boddy [10] reported for the first time the evolution of oxygen under illumination on n-TiO_2 (rutile) single crystals in 1968, preceding the well-known paper by Fujishima and Honda in 1972 [11]. The results of these experiments are, certainly, the basis for the heterogeneous

N. Alonso-Vante, *Chalcogenide Materials for Energy Conversion*, Nanostructure Science and Technology, https://doi.org/10.1007/978-3-319-89612-0_3

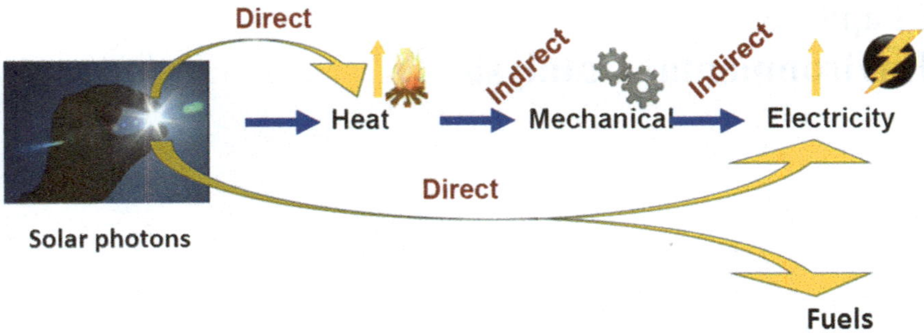

Fig. 3.1 A schematic representation of the direct and indirect conversions of solar photons

photocatalysis, accomplished in a photoelectrochemical cell (PEC) system. They also paved, since then, the long way up to the development of the artificial leaf [12–15], aiming at storing high-energy density chemical fuels for the future.

3.2 The Basics of Semiconductor Photoelectrochemistry

The understanding of semiconductor photoelectrochemistry goes back to the works devoted to the effect of light onto well-defined single-crystal materials in contact with electrolyte (solid/electrolyte junctions) since the 1960s and summarized in the 1980s by some pioneers [16–18]. Although some systems showed interesting performance (photon to electricity conversion), most of them failed due to the instability against (photo) corrosion and cost combined with a poor efficiency. The field of photoelectrochemistry was renewed thanks to the development of semiconducting materials in the nanoscale domain, as reported in the 1990s by O'Regan and Grätzel [19]. Nanostructured materials are now the object of research for light-driven water splitting and environmental remediation, linking the electrocatalysis at nanoparticulated semiconductors and semiconductor photoelectrochemistry.

3.2.1 Semiconductor Properties

The Gibbs free energy stored in the semiconductor, Fig. 3.2a, resulting from the photon absorption ($h\nu \geq E_g$), Fig. 3.2b, depends on the band gap of the material, E_g, to free the internal energy of electron–hole pairs, and further to their e^--h^+ pair concentration (n, p) release or recombination, Fig. 3.2c. The driving force is the photon absorption process that leads to the splitting of the Fermi level, i.e., the work per photon (photovoltage) $eV_{ph} = \Delta E_F$. This latter is equivalent to the quasi-Fermi levels (QFL) for electrons and holes, $E_{F,n}$ and $E_{F,p}$, respectively, Fig. 3.2b so that the incident photon energy $h\nu > E_g > \Delta E_F$. The QFL are given by the expressions:

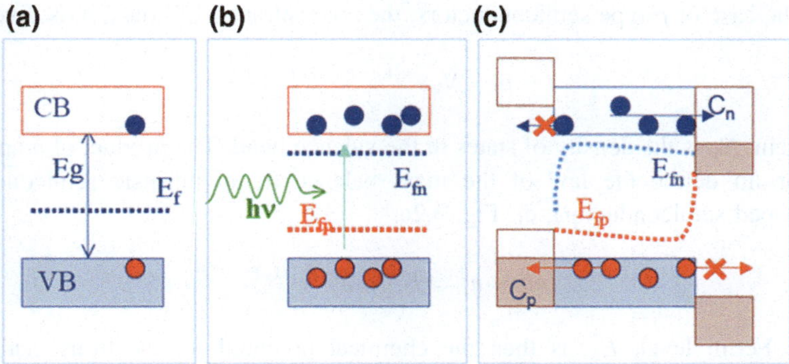

Fig. 3.2 Elementary step of photoabsorption by an intrinsic semiconductor (**a**), producing the work in (**b**) via the separation of the quasi-Fermi levels, a state of quasi-equilibrium or thermalization, and to charge separation or recombination in (**c**)

$$E_{F,n} = E_C + k_B T \ln \frac{n}{N_C} \tag{3.1a}$$

$$E_{F,p} = E_V - k_B T \ln \frac{p}{N_V} \tag{3.1b}$$

where n and p are electron and hole concentrations, and N_C and N_V are, respectively, the densities of states of the conduction and of the valence bands [20]. The shift of the quasi-Fermi level with absorption of light highlights the importance of this parameter, which defines the concentration of majority carriers (electrons and/or holes in a semiconductor) obtained by the effect of doping the material. If we consider an n-type semiconductor, the concentration of electrons, n, under equilibrium thermal conditions is given by the Fermi–Dirac equation:

$$n = N_C f(E) \tag{3.2}$$

where $f(E)$ is the Fermi–Dirac distribution probability, and N_C the density of states in the conduction band.

$$f(E) = \frac{1}{1 + e^{(E_C - E_F)/k_B T}} \tag{3.3}$$

For $(E_C - E_F)$ ca. 0.05 eV or a doping level of $\leq 10^{18}$ cm^{-3}, Eq. (3.2) can be approximated by the Boltzmann equation.

$$n = N_C e^{-(E_C - E_F)/k_B T} \tag{3.4}$$

In the case of *p*-type semiconductors, the concentration of holes, *p*, is given by:

$$p = N_V e^{-(E_F - E_V)/k_B T} \quad (3.5)$$

Herein, N_V is the density of states in the valence band. The product of *n* and *p* at equilibrium define the law of the mass balance in an intrinsic semiconductor (non-doped semiconductor), cf. Fig. 3.2a.

$$np = n_i^2 = N_C N_V e^{-(E_g)/k_B T} \quad (3.6)$$

The Fermi level, E_F, is then the chemical potential for electrons, acting as donors, N_D, in n-type semiconductors, and as an acceptor, N_A, levels in *p*-type semiconductors, and also known as the electrochemical potential for electrons, $\tilde{\mu}_{e^-}$. Equations (3.4) and (3.5) define the concept of the Fermi level at equilibrium in darkness and are similar to Eqs. (3.1a), and (3.1b). These latter define a system under a steady-state illumination, where photogenerated electrons and holes are a concentration of photostationary electrons and holes (*n*, *p*), far from their initial equilibrium, and in thermal equilibrium with the lattice. Such a concept, used to describe non-equilibrium processes at solid-state junctions, was applied for a quantitative analysis of charge transfer processes at *n*- and *p*-type GaAs electrodes in photoelectrochemistry [21, 22]. The bottom of the conduction band and the top of the valence band for *n*-type semiconductor are schematized in Fig. 3.3a compared to a metal in Fig. 3.3b, showing other usual energy parameters that allow to judge the properties of the semiconducting materials summarized in Fig. 3.4. Most of the oxide materials possess band gaps larger than 3 eV (< 413.3 nm), meaning that these materials are able to split water using photons in the ultraviolet (UV) region ($\lambda < 400$ nm). On the other hand, chalcogenides (e.g., CdS, MoS_2, CdSe) with gap energies (1.9–2.4 eV or 516.6–652.5 nm) within the solar spectrum are excellent photocatalyst materials for the hydrogen evolution.

The band edges of the oxide semiconductors usually follow the Nernst law with pH, i.e., a shift of −0.059 V/pH, at room temperature [23]. This means that the adsorption–desorption process of protons and hydroxides governs the double-layer capacitance at the SC/El junction, $C_H = q/V_H$. (q: the charge density $\mu C/cm^2$; V_H: the Helmholtz potential). For oxide surfaces, the two sites of a hydroxylated surface: M^+-OH^- should be equivalent to the neutral surface: M–OH, otherwise the site can give proton to the solution (3.7a), or accept a proton from the solution (3.7b):

$$\text{M-OH} = \text{M-O}^- + \text{H}^+ \quad (3.7a)$$

$$\text{H}^+ + \text{M-OH} = \text{M-OH}_2^+ \quad (3.7b)$$

The free energy of reaction (3.7b) must include the factor $-eV_H$ (Helmholtz potential) that should be overcome by the protons to be adsorbed, or $+eV_H$ if released (3.7a). Rearranging (3.7a) and (3.7b), we have, respectively:

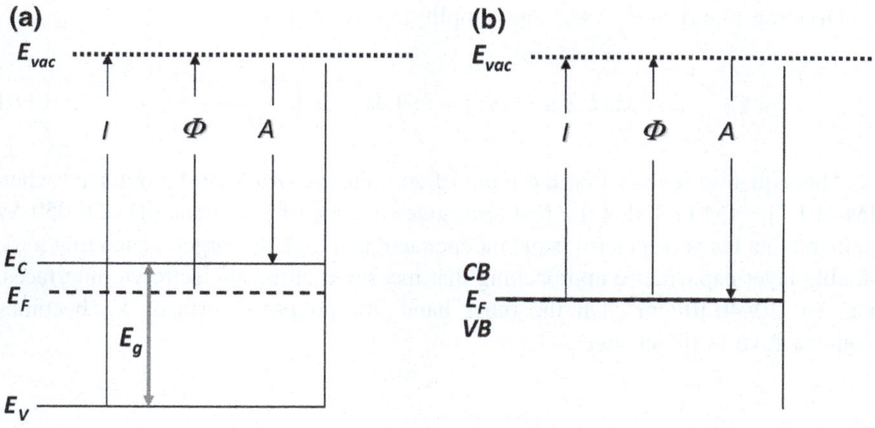

Fig. 3.3 **a** Energy diagram for an n-type semiconductor; E_{vac}—vacuum energy level. E_c—conduction band edge energy; E_v—valence band edge energy; E_g—energy gap; A—the electron affinity; I—the ionization energy; Φ—the work function. (**b**) Energy diagram for a metal with empty states (conduction band—CB) and filled states (valence band–VB). (E_F is the Fermi energy)

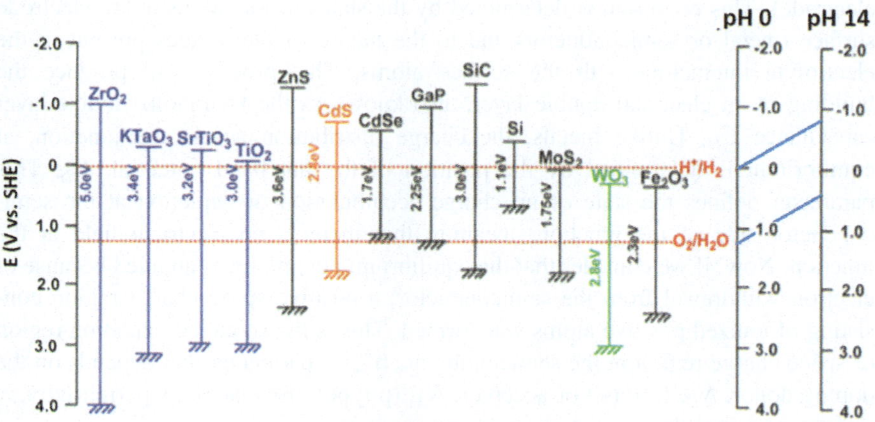

Fig. 3.4 Band–edge positions of oxide, carbide, chalcogenide semiconductors with respect to the redox potentials of water, at pH = 0, and pH = 014. Figure adapted with permission from reference [32] Copyright © 2008, Royal Society of Chemistry

$$\frac{[M-O^-][H^+]}{[M-OH]} = e^{-\frac{(\Delta G_a - eV_H)}{k_B T}} \tag{3.8a}$$

$$\frac{[M-OH_2^+]}{[H^+][M-OH]} = e^{-\frac{(\Delta G_b + eV_H)}{k_B T}} \tag{3.8b}$$

Dividing (3.8a) over (3.8b) and simplifying, we obtain:

$$eV_H = 2.303k_BT(pzc - pH) + 2.303k_BT\ln\left\{\frac{[M - OH_2^+]}{[M - O^-]}\right\}^{1/2} \tag{3.9}$$

This equation reveals that the point of zero charge (*pzc*) can be obtained when $[M\text{-}OH_2^+] = [M\text{-}O^-]$, that the first term gives a slope of V_H versus pH of 0.059 V/pH, and that the second term is pH independent and highly charged generating a C_H double-layer capacitance approaching that measured at metal/electrolyte interfaces, i.e., ca. 20–40 μF/cm^2. On the other hand, for nonpolar surfaces V_H becomes non-sensitive to pH changes.

3.2.2 Semiconductor/Electrolyte Junction

When a semiconductor is immersed in an electrolyte, it will necessarily reach (similar to metals) the establishment of a potential until equilibrium between both phases is attained (equilibrium potential measured with respect to a reference electrode). This condition is determined by the states of the atoms at the electrode surface (metal or semiconductor) and to the nature of the species present in the electrolyte interacting with the surface atoms. This process will produce the building of an electrical double layer, also known as the Helmholtz double-layer capacitance C_H. Unlike metals, the charge distribution across the junction, at equilibrium, is determined by the position of the flat band potential, E_{fb}. This parameter defines the state of no charge accumulation or depletion at the semiconductor side of the junction, meaning that there is no electrical field at the junction. Now, if we consider that the equilibrium potential was attained because of electron withdrawal from the semiconductor, a positive space charge region consisting of ionized positive atoms was formed. This is the so-called depletion region or space charge region in the semiconductor, W_{SC}, a parameter that depends on the doping donor, N_D, (n-type) or acceptor, N_A, (p-type), the material's permittivity, ε, and the potential drop at the space charge region.

$$W_{SC} = \left(\frac{2\varepsilon\varepsilon_0\Delta\phi_{SC}}{eN_D}\right)^{1/2} \tag{3.10}$$

The space charge region generates a linear electric field variation, E, from the surface (x = 0) to a distance x = W_{SC}. The field decreases to zero at x=W_{SC}. Therefore,

$$E_{\max} = \frac{eN_D.W_{SC}}{2\varepsilon\varepsilon_0} \tag{3.11}$$

Furthermore, the charge density at the space charge region, ρ, can be calculated by:

$$\rho = eN_D W_{SC} = (2\varepsilon\varepsilon_0 eN_D \Delta\phi_{SC})^{1/2} \tag{3.12}$$

The donor charge density, ρ, the electric field, E, and the electric potential $\Delta\phi$, are contrasted with the semiconductor/electrolyte junction in Fig. 3.5. The linear variation of the electric field corresponds to a parabolic variation of the electrical potential $\Delta\phi_{SC}$. This variation reflects the band bending at the semiconductor/electrolyte junction. If we assume, as an example, ε=10, and $\Delta\phi$ = 0.7 V, W_{SC}, E_{max}, and ρ can be calculated from Eqs. (3.10)–(3.12). The data are summarized in Table 3.1.

The positive charge in the space charge region in the semiconductor must be balanced by the opposite negative charge in the electrolyte, located at the outer Helmholtz plane of the electrical double layer, C_H. The variation of the space charge region (SCR) further defines the space charge capacitance, i.e.,

Fig. 3.5 A general scheme showing thermal equilibrium of (**a**) the space charge distribution; **b** the corresponding electric field distribution; **c** the electrostatic potential distribution; **d** the interface of the SC/El junction at band bending condition in darkness showing the ionized charges in the space charge region (SCR) with corresponding counter-ionic charge from the electrolyte; and **e** under illumination condition

Table 3.1 Variation of W_{SC}, $E_{\max}$, and ρ, as a function of the donor concentration, N_D

N_D (cm^{-3})	10^{14}	10^{16}	10^{18}	10^{20}
W_{SC} (nm)	2780	278	27.8	2.78
E_{max} (V/cm)	5×10^3	5×10^4	5×10^5	5×10^6
ρ (e/cm^2)	2.8×10^8	2.8×10^9	2.8×10^{10}	2.8×10^{11}

$$C_{SC} = \frac{d\rho}{d\Delta\phi_{SC}} \tag{3.13}$$

The dependence of which with the applied electrode potential is given by the Mott–Schottky equation [23], e.g., for *n*-type semiconductors:

$$\frac{1}{C_{SC}^2} = \frac{2}{eN_D\varepsilon\varepsilon_0}\left(\Delta\phi_{SC} - \frac{k_BT}{e}\right) \tag{3.14}$$

The applied electrode potential at the semiconductor/electrolyte junction will vary Δ_ϕSC at will, so that any change in the applied electrode potential should appear across the SCR. The capacity will extrapolate to infinity, when Δ_ϕSC = k_BT/e. The flat band potential E_{fb} will be read extrapolated to $C_{SC}^2 = 0$. E_{fb} means the actual position of the Fermi level of the semiconductor with respect to a reference electrode before equilibrium, and it provides the conduction and valence band energies' position, providing in this manner the thermodynamic feasibility of the semiconductor to reduce and/or oxidize a given species in the electrolyte under illumination, cf. Fig. 3.3. The additional benefit of this experiment is the determination of the majority carrier concentration, N_D, directly from the slope. It is important to recall that at the SC/El junction the capacitance of the semiconductor, C_{SC}, is in series with the capacitance of the double layer, C_H. This latter parameter is usually of the order of 20–40 μF/cm^2 and thus much higher (one- to two-orders of magnitude) than the values expected for a reasonable moderately doped semiconductor [16].

The photons impinging on the n-type semiconductor with energy of $h\nu \geq E_g$ produce electron–hole pairs separated by the electric field in the SCR, Fig. 3.5e. Electrons are transported to the rear ohmic contact in the way to the external circuit, whereas holes are transported to the semiconductor/electrolyte junction. Under illumination, the Fermi level rises producing a photopotential, eV_{ph}, and a concomitant splitting of the original Fermi level into two quasi-Fermi levels (QFL) related to the concentration of electrons, $E_{F,n}$, and the other to the concentration of holes, $E_{F,p}$, cf. Equation (3.1a), (3.1b). As noted in Sect. 3.2.1, if this free energy of photoexcited carriers is suitable, reduction via electrons and oxidation via holes will be driven separately. Indeed, the oxidation reaction, prevented in darkness Fig. 3.5d, takes easily place under illumination: *Red* + h^+ = *Ox* in the valence band, as far as the process of electron–hole recombination in the SCR is not favored. As also noted in Chap. 1, oxidation and reduction reactions are generally submitted to an overpotential, η, to efficiently drive the process due to kinetics constraints.

The lifetime of the minority carriers (the holes) limits the quantum efficiency, or the incident photon-to-current conversion efficiency (*IPCE*). This parameter is the ratio of the measured electron flux (current density, j_{ph}) in the external circuit to the incident photon flux, I_0 (at a wavelength, λ). As shown in Fig. 3.2c, if there is no vectorization (charge transfer) of the photogenerated charges, recombination will be the preferred path, otherwise electrons and holes will serve to promote reduction and oxidation processes of species in intimate contact with the semiconductor surface.

$$ICPE = \frac{j_{ph}}{eI_0} = \frac{j_{ph}(mA/cm^2) \times 1239.8(V/nm)}{P_\lambda(mW/cm^2) \times \lambda(nm)} = \frac{electrons\ cm^{-2}s^{-1}}{photons\ cm^{-2}s^{-1}} \quad (3.15)$$

It is worth to note that lower values of *ICPE* are obtained as those predicted by a model developed by Gärtner [24], and the reason is due to the e^-–h^+ pair recombination in the SCR, and additionally in the material's bulk (neutral region). Strictly speaking, the *ICPE* must be collected at zero bias conditions, i.e., two-electrodes in short circuit. Thus, all reported ICPE under a determined bias should be considered as a diagnostic tool to compare material properties or understand material performance [25]. Summing up, *ICPE* considers photon absorption, leading to the fraction of electron–hole pairs generated per incident photon [as defined in Eq. (3.15)], but additionally, one must also consider the charge transport to the interface and the charge efficiency, cf. Fig. 3.5e.

3.2.2.1 The Fluctuating Energy Model at the Semiconductor/Electrolyte Junction

Figure 3.3 contrasts the various materials' properties for a semiconductor and a metal prior to form a contact with, e.g., electrolyte (solid/electrolyte junction). In both cases, the electrochemical potential of the electron is defined by the position of the Fermi level, which states the boundary between the conduction band and valence band in a metal, and the doping level of the *n*-type or *p*-type semiconductor in the band gap. The magnitude of the latter separates the valence band from the conduction band. All these parameters are, naturally, defined by the electronic structure of the element and/or compound. A similar approach to this statement was made to the redox species in an electrolyte, doing likeness to the solid state with empty levels (conduction band) and filled levels (valence band) with the oxidized state (Ox) and the reduced state (Red), respectively [26], and with reference to the vacuum level, the energy of the redox couple can be given by:

$$E_{F,redox} = E^0_{F,redox} - \frac{k_B T}{ne} \ln \frac{C_{Ox}}{C_{Red}} = -4.5eV - eE_{redox} \quad (3.16)$$

The correspondence of the electrochemical potential versus the standard hydrogen electrode scale (SHE), cf. Fig. 2.4, on the vacuum scale is given as

−4.5 eV. From this concept, the electrons (which are not free to move in the liquid phase) are localized on the charged ions (anions and cations) with different energies as a consequence of their interaction with the solvent and change with time due to fluctuations of the dipoles of the solvents, i.e., the fluctuating energy model of the Marcus theory [27–29]; and Gerischer's [30, 31], known as the Marcus–Gerischer Model.

The density of states *D*(*E*) for the Ox and Red species is, respectively, given as:

$$D_{Ox}(E) = C_{Ox}W_{Ox}(E) \tag{3.17}$$

$$D_{\mathrm{Red}}(E) = C_{\mathrm{Red}}W_{\mathrm{Red}}(E) \tag{3.18}$$

C_{Ox} and C_{Red} are, respectively, the concentrations of the oxidized and reduced species; W_{Ox} and W_{Red} are, respectively, the probability to find an empty (oxidized) state or an occupied (reduced) state. The electronic level has a Gaussian distribution shape of *W*(*E*):

$$W(E) = (\pi 4k_BT)^{-1/2}\exp\left(-\frac{(E - E^0)^2}{4\lambda k_BT}\right) \tag{3.19}$$

In the exponential term, is the reorganization energy for the corresponding species. The density of states of the species in the electrolyte shape is contrasted for a semiconductor/electrolyte, and a metal/electrolyte junction, Fig. 3.6. This approach is ingenious, on the one hand, for semiconductor electrodes because the electrons can be transferred via the conduction band or the valence band, and is

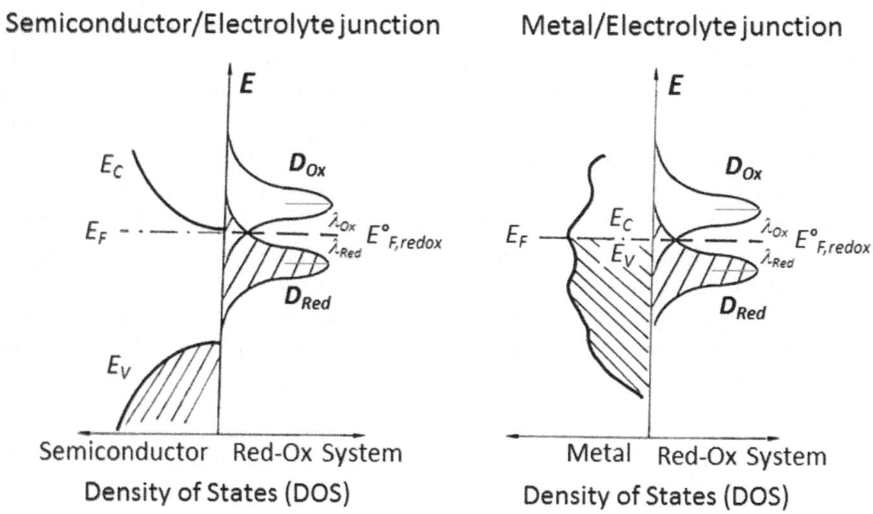

Fig. 3.6 Density of states at SC/electrolyte and M/electrolyte junctions deduced from reference [31]

valid for non-adiabatic processes—weak interactions; on the other hand, with the model it is possible to easily predict the band energy involved in the process of charge transfer.

3.2.3 Photoelectrolysis of Water

Putting together all the elements that constitute a photoelectrochemical cell (PEC), i.e., a semiconductor and a metal counter-electrode, for the water splitting process, it is clear that a photovoltage, $eV_{ph} < E_g$, will not be sufficient to release the stored free energy since the energy needed to split water into hydrogen and oxygen is $\Delta E_F > 1.229$ eV; therefore, the use of higher band gap materials to ensure an adequate photovoltage as schematically shown in Fig. 3.7 with a single *n*-type semiconductor is required. As assessed in Chap. 2, interfacial charge transfer kinetics for the water splitting reaction is sluggish (overpotential for holes ($e\eta_h$) and electrons ($e\eta_e$), particularly at the oxygen electrode. Besides, such losses are also associated with free energy losses within the semiconductor (recombination, surface states, low charge carriers' mobility, etc.). Such macroscopic phenomena are usually encountered in massif materials; therefore, the necessity to further engineer materials chemistry to tailor materials in the nanoscale range, in which such microscopic phenomena are further mitigated, is a strategy nowadays. One interesting example is the use of oxide semiconductors; however, due to the small crystalline domains and high defect and impurity concentrations, the metal oxides generally show an important energy loss. As cited in Sect. 2.2 (Table 2.1), the electrolysis of water requires energy of 237.1 kJ/mol, or 1.229 V. However, the production of hydrogen and oxygen in the gas phase goes together with an increased entropy change, so that the thermoneutral voltage is calculated from the standard enthalpy change, i.e., 285.8 kJ/mol or 1.48 V. Indeed, in commercial electrolyzers applied voltages, to overcome the associated kinetics and ohmic drop, range between 1.7 and 1.8 V. This consideration is also applied in

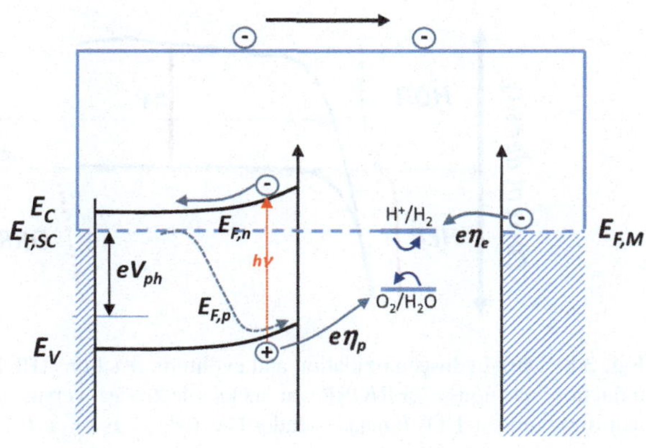

Fig. 3.7 Energy diagram of a single band gap n-type semiconductor in a photoelectrochemical cell for water splitting. Ohmic contact with the semiconductor is shown. A connection between the ohmic contact and the counter-electrode completes the circuit

photoelectrolysis, with the condition that a single semiconductor band edges straddle the redox energy of water: $E^0_{H^+/H_2}$ and $E^0_{O_2/H_2O}$, cf. Fig. 2.4. Following this latter scheme, Fig. 3.4 recaps some semiconductor materials reported so far [32] showing on the one hand, their relative band edge positions (E_C, and E_V), and on the other hand, their band-gap width (E_g). Few of them fulfill the requirement by having the bottom of the conduction band more negative than the (H^+/H_2) redox potential, and the top of the valence band more positive than the (O_2/H_2O) redox potential.

Putting together the dark process on metals (electrocatalysis) with the light process (photoelectrocatalysis), using nanostructured n-TiO_2 (anatase) under UV-illumination on the same electrochemical potential scale, see Fig. 3.8, we can appreciate the important potential shift of the onset oxidation potential of water to molecular oxygen in acid medium compared to, say the best OER catalyst RuO_2, cf. Fig. 2.12, as a result of the hole energetics at the titanium dioxide valence band. Although reported in 1972 [11], the product was detected online in the 1990s using the mass spectrometry (DEMS) coupled with the PEC system [33].

In the field of solar water splitting, the solar-to-hydrogen (STH) efficiency has been estimated for single absorbers [25], as shown in Fig. 3.9. The STH describes the overall efficiency in a PEC water splitting device exposed to a solar Air-Mass 1.5 (AM 1.5) irradiation in the absence of any applied bias (i.e., two-electrode systems short circuited) in the same pH electrolyte. Under this circumstance, the STH efficiency is defined as the chemical energy (rate of hydrogen—mmol/s) produced with respect to the solar energy input, P, closely matched to the intensity of Air-Mass 1.5 Global (AM 1.5 G) standard [34].

$$STH = \left[\frac{(mmolH_2/s) \times (237.1kJ/mol)}{P(mW/cm^2) \times A(cm^2)}\right]_{\mathrm{AM1.5G}} \tag{3.20}$$

It is recalled that in Eq. (3.20), the free energy at 298 K is used instead of the standard enthalpy change. Otherwise, if an applied bias potential, E_{bias}, is used, one

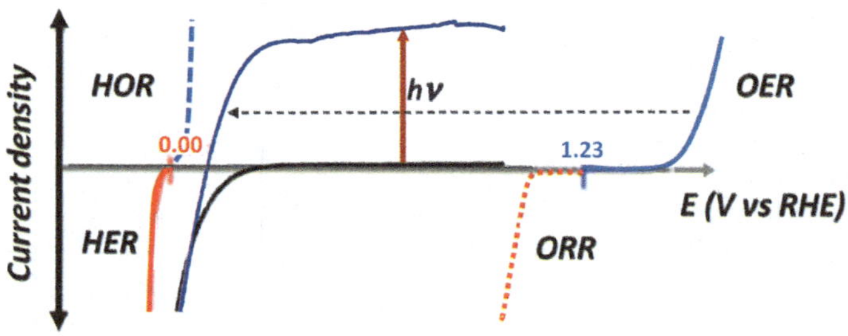

Fig. 3.8 Dark hydrogen oxidation and evolution reactions (HOR/HER) and oxygen evolution and reduction reactions (OER/ORR) at metal electrodes versus light processes. The OER, at the nanostructured n-TiO_2 (anatase) under UV-light, sets on at 0.1 V/RHE in 0.5 M H_2SO_4

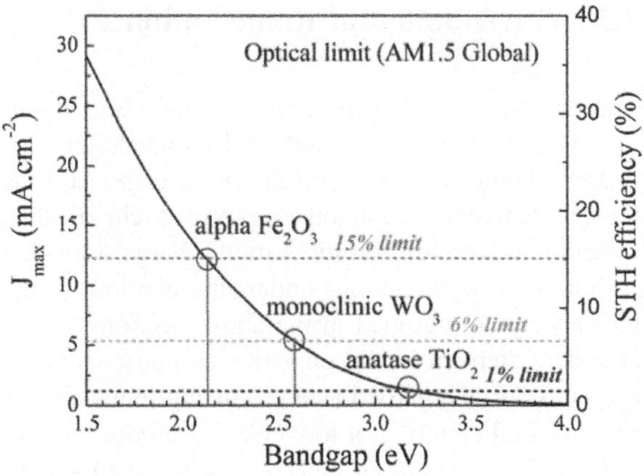

Fig. 3.9 Theoretical maximal photocurrent j_{max}, and solar to hydrogen (STH) as a function of material band gap, for various oxides: α-Fe_2O_3, monoclinic WO_3, and anatase TiO_2, with permission from reference [25] Copyright © Materials Research Society 2010

exceeds the thermodynamic water splitting potential of 1.229 V, and the photo-generated current is directly related to the production of hydrogen. In this sense, Eq. (3.20) can be rewritten as:

$$STH_{bias} = \left[\frac{j_{ph}(mA/cm^2) \times (1.229 - E_{bias})(V)}{P(mW/cm^2)} \right]_{\text{AM1.5G}} \tag{3.21}$$

Errors to avoid when reporting STH have been discussed [25]. Under these premises, and for the sake of comparing such a parameter with one single absorber (semiconductor) Fig. 3.9 contrasts the theoretical STH as a function of its optical band gap. The trend is to have narrower band gap materials with a concomitant photovoltage and energy band position to enhance the STH factor, cf. Fig. 3.4. From the data collected in the vast literature, materials having a gap energy between 1.5 and 2.5 eV can be considered as interesting candidates.

Although water splitting is a promising path for a sustainable hydrogen production and solar energy storage, the hurdle to overcome is the cost of hydrogen production. Furthermore, most single photoelectrodes cannot split water, and adding complexity to the system, one must provide an additional bias to split water [35]. Besides, complications will be further encountered to separate hydrogen and oxygen produced in a classical PEC. This problem is in part solved in conventional electrolyzer structures, which use ion exchange membranes to avoid H_2/O_2 cross-over. A feasible approach consists of separating anode and cathode compartments, coupling a redox mediator in each compartment, and adding a photovoltaic (PV) device [36]. The key issue with the redox mediators (working as inner battery-like system) in each separate compartment is the fact that they provide the chemical potential to generate H_2 and O_2 evolution processes at the cathode and anode. This is, in fact, wiring PV with an electrolysis cell and/or photoelectrolysis cell [37].

3.3 Environmental Remediation Principle

The general definition of environmental remediation is to reduce all kinds of pollutants present in the air, soil, and groundwater or surface water. The word remediation should not be confused with a response to an accident. The issue is huge; thus, we will concentrate on the physical–chemical aspects using the current state of materials research to promote multi-charge transfer reactions (e.g., water splitting, carbon dioxide reduction) under illumination using the physical–chemical principles succinctly outlined in the above sections of this chapter. Hydrogen, the most abundant element in the universe, combusts with oxygen to deliver water, and in this sense, it is regarded as a promising renewable energy source, which should be produced efficiently at a low cost. Its production nowadays is via reforming fossil fuels with water vapor ($CH_4 + H_2O \rightarrow CO + 3H_2$), or via gasification of coal under oxygen control ($3C + O_2 + H_2O \rightarrow 3CO + H_2$) [38]. However, the excessive use of fossil fuels perturbs the environment through the production of toxic species: carbon oxides (CO_x), nitrogen oxides (NO_x), sulfur oxides (SO_x), organic compounds (C_xH_y), particulate matter (soot), and metals [39]. So far, burning fossil fuels is a predominant activity for the hydrogen production. However, an alternative, as discussed in the sections above, is the use of photons as chemical precursors and the use of abundant materials to drive the hydrogen production. This is the inevitable energy transition to the renewable sources of energy like, wind, water, sun, and biomass to be used for thermal, electrical, photonic, and biochemical energy for the hydrogen generation via the water splitting process. The major driving force to generate hydrogen is the thermal energy through the steam-reforming method. In comparison, the water electrolysis through electrical energy represents only 4% of the hydrogen production [40]. Here, again the electricity production powered by renewables means that electrolysis of water will be, environmentally speaking, a benign process for the future hydrogen production.

3.4 Energy-Driving Water Splitting Systems

Most photocatalysts are active in the UV-region (mostly UV-A) of the solar spectrum. This region accounts for only 4% of the total visible solar energy, whereas the visible (400–800 nm, or 3.09–1.55 eV) and the infrared (>800 nm, <1.55 eV) account, respectively, for 53 and 43% of the total solar energy [41, 42]. It seems clear that less efficient photocatalysts that absorb visible light can be more useful than more efficient photocatalysts absorbing uniquely ultraviolet light. The materials design leads to photocatalytic samples in powder or colloidal form able to integrate dopant elements as co-catalysts, necessary for their optimization. Within this context, the link between photoelectrochemistry (see Sect. 3.2) and nanoscience (quantum size effect) comes across. The former reported by Brattain and Garrett [43] was born in 1955, the physical [44] in the 1970s, and the chemical [45, 46] aspects of quantum size effect in the 1980s.

Photocatalytic processes, by means of semiconducting nanopowders, have been used for the production of hydrogen either from water, or from organics present in wastewaters. In general, the scheme presented for nanoparticulated semiconductors, Fig. 3.10, shows the various approaches undertaken to enhance the water splitting reaction on such nano-objects (oxide or non-oxide). Based on the energetics of semiconductors, e.g., a single oxide nanoparticle can produce H_2 and O_2 via the photogenerated electron–hole pairs (e^--h^+)(see Fig. 3.10a). This phenomenon is feasible, provided the energetic requirements are fulfilled, cf. Fig. 3.4; with a further prerequisite, the semiconducting materials are stable in aqueous solutions under irradiation.

The process of photocatalysis implies at least three steps, namely (i) absorption of photons ($h\nu \geq E_g$), (ii) charge separation followed by migration of the photogenerated charge carriers, and (iii) surface chemical reactions between the charge carriers and the adsorbed species (e.g., water). Recombination of electrons and holes can occur in the bulk of nanoparticles reducing the efficiency of the heterogeneous photocatalytic process (iii). One single oxide particle cannot, however, perform under visible light both reactions if the flat band position is not sufficiently negative for the hydrogen production, as contrasted in Fig. 3.11 [47], and if the energy gap is >3 eV. Conversely, visible-light-sensitive oxide semiconductors, such as WO_3, Fe_2O_3, cannot produce H_2 (see Fig. 3.4). Under this circumstance,

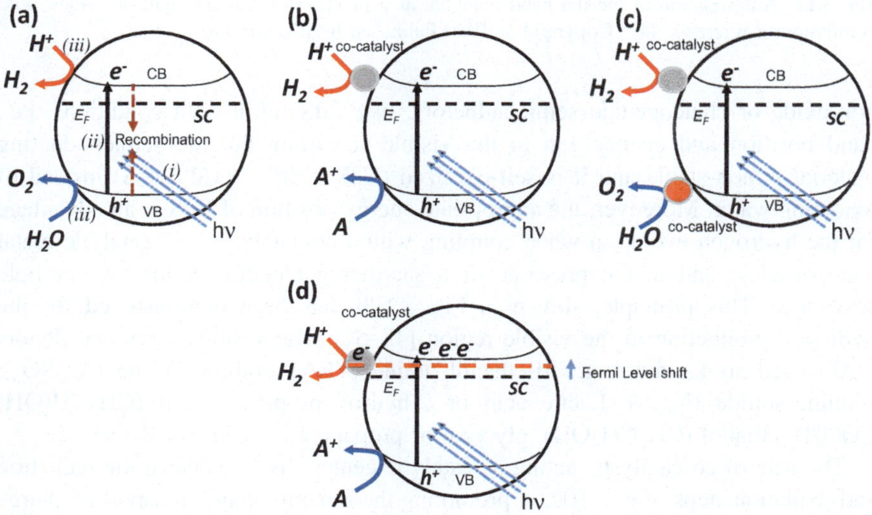

Fig. 3.10 Schematic illustration for the water photosplitting process using nanoparticulated semiconductors (SCs) either an oxide or a chalcogenide (**a**), (**c**), and (**d**). The hydrogen production is shown in (**b**) via sacrificial agents. The oxidation of water by positive holes in the valence band (VB) and reduction of proton ions by photogenerated electrons in the conduction band (CB) in the presence of selected co-catalysts. The effect of charge accumulation is shown in (**d**)

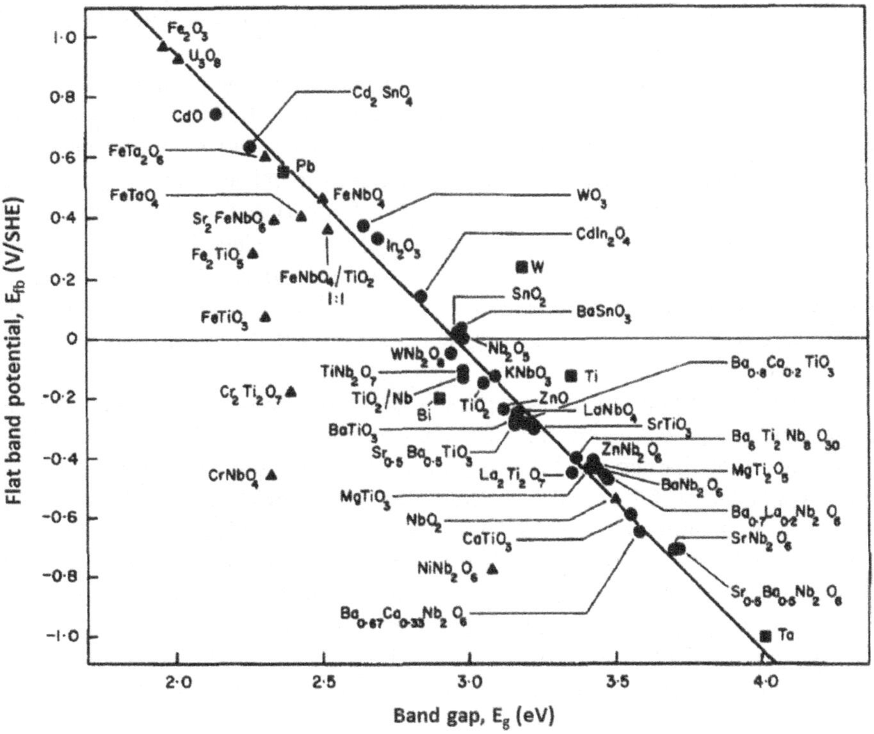

Fig. 3.11 A correlation of the flat band potential as a function of the band gap of oxides, with permission of reference [47] Copyright © 1980 Published by Elsevier Ltd

non-oxide or chalcogenide semiconductors, like CdS fulfill both conditions, i.e., band position and energy gap in the visible spectrum, but the semiconducting material is non-stable and it is self-oxidized ($CdS + 2h^+ \rightarrow Cd^{2+} + S$) instead of oxidizing water. Moreover, the appropriate energy position of CdS is an advantage for the hydrogen evolution when coupling with a co-catalyst (e.g., catalytic metal nanoparticles) and in the presence of a sacrificing electron donor "A" or hole scavenger. This principle, shown in Fig. 3.10b, has been demonstrated for the hydrogen production in the visible region [48–52]. The various sacrificial donors ("A") used so far are, e.g., tri-ethanol amine (TEA), sodium sulfite (Na_2SO_3), sodium sulfide (Na_2S), Lactic acid or 2-hydroxypropanoic acid ($CH_3CH(OH)COOH$), ethanol (CH_3CH_2OH), glycerol or propane-1,2,3-triol ($C_3H_8O_3$).

The role of co-catalysts, acting as catalytic centers, is to enhance the reduction and oxidation steps, Fig. 3.10c, in promoting the separation and transport of charge carriers: processes (ii) and (iii), Fig. 3.10a. With one co-catalyst, Fig. 3.10b, one has the advantage of using organic or inorganic species ("A") as sacrificial agents, or hole scavengers to generate the proton ions, that are reduced by the photogenerated electrons to form H_2 in the presence of an appropriate co-catalyst. For the photoreforming process, sacrificial organics are required [53]. In the frame of the environmental remediation principle, the use of wastewater streams containing organics should be considered, for the sake of water treatment and hydrogen generation [54, 55].

The physical–chemical properties of nanoparticulated semiconductor systems, compared to bulk semiconducting materials, are that such nano-objects possessing depletion width near their particle size do not show band bending, as just schematized in Fig. 3.10. Undeniably, the process of charge transfer is governed by diffusion, and the photopotential should be determined by the charge recombination rate rather than by the space charge region where an electrical field exists, cf., e.g., Fig. 3.5. There are various examples that assess such statements in the literature [56–58] and are summarized in Fig. 3.10d. This latter figure depicts the property of storing electrons (Fermi level shift) observed for gold nanoparticles on TiO_2 [56, 57], and for silver nanoparticles on ZnO [58]. In a photoelectrochemical cell system, this phenomenon is measured through the onset photopotential shift toward a negative electrode potential in the absence of an electron scavenger (e.g., oxygen). The photocatalytic system offers, in a simple way, opportunities to tune co-catalysts that interact (forming nanoheterojunctions, or decorating semiconducting nanoparticles' surface) with the semiconducting nanoparticles in order to improve the charge transfer rate of minority carriers to the adsorbed species without the necessity to improve the transfer of the majority carriers. Hence, for water splitting applications catalytic centers based on metallic nanoparticles are used. As discussed in Chap. 2, some existing metallic elements of the periodic table have been identified as "good" electrocatalysts for HER (cf. Fig. 2.8) and OER (cf. Fig. 2.12). Although, the actual trend on the quest for improving the water splitting reactions is to tailor novel materials, such as chalcogenides [49, 59–65], phosphides [66–74], perovskites [75–82], and layered double hydroxides [83–89] based on earth abundant elements. In the premises of novel materials, the water splitting photocatalytic system, depicted in Fig. 3.10c, must be designed with the best low-cost HER and OER co-catalysts.

3.4.1 Photocatalysis on Some Selected Oxides

An overview, similar to that made for S, Se and Te chalcogenides, cf. Fig. 1.4 [90], summarizing some selected oxides employed for water splitting is represented in Fig. 3.12. The left block shows elements that form oxides with d^0 configuration or empty *d*-orbitals of metal cations in the conduction band. Conversely, their valence band is composed of O *2p* orbitals.

The representative oxides of group 4B are TiO_2 (anatase, rutile, brookite), and ZrO_2 (monoclinic). The band gap of TiO_2 ranges from 3 eV (rutile), 3.2 eV (anatase) to 3.5 eV (brookite) [90] and that of ZrO_2 of 5.0 eV [91]. The most active photocatalyst for water splitting with the help of an external bias is the anatase phase, because of its more negative conduction band position, as compared to the rutile phase. Their photon absorption is in the UV-region, and one of the driving

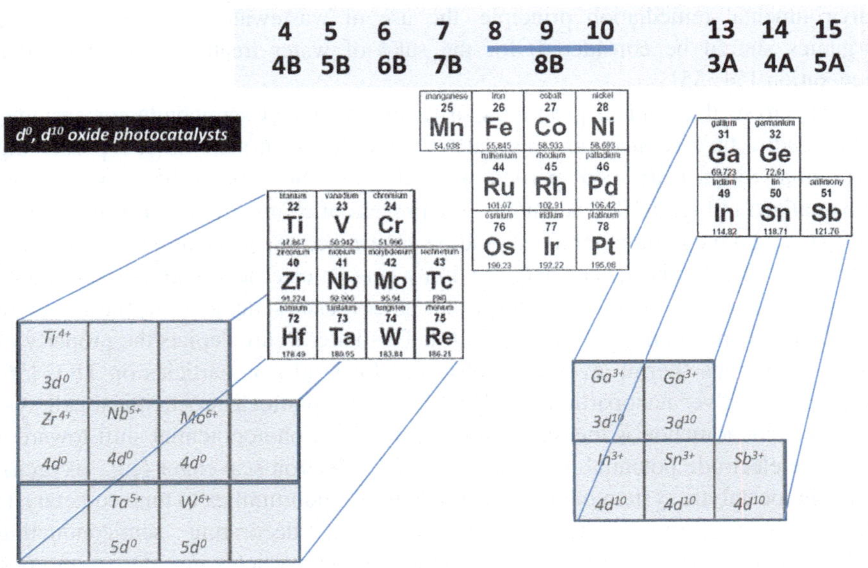

Fig. 3.12 A portion of the periodic showing some transition metal oxides (M_xO_y). The elements of the right block are essentially d^{10} structures, the elements of block left are d^0 structures

forces to shift the absorption threshold to the visible solar region is obtained by doping TiO_2 either with non-metals (*p*-block elements of the periodic table), such as, B [92–97], C [95, 97–105], N [94–97, 100, 106–111], S [112], P [107, 113, 114], or by introducing metal ions, such as, e.g., V^{4+} [115–118], Mo^{6+} [119, 120], Ru^{3+}, [121, 122], etc. Among the metal ions, the latter ones increase the oxide photoactivity, whereas others, such as Al^{3+}, Co^{3+}, decrease the oxide photoactivity [123]. However, the ultimate goal of the dopant is to trap and to favor the transfer of electrons or holes. *In fine*, the judicious incorporations into the oxide lattice of dopants can extend the photon absorption edge into the visible region (see Fig. 3.13a [124]). The base photocatalysts (TiO_2) can be combined with other oxides (e.g., $SrTiO_3$ [125] or ZrO_2 [126]) to form heterojunctions and improve the photocatalysis of hydrogen evolution.

Hydrogen production by water splitting was also examined on oxides of group 5B such as Nb_2O_5 (tetragonal phase [127], avg. band gap of 3.4 eV [128]), and Ta_2O_5 (tetragonal phase [129], $E_g = 3.8$ eV) in which sacrificial donors such as methanol were employed for the hydrogen production under UV-irradiation [130], for the former, and for the latter, forming a composite with Bi_2O_3 in order to shift its spectral response into the visible region [131]. The niobium [130, 132] and tantalum pentoxides [131, 133] were also modified with metals (e.g., Pt, Au, Cu) and/or metallic oxides (e.g., NiO, RuO_2) co-catalysts to enhance their water splitting activity under UV-illumination. Under the same purpose of water splitting, the oxides, in the group 6B, considered were WO_3 (monoclinic phase, $E_g = 2.8$ eV) [134–157] and MoO_3 (orthorhombic phase, $E_g = 2.45$ eV [158]) [159–163]. Again,

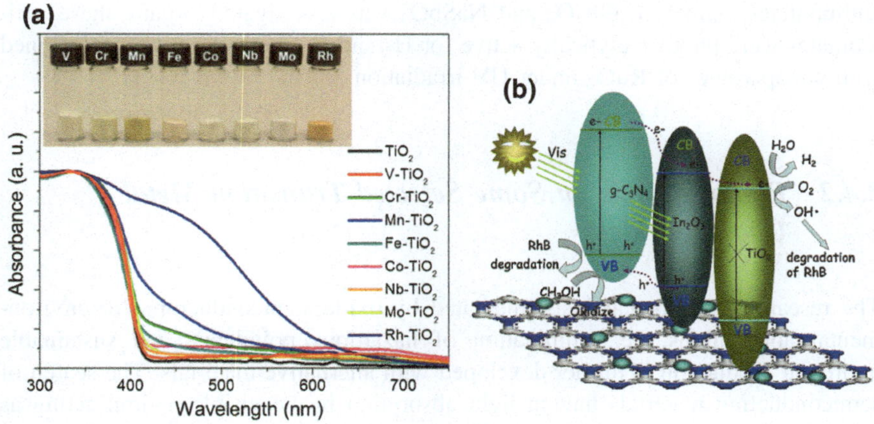

Fig. 3.13 **a** UV–vis spectra of various transition metal (TM)-doped TiO_2 nanowires. Inset show the pictures of various TM-doped TiO_2 nanowires, with permission from reference [124] Copyright © 2013, American Chemical Society. **b** The photocatalytic process of charge separation toward hydrogen production and dye degradation, with permission of reference [173] Copyright © 2015 Elsevier B.V

a similar approach of modifying the reactivity and photon absorption threshold of WO_3 and MoO_3 was done by admixing co-catalysts based on g-C_3N_4, metals (e.g., Pt, Au), oxides (e.g., TiO_2, NiO, V_2O_5, CeO_2, CoO_x, Fe_2O_3) chalcogenides (e.g., MoS_2, WS_2) [155, 162, 164–168].

In the group 3A, Fig. 3.12, the most stable phase of Ga_2O_3 is the β-Ga_2O_3 (monoclinic phase [169], E_g = 4.59 eV [170]). Ga_2O_3 has five different polytypes (α, β, γ, δ, ε). Interestingly, using DFT it was found that the band alignments for Se-doped and I-doped β-Ga_2O_3 are well positioned allowing both photo-oxidation and photoreduction of water in the visible region [170]. It was, however, experimentally demonstrated that the addition of ions, such as, Ca, Cr, Zn, Sr, and Ta, boosted remarkably the water splitting photocatalytic activity [171]. On the other hand, In_2O_3 (bixbyite (bcc) phase, E_g = 2.63 eV [172]) was investigated with other oxides forming ternary composites, e.g., TiO_2–In_2O_3@g–C_3N_4. It was found, indeed, that such composites enhance the hydrogen generation rate [173] and the degradation rate of a dye (Rhodamine B). This phenomenon, evidenced by photoluminescence spectroscopy, electrochemical impedance spectroscopy, and photocurrent analysis, was attributed to the interfacial transfer of photogenerated electrons and holes (in the Vis-region), among TiO_2, In_2O_3, and g–C_3N_4, leading to the effective charge separation on these semiconductors, Fig. 3.13b. Finally, photocatalysts of groups 4A and 5A, sensitive to UV-light irradiation, are mixed oxides containing Ge^{4+} and Sb^{5+}. Indeed, RuO_2-decorated $LiInGeO_4$ showed a high photocatalytic activity for the overall splitting of water to produce H_2 and O_2 under UV-irradiation. The combination of highly crystallized $LiInGeO_4$ and highly dispersed nanoparticles of RuO_2 enhanced the photocatalytic activity [174]. In a similar context, the photocatalytic activity for water decomposition of metal antimoniates, namely $CaSb_2O_6$ and $NaSbO_3$ was investigated. Again, these antimoniates were photocatalytically active for H_2 and O_2 production when combined with nanoparticles of RuO_2 under UV-irradiation [175].

3.4.2 Photocatalysis on Some Selected Transition Metal Chalcogenides

The research on photocatalysts, mediated by oxide semiconductors, for environmental applications, i.e., elimination of hazardous pollutants, and sustainable hydrogen production is further developed with alternative materials. The search of semiconducting materials having light absorption in the visible region; acting as catalytic centers and as co-catalysts is important for an efficient charge separation and photocatalytic activity. The transition metal dichalcogenide appears to fulfill this expectation, since most of them possess a relatively favorable energy level for hydrogen evolution and oxygen activation [176] (see Fig. 3.10). They are recognized as alternatives for noble-metal-based catalysts. Though, popular metal cations d^{10} chalcogenide materials such as: ZnS (zincblende phase, E_g= 3.6 eV), ZnSe (zincblende phase, E_g = 2.7 eV) [177], CdS (wurtzite phase, E_g = 2.4 eV), CdSe (wurtzite phase, E_g = 1.7 eV) [49, 52, 128, 178–183] have also been explored. Most of these materials are, however, sensitive to corrosion in aqueous medium, but could be kinetically stabilized using sulfide as sacrificial agents Na_2S and Na_2SO_3 [184, 185]. This is certainly the reason to explore 2D transition metal dichalcogenides (2D-TMD) in order to expose the highly active centers at the edges (planes perpendicular to the Van der Waals surface); see, e.g., MoX_2 depicted in Fig. 1.5, and take advantage of the tunable band gap and the energy band positions.

3.4.2.1 Photocatalytic Hydrogen Evolution

MoS_2 (trigonal prismatic phase—2H, E_g = 1.75 eV) and WS_2 (trigonal prismatic phase, E_g = 2.1 eV) can be considered as representative of the 2D-TMD compounds, cf. Figs. 1.4 and 1.5. Indeed, aiming at enhancing the hydrogen evolution rate, MoS_2 has been employed to modify the photocatalytic activity of other 3D-chalcogenides, e.g., CdS [186, 187] and $ZnIn_2S_4$ [188], as well as oxides, e.g., TiO_2 [189–192], ZnO [193], $SrZrO_3$ [194]. It is clear that the junctions of MS_2 with other materials mentioned above constitute the so-called multi-component or composite photocatalysts. The intimate contact established between two components (semiconductor-semiconductor heterojunction, SC/SC), say, MoS_2 and CdS, using an optimized chemical synthetic route in aqueous solution, is shown in the inset of Fig. 3.14a responsible for the enhanced photocatalytic hydrogen production [186]. Other examples of SC/SC photocatalysts, based on MoS_2 with TiO_2, ZnO and $SrZrO_3$, are depicted in Figs. 3.14b–d. A common parameter in this kind of

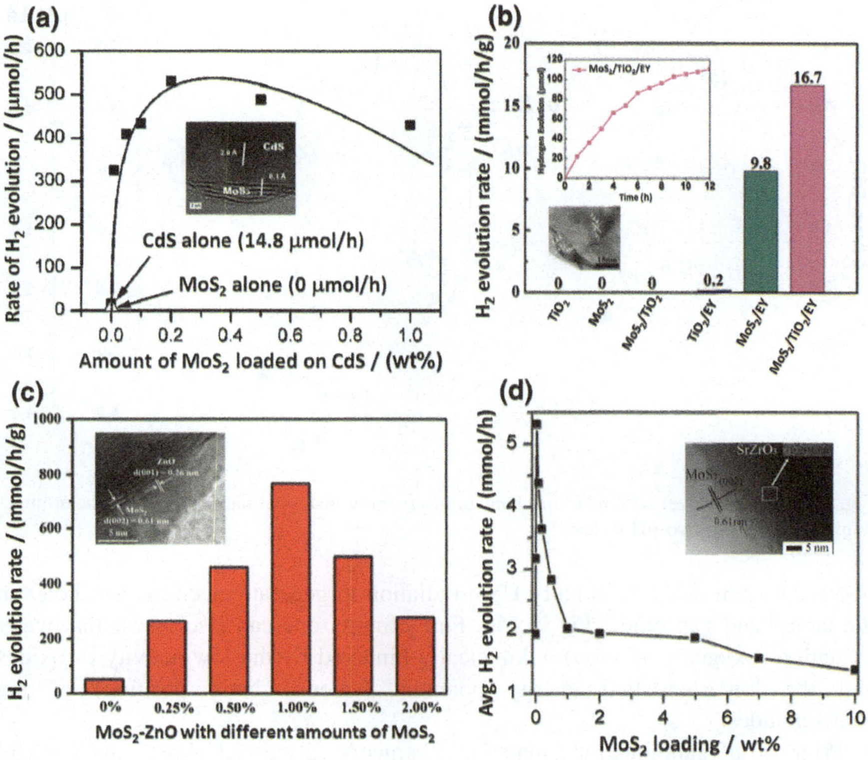

Fig. 3.14 **a** Rate of H_2 evolution on MoS_2/CdS photocatalysts loaded with different amounts of MoS_2 under visible light (λ > 420 nm), the inset shows the HRTEM image of 1 wt% MoS_2/CdS. Adapted with permission from reference [186] Copyright © 2008, American Chemical Society. **b** The hydrogen evolution rate on 1D MoS_2 nanosheet–TiO_2 nanowire hybrid nanostructures. Inset is a high-resolution TEM image of as-synthesized hybrid structures. EY means: dye eosin-Y (2-(2,4,5,7-tetrabromo-6-oxido-3-oxo-3H-xanthen-9-yl) benzoate), adapted with permission from reference [189] Copyright © 2014, Royal Society of Chemistry. **c** The dependence of the steady rate of H_2 production by MoS_2–ZnO on the loading amounts of MoS_2. Inset TEM image of 2.00 wt% MoS_2–ZnO composite photocatalyst. Adapted with permission from reference [193] Copyright © 2015, Royal Society of Chemistry. **d** Photocatalytic activity of $MoS_2/SrZrO_3$ catalysts with different amount of MoS_2. Inset: HRTEM images of (SM8, = 5%MoS_2). Figure adapted with permission of reference [194] Copyright © 2014, Royal Society of Chemistry

heterojunction (as demonstrated in the TEM pictures) is the role that the amount of MoS_2 plays toward the hydrogen evolution efficiency. The analysis of hydrogen evolution efficiency is hard due to the different experimental conditions reported so far. However, a rationalization of the various heterojunctions can be summarized in Fig. 3.15, considering the band positions of the semiconductors, reported in Fig. 3.4. A sacrificial agent (A) or hole scavenger is necessary in the photocatalytic system. The feasibility for hydrogen to evolve, under visible light, for CdS/MoS_2 system, is shown in Fig. 3.15a. It is clear that this process is less favorable under visible light in the oxides (TiO_2, ZnO)/MoS_2. A similar phenomenon can be also

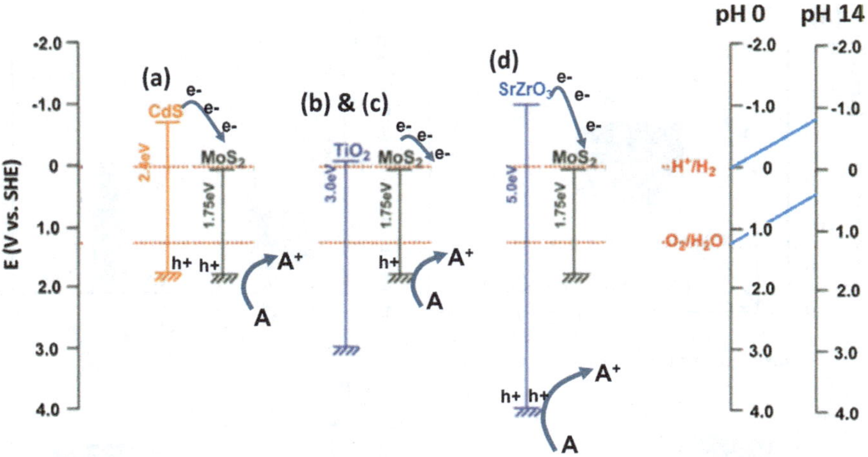

Fig. 3.15 Energy level schematic mechanism of charge transfers at nanoparticulated heterojunctions, compared to two pH 0, and 14

observed when using essentially UV-irradiation to generate electrons and holes in the large-band gap oxide, Fig. 3.15d. Furthermore, one can also assess that water oxidation (oxygen evolution) is kinetically hindered by the low activity of oxides and the low oxidation energy potential (valence band position) of the chalcogenides.

Prior to all nanojunctions materials chemistry, discussed above, the Chevrel phase ($Mo_4Ru_2Se_8$, cf. Fig. 1.9) in powder (crystallites) form was fixed onto a well-defined p-type semiconductor (p-GaP) surface via an ionomer Nafion® thin film. In spite of the physical contact between the chalcogenide crystallites and GaP, it was demonstrated that photogenerated charge carriers, in the visible region, could be conveyed through the chalcogenide material to kinetically enhance the hydrogen evolution process. This phenomenon was visible by the increase in the photocurrent in the region of charge recombination (flat band region) observed on the naked p-GaP, Fig. 3.16 [195].

3.4.2.2 Photocatalysis and Charge Storage and Transfer via Chalcogenides in Colloidal Solution

Transition metal chalcogenides based on ruthenium centers can be prepared in colloidal solution. The synthetic route is shown in Figs. 1.11 and 1.13. As reported some time ago [196], the addition of a surfactant serving as stabilizer prior to the formation of the heteronuclear-complex $Ru_4Se_2(CO)_{11}$ permits, after pyrolysis of the heteronuclear-complex, in an organic solvent (e.g., Xylene or di-chlorobenzene), to obtain colloidal Ru_xSe_y nanoparticles [196–198]. For Ru_xSe_y material 1-octa-decanthiol ($CH_3(CH_2)_{17}$-SH) was employed as the stabilizer agent

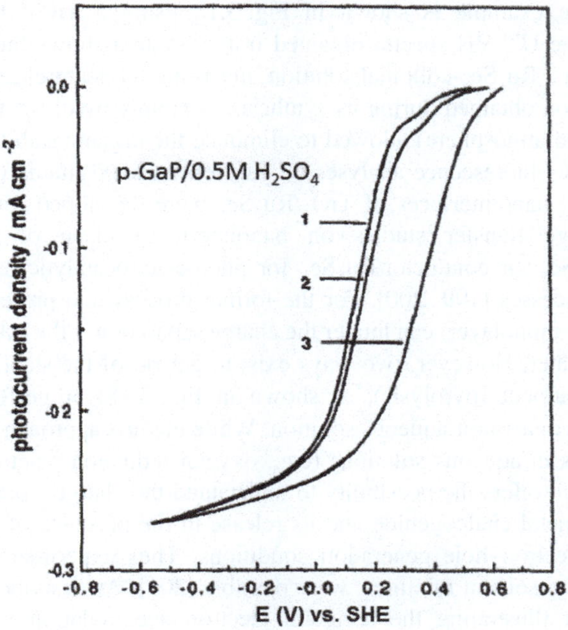

Fig. 3.16 Photoevolution of hydrogen of the Nafion-fixed Chevrel-phase chalcogenide ($Mo_4Ru_2Se_8$). Curves: (1) naked p-GaP; (2) Nafion-modified p-GaP; and (3) Nafion®-attached chalcogenide crystallites. With permission of reference [195] Copyright © 1987 Published by Elsevier B.V

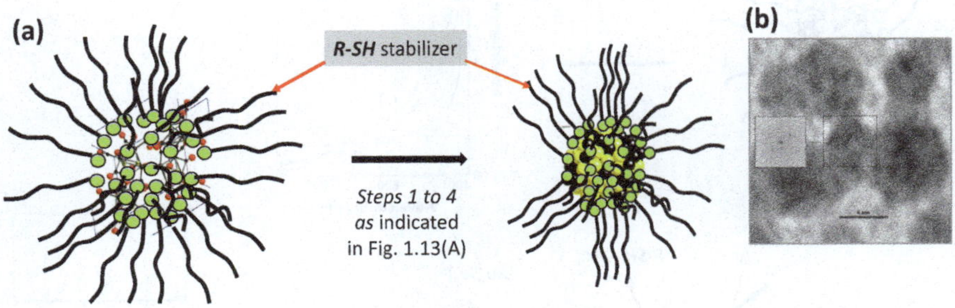

Fig. 3.17 **a** Schematics of the structural dynamics of Ru_xSe_y-ligand stabilized ($CH_3(CH_2)$ 17-SH) nanoclusters in organic solvent (Xylene, or Di-chlorobenzene), cf. Fig. 1.14. The TEM image illustrates the crystallinity nature of a Ru_xSe_y cluster

[196]. The resulting product, after steps 1 to 4 (Fig. 1.13), was identified, as shown in Fig. 3.17. The synthesis of the colloidal solution could be followed by UV-Vis resulting in a featureless curve. It is interesting to notice that the colloidal solution containing small Ru_xSe_y clusters with a narrow-size envelope shown in the TEM image, Fig. 3.17b, was stable in ambient conditions for months. The colloidal solution allowed to modify by simply dipping different metallic or semiconducting

substrates. One example is shown in Fig. 3.18 using a TiO_2 thin layer. The follow-up of the UV-Vis spectra obtained in this system shows this effect on the dipping time in a Ru_xSe_y-colloidal solution, mirroring the featureless spectra of the colloidal solution obtained during its synthesis. A simply pyrolysis thereafter (e.g., 230 °C in argon atmosphere) allowed to eliminate the organic stabilizer, as proven by TEM and X-Fluorescence analyses on Ru_xSe_y-modified titania (see Fig. 3.18). In this manner, nanointerfaces of TiO_2/Ru_xSe_y were developed for photoelectrochemical charge transfer studies on nanointerfaces made of semiconductor substrates/Ru_xSe_y, or conductor/Ru_xSe_y for photoelectrocatalytic as well as electrocatalytic processes [199, 200]. For the former process, the presence of the stabilizer (organic monolayer) can hinder the charge separation if the system is put into contact with water. However, two ways exist to get rid of the stabilizer molecule, using heat treatment (pyrolysis), as shown in Fig. 3.18, or performing experimentations in a quasi-non-aqueous solution. While the first approach is necessary to perform studies in aqueous solutions (e.g., oxygen reduction reaction [200]). The second approach offers the possibility to understand the charge storage capacity of the transition metal chalcogenide and its release in the presence of a proper scavenger under electron–hole generation conditions. Thus experiments, in colloidal suspension using solvent mixtures, were possible [201]. As described in Fig. 3.19, the experiment illustrating the temporal electron accumulation via UV-Vis on titania nanoparticles was done using colloidal TiO_2. Figure 3.19 shows that a

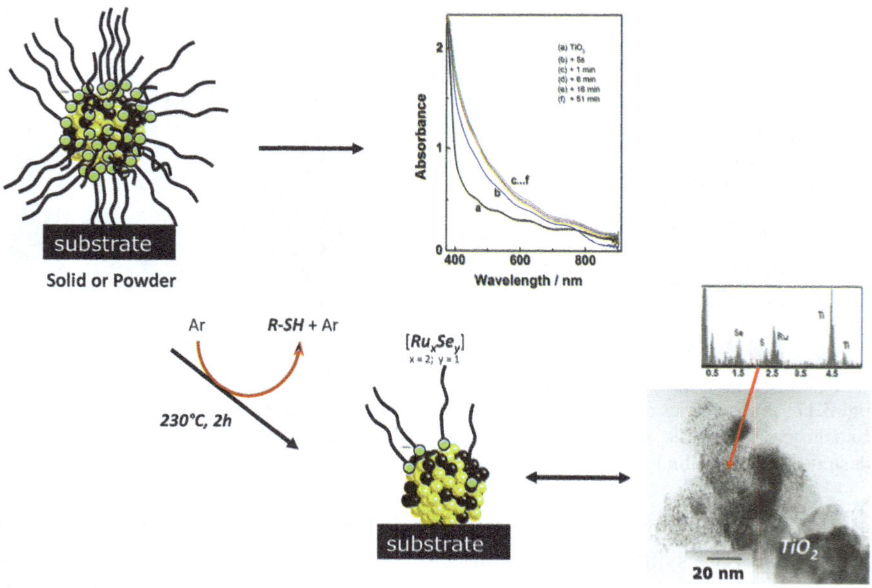

Fig. 3.18 Absorption spectra for a TiO_2 nanostructured thin film before thermal treatment and after immersion in the colloidal Ru_xSe_y solution for different times: (1) before immersion; (2) after 5 s; (3) after 1 min; 4. after 6 min; 5. after 16 min; 6. after 51 min. The pyrolysis process after adsorption of the Ru_xSe_y-colloidal solution onto a substrate is illustrated in the TEM image, using as a substrate titania thin layer

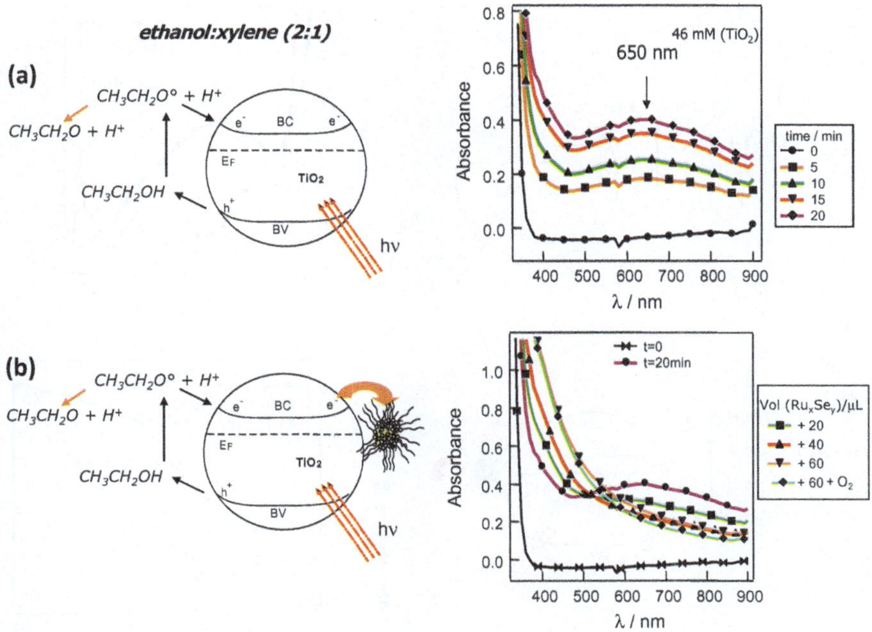

Fig. 3.19 **a** Absorption spectra evolution of TiO_2 (0.023 M solution purged with nitrogen), **b** Absorption evolution for different previously illuminated samples: TiO_2, TiO_2 + (20 µl; 40 µl, 60 µl Ru_xSe_y)

relative broad peak centered at 650 nm grows as a function of increasing illumination time. Electron–hole pairs are generated by the band-to-band excitation using the UV light. The holes diffuse to the particle surface where they react with reduced dissolved species, such as ethanol, as shown in the upper scheme at the left of Fig. 3.19. In contrast, the electrons remain in the particles, leading to a transitory charge. This negative charge induces recombination and accelerates the electron charge transfer to the solution. The accumulation of electrons in the TiO_2 colloidal solution can be detected by monitoring the increase in the absorbance in the visible region; i.e., the semiconductor particles become blue colored [202]. The release of the accumulated electrons can be favored via an electron scavenger, and/or via a mediator to speed up such a charge release (increasing charge transfer kinetics). To illustrate this approach, the UV-Vis spectra of the semiconductor colloidal solution, purged with nitrogen after 20 min under illumination, were measured using different amounts of the Ru_xSe_y colloidal solution distributed in various quartz cuvettes with a certain volume of TiO_2 colloidal solution and each illuminated during 20 min. As the system TiO_2/Ru_xSe_y is illuminated, electron–hole pairs are generated. The holes will diffuse to the semiconductor surface as mentioned above, where they will react with ethanol species present in solution. On the one hand, the photogenerated electrons can be trapped at TiO_2 surface states or can move to the Ru_xSe_y nanoparticles (see Fig. 3.19b). The chalcogenide is embedded in a sheath of

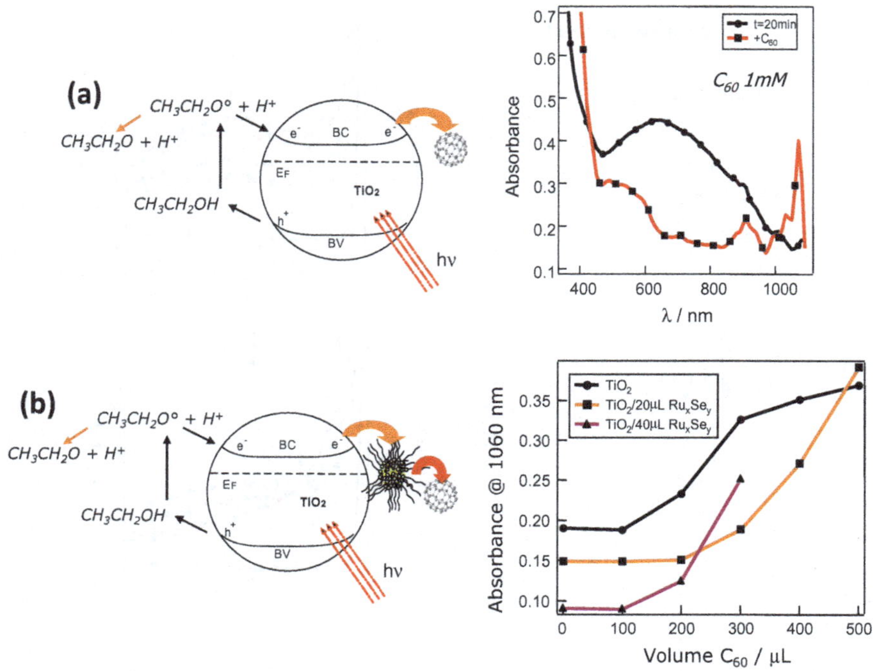

Fig. 3.20 **a** Absorption spectra evolution of TiO_2 (0.023 M solution purged with nitrogen), with thereafter addition of deaerated 1 mM C60 solution. **b** Accumulation of electrons. Bottom: transfer to C_{60} to produce the radical C_{60}. Absorption evolution at 1060 nm for different previously illuminated samples: **a** TiO_2, **b** TiO_2 + 20 µl Ru_xSe_y; and **c** TiO_2 + 40 µl Ru_xSe_y with the addition of C_{60}. Light was due to 150 W Xe lamp

the chemical stabilizer: 1-octadecanthiol. Moreover, it seems that charge transfer takes place in the presence of these thiol chains. This configuration gives a certain hydrophobic character to the chalcogenide clusters, which certainly remain covered with a monolayer of octadecanthiol, even in the presence of TiO_2. It can thus reasonably be assumed that the Fermi level of Ru_xSe_y nanoparticles must be placed below the conduction band of TiO_2 favoring the electrons to be transferred from the semiconductor to the chalcogenide and stored in the chalcogenide clusters. This process continues until a stationary state is attained (see Fig. 3.19b). Indeed, the decrease of the peak centered at 650 nm is visible with the increasing amount of the Ru_xSe_y colloidal. This is a clear evidence that Ru_xSe_y clusters accumulate (store) charges released by the oxide nanoparticles. The charges can be further transferred if an electron scavenger is added, such as the molecular oxygen (O_2). Thus, molecular oxygen annihilates the accumulation of electrons. This experiment can be considered as a photocatalytic reduction of molecular oxygen using as sacrificial agent: ethanol (cf. Fig. 3.10d). However, a more spectacular proof is to use another spectroscopic probe molecule: the fullerene C_{60}. The electron transfer to C_{60} leads to the radical formation of C_{60}: ($C_{60} + e^- \rightarrow C^{\bullet}_{60}$) with a characteristic peak centered at 1060 nm. This phenomenon takes place testifying that C_{60} can interact with the electrons accumulated on the TiO_2 semiconductor surface (see Fig. 3.20a). The electronic transfer between the TiO_2 particles and C_{60} with a quantum efficiency of 24% was observed some time ago [203]. The electron storage and transfer via the chalcogenide clusters to the C_{60} molecule was performed monitoring the 1060 nm peak. The scheme is presented in Fig. 3.20b. The experiment was performed in such a way as to keep, without altering the ratio xylene: ethanol, the addition of C_{60} solution (1 mM) in xylene introduced directly into the TiO_2 (cf. Fig. 3.20a), or TiO_2/Ru_xSe_y suspensions (Fig. 3.20b), previously illuminated. The change in absorbance with $C^{\bullet}_{60}$ took place under two different numbers of chalcogenide nanoparticles. As compared to TiO_2 alone, the highest concentration of ruthenium chalcogenide nanoparticles (40 μL) accelerate the increase in the 1060-nm signal of the $C^{\bullet}_{60}$ molecule, testifying again that C_{60} reacts with the accumulated electrons on the semiconductor via Ru_xSe_y to form $C^{\bullet}_{60}$. In other words, electrons can accumulate in both materials.

References

1. Farrauto RJ, Heck RM (2000) Environmental catalysis into the 21st century. Catal Today 55 (1–2):179–187. https://doi.org/10.1016/S0920-5861(99)00237-0
2. Litter MI, Candal RJ, Meichtry JM (eds) (2014) Advanced oxidation technologies: sustainable solutions for environmental treatments. Sustainable energy developemnts, vol 9. CRC Press, Leiden, The Netherlands
3. Kharisov BI, Kharissova OV, Dias HVR (eds) (2014) Nanomaterials for environmental protection. Wiley, Hoboken
4. Hernández-Ramírez A, Medina-Ramírez I (eds) (2015) Photocatalytic semiconductors. Synthesis, characterization, and environmental applications. Springer, Heidelberg
5. Anpo M, Kamat PV (eds) (2010) Environmentally benign photocatalysts. Applications of titanium oxide-based materials. Springer, New York. https://doi.org/10.1007/978-0-387-48444-0
6. Kim D, Sakimoto KK, Hong D, Yang P (2015) Artificial photosynthesis for sustainable fuel and chemical production. Angew Chem Int Ed 54(11):3259–3266. https://doi.org/10.1002/anie.201409116
7. Barber J (2009) Photosynthetic energy conversion: natural and artificial. Chem Soc Rev 38(1):185–196. https://doi.org/10.1039/B802262N
8. Andreiadis ES, Chavarot-Kerlidou M, Fontecave M, Artero V (2011) Artificial photosynthesis: from molecular catalysts for light-driven water splitting to photoelectrochemical cells. Photochem Photobiol 87(6):1478. https://doi.org/10.1111/j.1751-1097.2011.01000.x
9. Kärkäs MD, Verho O, Johnston EV, Åkermark B (2014) Artificial photosynthesis: molecular systems for catalytic water oxidation. Chem Rev 114(24):11863–12001. https://doi.org/10.1021/cr400572f
10. Boddy PJ (1968) Oxygen evolution on semiconducting TiO_2. J Electrochem Soc 115(2):199–203. https://doi.org/10.1149/1.2411080
11. Fujishima A, Honda K (1972) Electrochemical photolysis of water at a semiconductor electrode. Nature 238(5358):37–38
12. Nocera DG (2012) The artificial leaf. Acc Chem Res 45(5):767–776. https://doi.org/10.1021/ar2003013

13. Barber J, Tran PD (2013) From natural to artificial photosynthesis. J R Soc Interface 10(81):1–16. https://doi.org/10.1098/rsif.2012.0984
14. Marshall J (2014) Solar energy: springtime for the artificial leaf. Nature News 510:22–24
15. Esiner S, Willems REM, Furlan A, Li W, Wienk MM, Janssen RAJ (2015) Photoelectrochemical water splitting in an organic artificial leaf. J Mater Chem A 3:23936–23945. https://doi.org/10.1039/C5TA07325A
16. Gerischer H (1990) The impact of semiconductors on the concepts of electrochemistry. Electrochim Acta 35(11):1677–1699. https://doi.org/10.1016/0013-4686(90)87067-C
17. Bard AJ, Bocarsly AB, Fan F-RF, Walton EG, Wrighton MS (1980) The concept of fermi level pinning at semiconductor/liquid junctions. Consequences for energy conversion efficiency and selection of useful solution redox couples in solar devices. J Am Chem Soc 102(11):3671–3677
18. Memming R (1984) Processes at semiconductor electrodes. In: Conway BE, Bockris JOM, Yeager E, Khan SUM, White RE (eds) Comprehensive treatise of electrochemistry, vol 7 kinetics and mechanisms of electrode processes. Plenum Press, New York, pp 529–592
19. O'Regan B, Grätzel M (1991) A low-cost, high-efficiency solar cell based on dye-sensitized colloidal films. Nature 353:737–699
20. Sze SM (2002) Semiconductor devices. Physics and technology. Wiley, Hoboken
21. Reineke R, Memming R (1992) Comparability of redox reactions at n- and p-type semiconductor electrodes. 1. The quasi-Fermi level concept. J Phys Chem 96(3):1310–1317. https://doi.org/10.1021/j100182a054
22. Reineke R, Memming R (1992) Comparability of redox reactions at n- and p-type semiconductor electrodes. 2. Electrochemical overpotential and recombination in view of the quasi-Fermi level concept. J Phys Chem 96(3):1317–1323. https://doi.org/10.1021/j100182a055
23. Morrison SR (1990) The chemical physics of surfaces, 2nd edn. Springer, Berlin. https://doi.org/10.1007/978-1-4899-2498-8
24. Gärtner WW (1959) Depletion-layer photoeffects in semiconductors. Phys Rev 116(1):84–87
25. Chen Z, Jaramillo TF, Deutsch TG, Kleiman-Shwarsctein A, Forman AJ, Gaillard N, Garland R, Takanabe K, Heske C, Sunkara M, McFarland EW, Domen K, Miller EL, Turner JA, Dinh HN (2010) Accelerating materials development for photoelectrochemical hydrogen production: standards for methods, definitions, and reporting protocols. J Mater Res 25(1):3–16. https://doi.org/10.1557/JMR.2010.0020
26. Gerischer H (1969) Charge transfer processes at semiconductor-electrolyte interfaces in connection with problems of catalysis. Surf Sci 18(1):97–122. https://doi.org/10.1016/0039-6028(69)90269-6
27. Marcus RA (1963) On the theory of oxidation—Reduction reactions involving electron transfer. V. Comparison and properties of electrochemical and chemical rates constants. J Phys Chem 67(4):853–857. https://doi.org/10.1021/j100798a033
28. Marcus RA (2000) Tutorial on rate constants and reorganization energies. J Electroanal Chem 483(1–2):2–6. https://doi.org/10.1016/S0022-0728(00)00011-5
29. Marcus RA (1964) Chemical and electrochemical electron-transfer theory. Annu Rev Phys Chem 15(1):155–196. https://doi.org/10.1146/annurev.pc.15.100164.001103
30. Gerischer H (1960) Über den Ablauf von Redoxreaktionen an Metallen und an Halbleitern. Zeitschrift für Physikalische Chemie 26(5_6):325. https://doi.org/10.1524/zpch.1960.26.5_6.325
31. Gerischer H (1960) Über den Ablauf von Redoxreaktionen an Metallen und an Halbleitern. Zeitschrift für Physikalische Chemie 26(3_4):223. https://doi.org/10.1524/zpch.1960.26.3_4.223
32. Kudo A, Miseki Y (2009) Heterogeneous photocatalyst materials for water splitting. Chem Soc Rev 38(1):253–278. https://doi.org/10.1039/B800489G
33. Bogdanoff P, Alonso-Vante N (1993) On-line determination via differential electrochemical mass spectroscopy (DEMS) of chemical products formed in photoelectrocatalytic systems.

Berichte der Bunsen-Gesellschaft Phys Chem 97(7):940–942. https://doi.org/10.1002/bbpc.19930970716
34. Buhl ML, Bird RE, Bilchak RV, Connolly JS, Bolton JR (1984) Thermodynamic limits on conversion of solar energy to work or stored energy—Effects of temperature, intensity and atmospheric conditions. Sol Energy 32(1):75–84. https://doi.org/10.1016/0038-092X(84)90051-3
35. Seitz LC, Chen Z, Forman AJ, Pinaud BA, Benck JD, Jaramillo TF (2014) Modeling practical performance limits of photoelectrochemical water splitting based on the current state of materials research. Chemsuschem 7(5):1372–1385. https://doi.org/10.1002/cssc.201301030
36. Landman A, Dotan H, Shter GE, Wullenkord M, Houaijia A, Maljusch A, Grader GS, Rothschild A (2017) Photoelectrochemical water splitting in separate oxygen and hydrogen cells. Nat Mater. Advance online publication. https://doi.org/10.1038/nmat4876, http://www.nature.com/nmat/journal/vaop/ncurrent/abs/nmat4876.html#supplementary-information
37. Rothschild A, Dotan H (2017) Beating the efficiency of photovoltaics-powered electrolysis with tandem cell photoelectrolysis. ACS Energy Lett 2(1):45–51. https://doi.org/10.1021/acsenergylett.6b00610
38. Corneil HG, Heinzelmann FJ, Nicholson EWS (1977) Production economics for hydrogen, ammonia, and methanol during the 1980-2000 period. Exxon Research and Engineering Co., Linden, N.J. (USA). Government Research Lab
39. de Assuncao JV (2009) Environmental effects of toxic materials from oil and gas combustion. In: Goldenberg J (ed) Interactions: energy/environment. Eolss Publishers Co Ltd, Oxford, pp 98–114
40. Rashid MM, Al Mesfer MK, Naseem H, Danish M (2015) Hydrogen production by water electrolysis: a review of alkaline water electrolysis, PEM water electrolysis and high temperature water electrolysis. Int J Eng Adv Technol (TM) 4(3):80–93
41. Moan J (2001) Visible light and UV radiation. In: Brune D, Hellborg R, Persson BR, Pääkkönen R (eds) Radiation at home, outdoors and in the workplace. Scandinavian Science Publisher, Oslo, pp 69–85
42. Fleagle RG, Businger JA (eds) (1980) Solar and terrestrial radiation, vol 25. International geophysics, vol 25. Elsevier B.V., Amsterdam. https://doi.org/10.1016/S0074-6142(08)60506-9
43. Brattain WH, Garrett CGB (1955) Experiments on the interface between Germanium and an electrolyte. Bell Syst Tech J 34(1):129–176. https://doi.org/10.1002/j.1538-7305.1955.tb03766.x
44. Dingle R, Wiegmann W, Henry CH (1974) Quantum states of confined carriers in very thin $Al_xGa_{1-x}As$-GaAs-Al_xGa_{1-x} Heterostructures. Phys Rev Lett 33(14):827–830. https://doi.org/10.1103/PhysRevLett.33.827
45. Brus LE (1983) A simple model for the ionization potential, electron affinity, and aqueous redox potentials of small semiconductor crystallites. J Chem Phys 79(11):5566–5571. https://doi.org/10.1063/1.445676
46. Henglein A (1982) Photo-degradation and fluorescence of colloidal-cadmium sulfide in aqueous solution. Ber Bunsenges Phys Chem 86(4):301–305. https://doi.org/10.1002/bbpc.19820860409
47. Scaife DE (1980) Oxide semiconductors in photoelectrochemical conversion of solar energy. Sol Energy 25(1):41–54. https://doi.org/10.1016/0038-092X(80)90405-3
48. Kakuta N, Park KH, Finlayson MF, Ueno A, Bard AJ, Campion A, Fox MA, Webber SE, White JM (1985) Photoassisted hydrogen production using visible light and coprecipitated zinc sulfide+cadmium sulfide without a noble metal. J Phys Chem 89(5):732–734. https://doi.org/10.1021/j100251a002
49. Hong S, Kumar DP, Reddy DA, Choi J, Kim TK (2017) Excellent photocatalytic hydrogen production over CdS nanorods via using noble metal-free copper molybdenum sulfide

(Cu_2MoS_4) nanosheets as co-catalysts. Appl Surf Sci 396:421–429. https://doi.org/10.1016/j.apsusc.2016.10.171
50. Ma S, Xie J, Wen J, He K, Li X, Liu W, Zhang X (2017) Constructing 2D layered hybrid CdS nanosheets/MoS_2 heterojunctions for enhanced visible-light photocatalytic H_2 generation. Appl Surf Sci 391(Part B):580–591. https://doi.org/10.1016/j.apsusc.2016.07.067
51. Fan Y, Chen G, Li D, Li F, Luo Y, Meng Q (2011) Enhancement of photocatalytic H_2 evolution on hexagonal CdS by a simple calcination method under visible light irradiation. Mater Res Bull 46(12):2338–2341. https://doi.org/10.1016/j.materresbull.2011.08.040
52. Berr M, Vaneski A, Susha AS, Rodríguez-Fernández J, Döblinger M, Jäckel F, Rogach AL, Feldmann J (2010) Colloidal CdS nanorods decorated with subnanometer sized Pt clusters for photocatalytic hydrogen generation. Appl Phys Lett 97(9):093108. https://doi.org/10.1063/1.3480613
53. Clarizia L, Spasiano D, Di Somma I, Marotta R, Andreozzi R, Dionysiou DD (2014) Copper modified-TiO_2 catalysts for hydrogen generation through photoreforming of organics. A short review. Int J Hydrogen Energy 39(30):16812–16831. https://doi.org/10.1016/j.ijhydene.2014.08.037
54. Zhang W, Li Y, Wang C, Wang P, Wang Q (2013) Energy recovery during advanced wastewater treatment: simultaneous estrogenic activity removal and hydrogen production through solar photocatalysis. Water Res 47(3):1480–1490. https://doi.org/10.1016/j.watres.2012.12.019
55. Priya R, Kanmani S (2013) Design of pilot-scale solar photocatalytic reactor for the generation of hydrogen from alkaline sulfide wastewater of sewage treatment plant. Environ Technol 34(20):2817–2823. https://doi.org/10.1080/09593330.2013.790081
56. Jakob M, Levanon H, Kamat PV (2003) Charge Distribution between UV-Irradiated TiO_2 and Gold Nanoparticles: determination of shift in the Fermi level. Nano Lett 3(3):353–358. https://doi.org/10.1021/nl0340071
57. Subramanian V, Wolf EE, Kamat PV (2004) Catalysis with TiO_2/Gold nanocomposites. effect of metal particle size on the fermi level equilibration. J Am Chem Soc 126(15):4943–4950. https://doi.org/10.1021/ja0315199
58. Wood A, Giersig M, Mulvaney P (2001) Fermi level equilibration in quantum dot– metal nanojunctions. J Phys Chem B 105(37):8810–8815. https://doi.org/10.1021/jp011576t
59. Karfa P, Madhuri R, Sharma PK (2017) Multifunctional fluorescent chalcogenide hybrid nanodots ($MoSe_2$:CdS and WSe_2:CdS) as electro catalyst (for oxygen reduction/oxygen evolution reactions) and sensing probe for lead. J Mater Chem A. https://doi.org/10.1039/C6TA08172J
60. Zhang J, Wang T, Pohl D, Rellinghaus B, Dong R, Liu S, Zhuang X, Feng X (2016) Interface engineering of MoS_2/Ni_3S_2 heterostructures for highly enhanced electrochemical overall-water-splitting activity. Angew Chem Int Ed. https://doi.org/10.1002/anie.201602237
61. Xu J, Cui J, Guo C, Zhao Z, Jiang R, Xu S, Zhuang Z, Huang Y, Wang L, Li Y (2016) Ultrasmall Cu_7S_4@MoS_2 hetero-nanoframes with abundant active edge sites for ultrahigh-performance hydrogen evolution. Angew Chem Inter Ed. https://doi.org/10.1002/anie.201600686
62. Li DJ, Maiti UN, Lim J, Choi DS, Lee WJ, Oh Y, Lee GY, Kim SO (2014) Molybdenum sulfide/N-doped CNT forest hybrid catalysts for high-performance hydrogen evolution reaction. Nano Lett 14(3):1228–1233. https://doi.org/10.1021/nl404108a
63. Vrubel H, Merki D, Hu X (2012) Hydrogen evolution catalyzed by MoS_3 and MoS_2 particles. Energy Environ Sci
64. Benck JD, Chen Z, Kuritzky LY, Forman AJ, Jaramillo TF (2012) Amorphous molybdenum sulfide catalysts for electrochemical hydrogen production: insights into the origin of their catalytic activity. ACS Catal 2(9):1916–1923. https://doi.org/10.1021/cs300451q
65. Ramakrishna Matte HSS, Gomathi A, Manna AK, Late DJ, Datta R, Pati SK, Rao CNR (2010) MoS_2 and WS_2 analogues of graphene. Angew Chem Int Ed 49(24):4059–4062. https://doi.org/10.1002/anie.201000009

66. You B, Sun Y (2016) Chalcogenide and phosphide solid-state electrocatalysts for hydrogen generation. ChemPlusChem 81(10):1045–1055. https://doi.org/10.1002/cplu.201600029
67. Li D, Baydoun H, Verani CN, Brock SL (2016) Efficient water oxidation using CoMnP nanoparticles. J Am Chem Soc 138(12):4006–4009. https://doi.org/10.1021/jacs.6b01543
68. Kucernak ARJ, Fahy KF, Sundaram VNN (2016) Facile synthesis of palladium phosphide electrocatalysts and their activity for the hydrogen oxidation, hydrogen evolutions, oxygen reduction and formic acid oxidation reactions. Catal Today 262:48–56. https://doi.org/10.1016/j.cattod.2015.09.031
69. Ha D-H, Han B, Risch M, Giordano L, Yao KPC, Karayaylali P, Shao-Horn Y (2016) Activity and stability of cobalt phosphides for hydrogen evolution upon water splitting. Nano Energy 29:37–45. https://doi.org/10.1016/j.nanoen.2016.04.034
70. Zhuo J, Cabán-Acevedo M, Liang H, Samad L, Ding Q, Fu Y, Li M, Jin S (2015) High-Performance electrocatalysis for hydrogen evolution reaction using Se-doped pyrite-phase nickel diphosphide nanostructures. ACS Catal 5(11):6355–6361. https://doi.org/10.1021/acscatal.5b01657
71. Zeng M, Li Y (2015) Recent advances in heterogeneous electrocatalysts for the hydrogen evolution reaction. J Mater Chem A 3(29):14942–14962. https://doi.org/10.1039/C5TA02974K
72. Jiang N, You B, Sheng M, Sun Y (2015) Electrodeposited cobalt-phosphorous-derived films as competent bifunctional catalysts for overall water splitting. Angew Chem Int Ed 54(21):6251–6254. https://doi.org/10.1002/anie.201501616
73. Xiao P, Sk MA, Thia L, Ge X, Lim RJ, Wang J-Y, Lim KH, Wang X (2014) Molybdenum phosphide as an efficient electrocatalyst for the hydrogen evolution reaction. Energy Environ Sci 7(8):2624–2629. https://doi.org/10.1039/C4EE00957F
74. Tian J, Liu Q, Asiri AM, Sun X (2014) Self-supported nanoporous cobalt phosphide nanowire arrays: an efficient 3D hydrogen-evolving cathode over the wide range of pH 0–14. J Am Chem Soc 136(21):7587–7590. https://doi.org/10.1021/ja503372r
75. Vignesh A, Prabu M, Shanmugam S (2016) Porous $LaCo_{1-x}Ni_xO_{3-\delta}$ nanostructures as an efficient electrocatalyst for water oxidation and for a zinc-air battery. ACS Appl Mater Interfaces 8(9):6019–6031. https://doi.org/10.1021/acsami.5b11840
76. Petrie JR, Cooper VR, Freeland JW, Meyer TL, Zhang Z, Lutterman DA, Lee HN (2016) Enhanced bifunctional oxygen catalysis in strained $LaNiO_3$ perovskites. J Am Chem Soc 138(8):2488–2491. https://doi.org/10.1021/jacs.5b11713
77. . Mefford JT, Rong X, Abakumov AM, Hardin WG, Dai S, Kolpak AM, Johnston KP, Stevenson KJ (2016) Water electrolysis on $La_{1-x}Sr_xCoO_{3-\delta}$ perovskite electrocatalysts. Nat Commun 7. https://doi.org/10.1038/ncomms11053
78. Hong WT, Welsch RE, Shao-Horn Y (2016) Descriptors of oxygen-evolution activity for oxides: a statistical evaluation. J Phys Chem C 120(1):78–86. https://doi.org/10.1021/acs.jpcc.5b10071
79. Zhu Y, Zhou W, Chen Z-G, Chen Y, Su C, Tadé MO, Shao Z (2015) $SrNb_{0.1}Co_{0.7}Fe_{0.2}O_{3-\delta}$ perovskite as a next-generation electrocatalyst for oxygen evolution in alkaline solution. Angew Chem Int Ed 54(13):3897–3901. https://doi.org/10.1002/anie.201408998
80. Misono M (2013) Catalysis of perovskite and related mixed oxides. In: Makoto M (ed) Studies in surface science and catalysis, vol 176. Elsevier, Amsterdam, pp 67–95. https://doi.org/10.1016/B978-0-444-53833-8.00003-X
81. Vojvodic A, Nørskov JK (2011) Optimizing perovskites for the water-splitting reaction. Science 334(6061):1355–1356. https://doi.org/10.1126/science.1215081
82. Bursell M, Pirjamali M, Kiros Y (2002) $La_{0.6}Ca_{0.4}CoO_3$, $La_{0.1}Ca_{0.9}MnO_3$ and $LaNiO_3$ as bifunctional oxygen electrodes. Electrochim Acta 47(10):1651–1660. https://doi.org/10.1016/s0013-4686(02)00002-6
83. Zhong H, Tian R, Li D, Tang P, Alonso-Vante N, Feng Y (2017) Tuning the adsorption properties of layered double hydroxides to tailor highly active oxygen bifunctional electrocatalysts. J Electrochem Soc 164(6):F491–F498. https://doi.org/10.1149/2.0341706jes

84. Li X, Hao X, Abudula A, Guan G (2016) Nanostructured catalysts for electrochemical water splitting: current state and prospects. J Mater Chem A 4(31):11973–12000. https://doi.org/10.1039/C6TA02334G
85. Shao M, Ning F, Wei M, Evans DG, Duan X (2014) Hierarchical nanowire arrays based on ZnO core−layered double hydroxide shell for largely enhanced photoelectrochemical water splitting. Adv Func Mater 24(5):580–586. https://doi.org/10.1002/adfm.201301889
86. Feng J, He Y, Liu Y, Du Y, Li D (2015) Supported catalysts based on layered double hydroxides for catalytic oxidation and hydrogenation: general functionality and promising application prospects. Chem Soc Rev 44(15):5291–5319. https://doi.org/10.1039/C5CS00268K
87. Fan G, Li F, Evans DG, Duan X (2014) Catalytic applications of layered double hydroxides: recent advances and perspectives. Chem Soc Rev 43(20):7040–7066. https://doi.org/10.1039/C4CS00160E
88. Lu Z, Qian L, Tian Y, Li Y, Sun X, Duan X (2016) Ternary NiFeMn layered double hydroxides as highly-efficient oxygen evolution catalysts. Chem Commun 52(5):908–911. https://doi.org/10.1039/C5CC08845C
89. Gong M, Li Y, Wang H, Liang Y, Wu JZ, Zhou J, Wang J, Regier T, Wei F, Dai H (2013) An advanced Ni–Fe layered double hydroxide electrocatalyst for water oxidation. J Am Chem Soc 135(23):8452–8455. https://doi.org/10.1021/ja4027715
90. Zallen R, Moret MP (2006) The optical absorption edge of brookite TiO_2. Solid State Commun 137(3):154–157. https://doi.org/10.1016/j.ssc.2005.10.024
91. Agorku ES, Pandey AC, Mamba BB, Mishra AK (2015) Gd, C, N, S multi-doped ZrO_2 for photocatalytic degradation of indigo carmine dye from synthetic water under simulated solar light. Mater Today: Proc 2(7):3909–3920. https://doi.org/10.1016/j.matpr.2015.08.020
92. Xue X, Wang Y, Yang H (2013) Preparation and characterization of boron-doped titania nano-materials with antibacterial activity. Appl Surf Sci 264:94–99. https://doi.org/10.1016/j.apsusc.2012.09.128
93. Dozzi MV, Selli E (2013) Doping TiO_2 with p-block elements: effects on photocatalytic activity. J Photochem Photobiol, C 14:13–28. https://doi.org/10.1016/j.jphotochemrev.2012.09.002
94. Di Valentin C, Pacchioni G (2013) Trends in non-metal doping of anatase TiO_2: B, C, N and F. Catal Today 206:12–18. https://doi.org/10.1016/j.cattod.2011.11.030
95. Uddin MN, Shibly SUA, Ovali R, Saiful I, Mazumder MMR, Islam MS, Uddin MJ, Gulseren O, Bengu E (2013) An experimental and first-principles study of the effect of B/N doping in TiO_2 thin films for visible light photo-catalysis. J Photochem Photobiol, A 254:25–34. https://doi.org/10.1016/j.jphotochem.2012.12.024
96. F-t Li, X-j Wang, Zhao Y, J-x Liu, Y-j Hao, R-h Liu, D-s Zhao (2014) Ionic-liquid-assisted synthesis of high-visible-light-activated N-B–F-tri-doped mesoporous TiO_2 via a microwave route. Appl Catal B: Environ 144:442–453. https://doi.org/10.1016/j.apcatb.2013.07.050
97. T-h Xu, C-l Song, Liu Y, G-r Han (2006) Band structures of TiO_2 doped with N, C and B. J Zheijang Univ Sci B 7(4):299–303. https://doi.org/10.1631/jzus.2006.B0299
98. He D, Li Y, Wang I, Wu J, Yang Y, An Q (2017) Carbon wrapped and doped TiO_2 mesoporous nanostructure with efficient visible-light photocatalysis for NO removal. Appl Surf Sci 391(Part B):318–325. https://doi.org/10.1016/j.apsusc.2016.06.186
99. Neumann B, Bogdanoff P, Tributsch H, Sakthivel S, Kisch H (2005) Electrochemical mass spectroscopic and surface photovoltage studies of catalytic water photooxidation by undoped and carbon-doped titania. J Phys Chem B 109(35):16579–16586
100. Chen D, Jiang Z, Geng J, Wang Q, Yang D (2007) Carbon and nitrogen Co-doped TiO_2 with enhanced visible-light photocatalytic activity. Ind Eng Chem Res 46(9):2741–2746
101. Wu X, Yin S, Dong Q, Guo C, Li H, Kimura T, Sato T (2013) Synthesis of high visible light active carbon doped TiO_2 photocatalyst by a facile calcination assisted solvothermal method. Appl Catal B: Environ 142–143:450–457. https://doi.org/10.1016/j.apcatb.2013.05.052

102. Huang K, Li Y, Xing Y (2013) Carbothermal synthesis of titanium oxycarbide as electrocatalyst support with high oxygen evolution reaction activity. J Mater Res 28(03):454–460. https://doi.org/10.1557/jmr.2012.353
103. Tang Z-R, Li F, Zhang Y, Fu X, Xu Y-J (2011) Composites of titanate nanotube and carbon nanotube as photocatalyst with high mineralization ratio for gas-phase degradation of volatile aromatic pollutant. J Phys Chem C 115(16):7880–7886. https://doi.org/10.1021/jp1115838
104. Khan SUM, Al-Shahry M, Ingler WB Jr (2002) Efficient photochemical water splitting by a chemically modified n-TiO_2. Science 297(5590):2243–2245
105. Mohapatra SK, Misra M, Mahajan VK, Raja KS (2007) Design of a highly efficient photoelectrolytic cell for hydrogen generation by water splitting: application of $TiO_{2-x}C_x$ nanotubes as a photoanode and Pt/TiO_2 nanotubes as a cathode. J Phys Chem C 111(24):8677–8685
106. Le NT, Konstantinova EA, Kokorin AI, Kodom T, Alonso-Vante N (2015) Recharge processes of paramagnetic centers during illumination in nitrogen-doped nanocrystalline titanium dioxide. Chem Phys Lett 635:241–244. https://doi.org/10.1016/j.cplett.2015.06.062
107. Zheng P, Wu H, Guo J, Dong J, Jia S, Zhu Z (2014) P-N co-doping induced structural recovery of TiO_2 for overall water splitting under visible light irradiation. J Alloys Compd 615:79–83. https://doi.org/10.1016/j.jallcom.2014.06.003
108. Liu S-H, Syu H-R (2012) One-step fabrication of N-doped mesoporous TiO_2 nanoparticles by self-assembly for photocatalytic water splitting under visible light. Appl Energy 100:148–154. https://doi.org/10.1016/j.apenergy.2012.03.063
109. Mrowetz M, Balcerski W, Colussi AJ, Hoffmann MR (2004) Oxidative power of nitrogen-doped TiO_2 photocatalysts under visible illumination. J Phys Chem B 108(45):17269–17273. https://doi.org/10.1021/jp0467090
110. Li D, Ohashi N, Hishita S, Kolodiazhnyi T, Haneda H (2005) Origin of visible-light-driven photocatalysis: a comparative study on N/F-doped and N-F-codoped TiO_2 powders by means of experimental characterizations and theoretical calculations. J Solid State Chem 178(11):3293–3302
111. Yates HM, Nolan MG, Sheel DW, Pemble ME (2006) The role of nitrogen doping on the development of visible light-induced photocatalytic activity in thin TiO_2 films grown on glass by chemical vapour deposition. J Photochem Photobiol, A 179(1–2):213–223
112. Aranda P, Kun R, Martín-Luengo MA, Letaïef S, Dékány I, Ruiz-Hitzky E (2007) Titania–sepiolite nanocomposites prepared by a surfactant templating colloidal route. Chem Mater 20(1):84–91. https://doi.org/10.1021/cm702251f
113. Liu Z, Yan X, Chu W, Li D (2010) Effects of impurities containing phosphorus on the surface properties and catalytic activity of TiO_2 nanotube arrays. Appl Surf Sci 257(4):1295–1299. https://doi.org/10.1016/j.apsusc.2010.08.050
114. Kim J, Kim H-R, Choung S-J (2011) The characterization and hydrogen production from water decomposition with methanol in a semi-batch type reactor using in, P-TiO_2s. Int J Photoenergy 2011:7p. https://doi.org/10.1155/2011/359161
115. Zhou J, Takeuchi M, Zhao XS, Ray AK, Anpo M (2006) Photocatalytic decomposition of formic acid under visible light irradiation over V-ion-implanted TiO_2 thin film photocatalysts prepared on quartz substrate by ionized cluster beam (ICB) deposition method. Catal Lett 106(1–2):67–70
116. Kubacka A, Fuerte A, Martínez-Arias A, Fernández-García M (2007) Nanosized Ti–V mixed oxides: effect of doping level in the photo-catalytic degradation of toluene using sunlight-type excitation. Appl Catal B: Environ 74(1–2):26–33. https://doi.org/10.1016/j.apcatb.2007.01.011
117. Bhattacharyya K, Varma S, Tripathi AK, Bhattacharyya D, Mathon O, Tyagi AK (2009) Study of the oxidation state and the structural aspects of the V-doped TiO_2. J Appl Phys 106(9):093503. https://doi.org/10.1063/1.3253761

118. Dholam R, Patel N, Miotello A (2011) Efficient H_2 production by water-splitting using indium–tin-oxide/V-doped TiO_2 multilayer thin film photocatalyst. Int J Hydrogen Energy 36(11):6519–6528. https://doi.org/10.1016/j.ijhydene.2011.03.028
119. Luo S-Y, Yan B-X, Shen J (2012) Enhancement of photoelectric and photocatalytic activities: Mo doped TiO_2 thin films deposited by sputtering. Thin Solid Films 522:361–365. https://doi.org/10.1016/j.tsf.2012.07.121
120. Wang S, Bai LN, Sun HM, Jiang Q, Lian JS (2013) Structure and photocatalytic property of Mo-doped TiO_2 nanoparticles. Powder Technol 244:9–15. https://doi.org/10.1016/j.powtec.2013.03.054
121. Cao B, Li G, Li H (2016) Hollow spherical RuO_2@TiO_2@Pt bifunctional photocatalyst for coupled H_2 production and pollutant degradation. Appl Catal B: Environ 194:42–49. https://doi.org/10.1016/j.apcatb.2016.04.033
122. Tanaka K, Murata S, Harada K (1985) Oxygen evolution by the photo-oxidation of water. Sol Energy 34(4):303–308. https://doi.org/10.1016/0038-092X(85)90043-X
123. Choi W, Termin A, Hoffmann MR (1994) The role of metal ion dopants in quantum-sized TiO_2: correlation between photoreactivity and charge carrier recombination dynamics. J Phys Chem 98(51):13669–13679. https://doi.org/10.1021/j100102a038
124. Liu B, Chen HM, Liu C, Andrews SC, Hahn C, Yang P (2013) Large-scale synthesis of transition-metal-doped TiO_2 nanowires with controllable overpotential. J Am Chem Soc 135(27):9995–9998. https://doi.org/10.1021/ja403761s
125. Zielińska B, Borowiak-Palen E, Kalenczuk RJ (2008) Photocatalytic hydrogen generation over alkaline-earth titanates in the presence of electron donors. Int J Hydrogen Energy 33(7):1797–1802. https://doi.org/10.1016/j.ijhydene.2008.02.001
126. Kim JY, Kim CS, Chang HK, Kim TO (2011) Synthesis and characterization of N-doped TiO_2/ZrO_2 visible light photocatalysts. Adv Powder Technol 22(3):443–448. https://doi.org/10.1016/j.apt.2010.06.014
127. Mertin W, Andersson S, Gruehn R (1970) Über die Kristallstruktur von M Nb_2O_5. J Solid State Chem 1(3):419–424. https://doi.org/10.1016/0022-4596(70)90124-6
128. Kang JH, Myung Y, Choi JW, Jang DM, Lee CW, Park J, Cha EH (2012) Nb_2O_5 nanowire photoanode sensitized by a composition-tuned CdS_xSe_{1-x} shell. J Mater Chem 22(17):8413–8419. https://doi.org/10.1039/C2JM16008K
129. Liu XQ, Han XD, Zhang Z, Ji LF, Jiang YJ (2007) The crystal structure of high temperature phase Ta_2O_5. Acta Mater 55(7):2385–2396. https://doi.org/10.1016/j.actamat.2006.11.031
130. Lin H-Y, Yang H-C, Wang W-L (2011) Synthesis of mesoporous Nb_2O_5 photocatalysts with Pt, Au, Cu and NiO cocatalyst for water splitting. Catal Today 174(1):106. https://doi.org/10.1016/j.cattod.2011.01.052
131. Adhikari SP, Hood ZD, More KL, Ivanov I, Zhang L, Gross M, Lachgar A (2015) Visible light assisted photocatalytic hydrogen generation by Ta_2O_5/Bi_2O_3, TaON/Bi_2O_3, and Ta_3N_5/Bi_2O_3 composites. RSC Advances 5(68):54998–55005. https://doi.org/10.1039/C5RA06563A
132. Pai Y-H, Fang S-Y (2013) Preparation and characterization of porous Nb_2O_5 photocatalysts with CuO, NiO and Pt cocatalyst for hydrogen production by light-induced water splitting. J Power Sources 230:321. https://doi.org/10.1016/j.jpowsour.2012.12.078
133. Xu L, Guan J, Gao L, Sun Z (2011) Preparation of heterostructured mesoporous In_2O_3/Ta_2O_5 nanocomposites with enhanced photocatalytic activity for hydrogen evolution. Catal Commun 12(6):548. https://doi.org/10.1016/j.catcom.2010.11.027
134. Gao H, Zhang P, Hu J, Pan J, Fan J, Shao G (2017) One-dimensional Z-scheme TiO_2/WO_3/Pt heterostructures for enhanced hydrogen generation. Appl Surf Sci 391(Part B):211–217. https://doi.org/10.1016/j.apsusc.2016.06.170
135. Higashi M, Abe R, Teramura K, Takata T, Ohtani B, Domen K (2008) Two step water splitting into H_2 and O_2 under visible light by $ATaO_2N$ (A=Ca, Sr, Ba) and WO_3 with shuttle redox mediator. Chem Phys Lett 452(1–3):120–123. https://doi.org/10.1016/j.cplett.2007.12.021

136. Sivula K, Formal FL, Grätzel M (2009) WO_3−Fe_2O_3 photoanodes for water splitting: a host scaffold, Guest Absorber Approach. Chem Mater 21(13):2862–2867. https://doi.org/10.1021/cm900565a
137. Coridan RH, Shaner M, Wiggenhorn C, Brunschwig BS, Lewis NS (2013) Electrical and photoelectrochemical properties of WO_3/Si tandem photoelectrodes. J Phys Chem C. https://doi.org/10.1021/jp311947x
138. Kominami H, J-i Kato, S-y Murakami, Ishii Y, Kohno M, K-i Yabutani, Yamamoto T, Kera Y, Inoue M, Inui T, Ohtani B (2003) Solvothermal syntheses of semiconductor photocatalysts of ultra-high activities. Catal Today 84(3–4):181–189. https://doi.org/10.1016/S0920-5861(03)00272-4
139. Miecznikowski K, Ramírez A, Fiechter S, Bogdanoff P, Szaniawska E, Wadas A, Kulesza PJ Development of hybrid tungsten oxide photoanodes admixed with borododecatungstate-polyanion modified-hematite: enhancement of water oxidation upon irradiation with visible light. Electrochim Acta. http://dx.doi.org/10.1016/j.electacta.2015.05.001
140. Wang H, Lindgren T, He J, Hagfeldt A, Lindquist SE (2000) Photoelectrochemistry of nanostructured WO_3 thin film electrodes for water oxidation: mechanism of electron transport. J Phys Chem B 104:5686–5696
141. Bamwenda GR, Arakawa H (2001) The visible light induced photocatalytic activity of tungsten trioxide powders. Appl Catal A-Gen 210(1–2):181–191. https://doi.org/10.1016/S0926-860X(00)00796-1
142. Bamwenda GR, Uesigi T, Abe Y, Sayama K, Arakawa H (2001) The photocatalytic oxidation of water to O_2 over pure CeO_2, WO_3, and TiO_2 using Fe^{3+} and Ce^{4+} as electron acceptors. Appl Catal A-Gen 205(1–2):117–128. https://doi.org/10.1016/S0926-860X(00)00549-4
143. Komornicki S, Radecka M, Sobas P (2004) Structural, electrical and optical properties of TiO_2-WO_3 polycrystalline ceramics. Mater Res Bull 39(13):2007–2017
144. Patil PS, Mujawar SH, Inamdar AI, Shinde PS, Deshmukh HP, Sadale SB (2005) Structural, electrical and optical properties of TiO_2 doped WO_3 thin films. Appl Surf Sci 252(5):1643–1650
145. Wang S, Zhang X, Cheng G, Jiang X, Li Y, Huang Y, Du Z (2005) Study on electronic transport properties of WO_3/TiO_2 nanocrystalline thin films by photoassisted conductive atomic force microscopy. Chem Phys Lett 405(1–3):63–67. https://doi.org/10.1016/j.cplett.2005.01.118
146. de Tacconi NR, Chenthamarakshan CR, Yogeeswaran G, Watcharenwong A, de Zoysa RS, Basit NA, Rajeshwar K (2006) Nanoporous TiO_2 and WO_3 films by anodization of titanium and tungsten substrates: influence of process variables on morphology and photoelectrochemical response. J Phys Chem B 110(50):25347–25355. https://doi.org/10.1021/jp064527v
147. Ke D, Liu H, Peng T, Liu X, Dai K (2008) Preparation and photocatalytic activity of WO_3/TiO_2 nanocomposite particles. Mater Lett 62(3):447–450. https://doi.org/10.1016/j.matlet.2007.05.060
148. Watcharenwong A, Chanmanee W, de Tacconi NR, Chenthamarakshan CR, Kajitvichyanukul P, Rajeshwar K (2008) Anodic growth of nanoporous WO_3 films: morphology, photoelectrochemical response and photocatalytic activity for methylene blue and hexavalent chrome conversion. J Electroanal Chem 612(1):112–120
149. Xiao M, Wang L, Huang X, Wu Y, Dang Z (2009) Synthesis and characterization of WO_3/titanate nanotubes nanocomposite with enhanced photocatalytic properties. J Alloys Compd 470(1–2):486–491. https://doi.org/10.1016/j.jallcom.2008.03.003
150. Chen Q, Li J, Li X, Huang K, Zhou B, Cai W, Shangguan W (2012) Visible-light responsive photocatalytic fuel cell based on WO_3/W photoanode and Cu_2O/Cu photocathode for simultaneous wastewater treatment and electricity generation. Environ Sci Technol 46(20):11451–11458. https://doi.org/10.1021/es302651q

151. Wang F, Di Valentin C, Pacchioni G (2012) Doping of WO_3 for photocatalytic water splitting: hints from density functional theory. J Phys Chem C 116(16):8901–8909. https://doi.org/10.1021/jp300867j
152. Janáky C, Rajeshwar K, de Tacconi NR, Chanmanee W, Huda MN (2013) Tungsten-based oxide semiconductors for solar hydrogen generation. Catal Today 199:53–64. https://doi.org/10.1016/j.cattod.2012.07.020
153. Lai CW, Sreekantan S (2013) Preparation of hybrid WO_3–TiO_2 nanotube photoelectrodes using anodization and wet impregnation: improved water-splitting hydrogen generation performance. Int J Hydrogen Energy 38(5):2156–2166. https://doi.org/10.1016/j.ijhydene.2012.12.025
154. Liu G, Han J, Zhou X, Huang L, Zhang F, Wang X, Ding C, Zheng X, Han H, Li C (2013) Enhancement of visible-light-driven O_2 evolution from water oxidation on WO_3 treated with hydrogen. J Catal 307:148–152. https://doi.org/10.1016/j.jcat.2013.06.024
155. Hou H, Gao F, Wang L, Shang M, Yang Z, Zheng J, Yang W (2016) Superior thoroughly mesoporous ternary hybrid photocatalysts of $TiO_2/WO_3/g\text{-}C_3N_4$ nanofibers for visible-light-driven hydrogen evolution. J Mater Chem A 4(17):6276–6281. https://doi.org/10.1039/C6TA02307J
156. Huang J, Zhang Y, Ding Y (2017) Rationally designed/constructed CoO_x/WO_3 anode for efficient photoelectrochemical water oxidation. ACS Catal 7(3):1841–1845. https://doi.org/10.1021/acscatal.7b00022
157. Karimi-Nazarabad M, Goharshadi EK (2017) Highly efficient photocatalytic and photo-electrocatalytic activity of solar light driven $WO_3/g\text{-}C_3N_4$ nanocomposite. Sol Energy Mater Sol Cells 160:484–493. https://doi.org/10.1016/j.solmat.2016.11.005
158. Ma C, Zhou J, Cui Z, Wang Y, Zou Z (2016) In situ growth MoO_3 nanoflake on conjugated polymer: an advanced photocatalyst for hydrogen evolution from water solution under solar light. Sol Energy Mater Sol Cells 150:102–111. https://doi.org/10.1016/j.solmat.2016.02.010
159. Chen Z, Cummins D, Reinecke BN, Clark E, Sunkara MK, Jaramillo TF (2011) Core–shell MoO_3–MoS_2 nanowires for hydrogen evolution: a functional design for electrocatalytic materials. Nano Lett 11(10):4168–4175. https://doi.org/10.1021/nl2020476
160. Huang L, Xu H, Zhang R, Cheng X, Xia J, Xu Y, Li H (2013) Synthesis and characterization of $g\text{-}C_3N_4/MoO_3$ photocatalyst with improved visible-light photoactivity. Appl Surf Sci 283:25–32. https://doi.org/10.1016/j.apsusc.2013.05.106
161. Ma BJ, Kim JS, Choi CH, Woo SI (2013) Enhanced hydrogen generation from methanol aqueous solutions over $Pt/MoO_3/TiO_2$ under ultraviolet light. Int J Hydrogen Energy 38(9):3582–3587. https://doi.org/10.1016/j.ijhydene.2012.12.142
162. Ghaffar I, Warsi MF, Shahid M, Shakir I (2016) Unprecedented photocatalytic activity of carbon coated/MoO_3 core–shell nanoheterostructurs under visible light irradiation. Physica E 79:1–7. https://doi.org/10.1016/j.physe.2015.11.040
163. Yang M, Zhang L, Jin B, Huang L, Gan Y (2016) Enhanced photoelectrochemical properties and water splitting activity of self-ordered MoO_3–TiO_2 nanotubes. Appl Surf Sci 364:410–415. https://doi.org/10.1016/j.apsusc.2015.12.157
164. de Tacconi NR, Chenthamarakshan CR, Wouters KL, MacDonnell FM, Rajeshwar K (2004) CompositeWO_3–TiO_2 films prepared by pulsed electrodeposition: morphological aspects and electrochromic behavior. J Electroanal Chem 566(2):249–256
165. Ohno T, Murakami N, Koyanagi T, Yang Y (2014) Photocatalytic reduction of CO_2 over a hybrid photocatalyst composed of WO_3 and graphitic carbon nitride ($g\text{-}C_3N_4$) under visible light. J CO_2 Utilization 6:17–25. https://doi.org/10.1016/j.jcou.2014.02.002
166. Solarska R, Bienkowski K, Zoladek S, Majcher A, Stefaniuk T, Kulesza PJ, Augustynski J (2014) Enhanced water splitting at thin film tungsten trioxide photoanodes bearing plasmonic gold-polyoxometalate particles. Angew Chem Int Ed 53(51):14196–14200. https://doi.org/10.1002/anie.201408374

167. Jayaraman S, Jaramillo TF, Baeck SH, McFarland EW (2005) Synthesis and characterization of Pt-WO_3 as methanol oxidation catalysts for fuel cells. J Phys Chem B 109(48): 22958–22966
168. Abadias G, Gago AS, Alonso-Vante N (2011) Structural and photoelectrochemical properties of $Ti_{1-x}W_xO_2$ thin films deposited by magnetron sputtering. Surf Coat Technol 205(2):S265–S270
169. Geller S (1960) Crystal Structure of β-Ga_2O_3. J Chem Phys 33(3):676–684. https://doi.org/10.1063/1.1731237
170. Guo W, Guo Y, Dong H, Zhou X (2015) Tailoring the electronic structure of β-Ga_2O_3 by non-metal doping from hybrid density functional theory calculations. Phys Chem Chem Phys 17(8):5817–5825. https://doi.org/10.1039/C4CP05637J
171. Sakata Y, Matsuda Y, Yanagida T, Hirata K, Imamura H, Teramura K (2008) Effect of metal ion addition in a Ni Supported Ga_2O_3 photocatalyst on the photocatalytic overall splitting of H_2O. Catal Lett 125(1):22–26. https://doi.org/10.1007/s10562-008-9557-7
172. Christoph J, Valentina S, Mansour M, Alica K, Helmut D, Recardo M, Zbigniew G, Reinhard U, Klaus I, Roberto F, Marcel M, Dieter S, Justin RW, Joel BV, Chris GVdW (2011) Experimental electronic structure of In_2O_3 and Ga_2O_3. New J Phys 13(8):085014
173. Jiang Z, Jiang D, Yan Z, Liu D, Qian K, Xie J (2015) A new visible light active multifunctional ternary composite based on TiO_2–In_2O_3 nanocrystals heterojunction decorated porous graphitic carbon nitride for photocatalytic treatment of hazardous pollutant and H2 evolution. Appl Catal B: Environ 170–171:195–205. https://doi.org/10.1016/j.apcatb.2015.01.041
174. Kadowaki H, Sato J, Kobayashi H, Saito N, Nishiyama H, Simodaira Y, Inoue Y (2005) Photocatalytic activity of the RuO_2-dispersed composite p-block metal oxide $LiInGeO_4$ with d^{10}–d^{10} configuration for water decomposition. J Phys Chem B 109(48):22995–23000. https://doi.org/10.1021/jp0544686
175. Sato J, Saito N, Nishiyama H, Inoue Y (2002) Photocatalytic water decomposition by RuO_2-loaded antimonates, $M_2Sb_2O_7$ (M=Ca, Sr), $CaSb_2O_6$ and $NaSbO_3$, with d^{10} configuration. J Photochem Photobiol, A 148(1–3):85–89. https://doi.org/10.1016/S1010-6030(02)00076-X
176. Lu Q, Yu Y, Ma Q, Chen B, Zhang H (2016) 2D transition-metal-dichalcogenide-nanosheet-based composites for photocatalytic and electrocatalytic hydrogen evolution reactions. Adv Mater 28(10):1917–1933. https://doi.org/10.1002/adma.201503270
177. Kaneko H, Minegishi T, Nakabayashi M, Shibata N, Domen K (2016) Enhanced hydrogen evolution under simulated sunlight from neutral electrolytes on $(ZnSe)_{0.85}$ $(CuIn_{0.7}Ga_{0.3}Se_2)_{0.15}$ photocathodes prepared by a bilayer method. Angew Chem 128 (49):15555–15559. https://doi.org/10.1002/ange.201609202
178. Chen J, Wu X-J, Yin L, Li B, Hong X, Fan Z, Chen B, Xue C, Zhang H (2015) One-pot synthesis of CdS nanocrystals hybridized with single-layer transition-metal dichalcogenide nanosheets for efficient photocatalytic hydrogen evolution. Angew Chem Int Ed 54(4):1210–1214. https://doi.org/10.1002/anie.201410172
179. Xu L, Shi W, Guan J (2012) Preparation of crystallized mesoporous CdS/Ta_2O_5 composite assisted by silica reinforcement for visible light photocatalytic hydrogen evolution. Catal Commun 25:54. https://doi.org/10.1016/j.catcom.2012.03.037
180. Yuan J, Wen J, Zhong Y, Li X, Fang Y, Zhang S, Liu W (2015) Enhanced photocatalytic H_2 evolution over noble-metal-free NiS cocatalyst modified CdS nanorods/g-C_3N_4 heterojunctions. J Mater Chem A 3(35):18244–18255. https://doi.org/10.1039/C5TA04573H
181. Tian F, Hou D, Hu F, Xie K, Qiao X, Li D (2017) Pouous TiO_2 nanofibers decorated CdS nanoparticles by SILAR method for enhanced visible-light-driven photocatalytic activity. Appl Surf Sci 391(Part B):295–302. https://doi.org/10.1016/j.apsusc.2016.07.010
182. Qi L, Yu J, Jaroniec M (2011) Preparation and enhanced visible-light photocatalytic H2-production activity of CdS-sensitized Pt/TiO_2 nanosheets with exposed (001) facets. Phys Chem Chem Phys

183. Park H, Choi W, Hoffmann MR (2008) Effects of the preparation method of the ternary CdS/TiO_2/Pt hybrid photocatalysts on visible light-induced hydrogen production. J Mater Chem 18(20):2379–2385
184. Zhou Q, Kang S-Z, Li X, Qin L, Mu J (2015) $AgGaS_2$ nanoplates loaded with CuS: an efficient visible photocatalyst for rapid H_2 evolution. Int J Hydrogen Energy 40(11):4119–4128. https://doi.org/10.1016/j.ijhydene.2015.01.143
185. Cheng F, Yin H, Xiang Q (2017) Low-temperature solid-state preparation of ternary CdS/g-C_3N_4/CuS nanocomposites for enhanced visible-light photocatalytic H_2-production activity. Appl Surf Sci 391(Part B):432–439. https://doi.org/10.1016/j.apsusc.2016.06.169
186. Zong X, Yan H, Wu G, Ma G, Wen F, Wang L, Li C (2008) Enhancement of photocatalytic H_2 evolution on CdS by loading MoS_2 as cocatalyst under visible light irradiation. J Am Chem Soc 130(23):7176–7177. https://doi.org/10.1021/ja8007825
187. Zong X, Wu G, Yan H, Ma G, Shi J, Wen F, Wang L, Li C (2010) Photocatalytic H_2 evolution on MoS_2/CdS catalysts under visible light irradiation. J Phys Chem C 114 (4):1963–1968. https://doi.org/10.1021/jp904350e
188. Chen G, Ding N, Li F, Fan Y, Luo Y, Li D, Meng Q (2014) Enhancement of photocatalytic H_2 evolution on $ZnIn_2S_4$ loaded with in-situ photo-deposited MoS_2 under visible light irradiation. Appl Catal B: Environ 160–161:614–620. https://doi.org/10.1016/j.apcatb.2014.05.028
189. Shen M, Yan Z, Yang L, Du P, Zhang J, Xiang B (2014) MoS_2 nanosheet/TiO_2 nanowire hybrid nanostructures for enhanced visible-light photocatalytic activities. Chem Commun 50 (97):15447–15449. https://doi.org/10.1039/C4CC07351G
190. Qin N, Xiong J, Liang R, Liu Y, Zhang S, Li Y, Li Z, Wu L (2017) Highly efficient photocatalytic H_2 evolution over MoS_2/CdS-TiO_2 nanofibers prepared by an electrospinning mediated photodeposition method. Appl Catal B: Environ 202:374–380. https://doi.org/10.1016/j.apcatb.2016.09.040
191. Han W, Zang C, Huang Z, Zhang H, Ren L, Qi X, Zhong J (2014) Enhanced photocatalytic activities of three-dimensional graphene-based aerogel embedding TiO_2 nanoparticles and loading MoS_2 nanosheets as Co-catalyst. Int J Hydrogen Energy 39(34):19502–19512. https://doi.org/10.1016/j.ijhydene.2014.09.043
192. Li H, Wang Y, Chen G, Sang Y, Jiang H, He J, Li X, Liu H (2016) Few-layered MoS_2 nanosheets wrapped ultrafine TiO_2 nanobelts with enhanced photocatalytic property. Nanoscale 8(11):6101–6109. https://doi.org/10.1039/C5NR08796A
193. Yuan Y-J, Wang F, Hu B, Lu H-W, Yu Z-T, Zou Z-G (2015) Significant enhancement in photocatalytic hydrogen evolution from water using a MoS_2 nanosheet-coated ZnO heterostructure photocatalyst. Dalton Trans 44(24):10997–11003. https://doi.org/10.1039/C5DT00906E
194. Tian Q, Zhang L, Liu J, Li N, Ma Q, Zhou J, Sun Y (2015) Synthesis of MoS_2/$SrZrO_3$ heterostructures and their photocatalytic H_2 evolution under UV irradiation. RSC Advances 5(1):734–739. https://doi.org/10.1039/C4RA11135D
195. Alonso-Vante N, Tributsch H (1987) Electrode surface modification through Nafion-attached transition-metal cluster chalcogenide particles. J Electroanal Chem 229(1–2):223–237
196. Alonso-Vante N (1998) Inert for selective oxygen reduction of oxygen and method for the production thereof. Germany Patent WO1997DE02453 19971016; DE19961044628 19961017
197. Fischer CH, Henglein A (1989) Photochemistry of colloidal semiconductors. 31. Preparation and photolysis of cadmium sulfide sols in organic solvents. J Phys Chem 93(14):5578–5581. https://doi.org/10.1021/j100351a049
198. Henglein A (1989) Small-particle research: physicochemical properties of extremely small colloidal metal and semiconductor particles. Chem Rev 89(8):1861–1873. https://doi.org/10.1021/cr00098a010

199. Lana Villarreal T, Bogdanoff P, Salvador P, Alonso-Vante N (2004) Photocatalytic oxidation on nanostructured chalcogenide modified titanium dioxide. Sol Energy Mater Sol Cells 83(4):347–362
200. Alonso-Vante N (2003) Physico-chemical properties of novel nanocrystalline ruthenium based chalcogenide materials. In: Kokorin AI, Bahnemann DW (eds) Chemical physics of nanostructured semiconductors VSP brill academic. Zeist, Neth, pp 135–152
201. Lana Villarreal T, Alonso-Vante N (2005) Photoinduced charge transfer between TiO_2 particles and Ru_xSe_y clusters: a photocatalytic performance improvement. In: Meeting Abstracts, ECS 207th—Quebec City, Canada. p 1375
202. Bahnemann D, Henglein A, Spanhel L (1984) Detection of the intermediates of colloidal TiO_2-catalysed photoreactions. Faraday Discuss Chem Soc 78:151–163
203. Kamat PV, Gevaert M, Vinodgopal K (1997) Photochemistry on semiconductor surfaces. Visible light induced oxidation of C_{60} on TiO_2 Nanoparticles. J Phys Chem B 101 (22):4422–4427. https://doi.org/10.1021/jp970047f

Chapter 4
Precious Versus Non-precious Electrocatalyst Centers

4.1 Introduction

Advances in earth-abundant materials is of paramount importance to promote reactions of technological interest, such as the hydrogen evolution reaction (HER), the hydrogen oxidation reaction (HOR), the oxygen reduction reaction (ORR), and the oxygen evolution reaction (OER) discussed in the previous chapters. To find a balance among the materials' activity, stability, and selectivity with an unlimited materials' availability represents challenging tasks in laboratories all over the world working in the rationalization of this problem. The gap to overcome is to justify the design of catalytic centers based on non-precious ones, aiming at approaching the activity of the so-called precious catalytic centers. However, most of the precious metals with respectable electrocatalytic properties have limited availability and high price, as in Fig. 4.1. An availability limitation also exists for some chalcogen elements. Among the chalcogens, sulfur is by far the most abundant (4 and 5.6 orders of magnitude higher than Se and Te, respectively), with a low average price of 2 orders of magnitude with respect to Au, Pt, and Pd. A similar observation with respect to the availability and price can be drawn for non-precious elements, such as W, Mo, Co, Ni, and Fe. In this chapter, we will concentrate on the role of chalcogenides (provided with precious and non-precious centers) as electrocatalysts.

4.2 Precious Metal Center Chalcogenides

Among these electrocatalytic materials, we can account on Pt- [1–3], Ru- [4–39], Pd- [40–43], Rh- [18, 44–46], and Ir- [47–49] based chalcogenide centers. The most fundamental studies on this kind of materials are devoted so far, to a large extent, to the oxygen reduction reaction in acid medium.

N. Alonso-Vante, *Chalcogenide Materials for Energy Conversion*, Nanostructure Science and Technology, https://doi.org/10.1007/978-3-319-89612-0_4

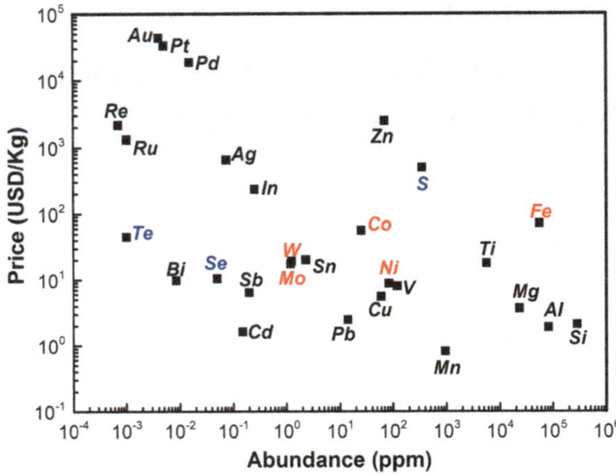

Fig. 4.1 Relationship between metal prices and relative abundance (ppm) of the chemical elements in Earth's upper continental crust. The corresponding metal price data on June 8, 2017, are from www.metalprices.com

4.2.1 Materials for HER/HOR

Few (or none) chalcogenide materials containing a precious catalytic center (e.g., Pt, Pd) are reported as electrocatalysts for the HER and/or HOR. Since platinum is one of the best electrocatalytic material for the HER/HOR in acid medium, cf. Figs. 2.8 and 2.17, the purpose is to find alternative materials [50, 51] to platinum-based ones to produce hydrogen from water at low overpotentials (see Sect. 4.3.1). In this connection, precious catalytic centers, such as Pt, Pd, and Ir, were incorporated into chalcogenide matrices [52], by modifying a chalcogenide layer to boost the HER electrocatalytically in dark [53] and under illumination (AM 1.5G) [54]. One interesting example to mention is the epitaxial growth of a monolayer of semiconducting $PtSe_2$ (E_g = 1.2 eV), cf. Figs. 1.4 and 1.7, performed by a single-step selenization on Pt (111). This material is a potential suitable visible-light photocatalysis for catalysis [1], cf. Chap. 3.

Chalcogels come into play (cf. Sect. 1.5.2) with the development of ternary chalcogels assimilating $[Fe_4S_4]$ biological redox-active centers bridging $[Sn_4S_{10}]$ units linked with metals ions forming the chalcogel framework: $[Fe_4S_4]_xM_y[Sn_4S_{10}]_z$ = M-ITS-cg3, where M (Pt, Zn, Co, Ni, Sn) [55]. As shown in Fig. 4.2a, the nature of the metal affects the electron charge transfer processes within the chalcogel framework, going from the more cathodic to the more anodic onset potential. It is apparent that platinum best reinforces kinetically the evolution of hydrogen, as compared to Co-ITS-cg3.

On the other hand, the electrocatalytic behavior of a series of composites made with palladium atoms deposited onto defect-rich (DR) molybdenum disulfide ($DR-MoS_2$) showed an electrocatalytic activity similar to that of platinum (see Fig. 4.2b [53]). In this figure, the TEM pictures of the molybdenum chalcogenide are contrasted with the HER current–potential characteristics, stressing that the way Pd atoms were grown onto the basal plane of the chalcogenide (TEM Images b–d), going from the less to the most active material, i.e., (b) Pd NPs/$DR-MoS_2$ (epitaxial),

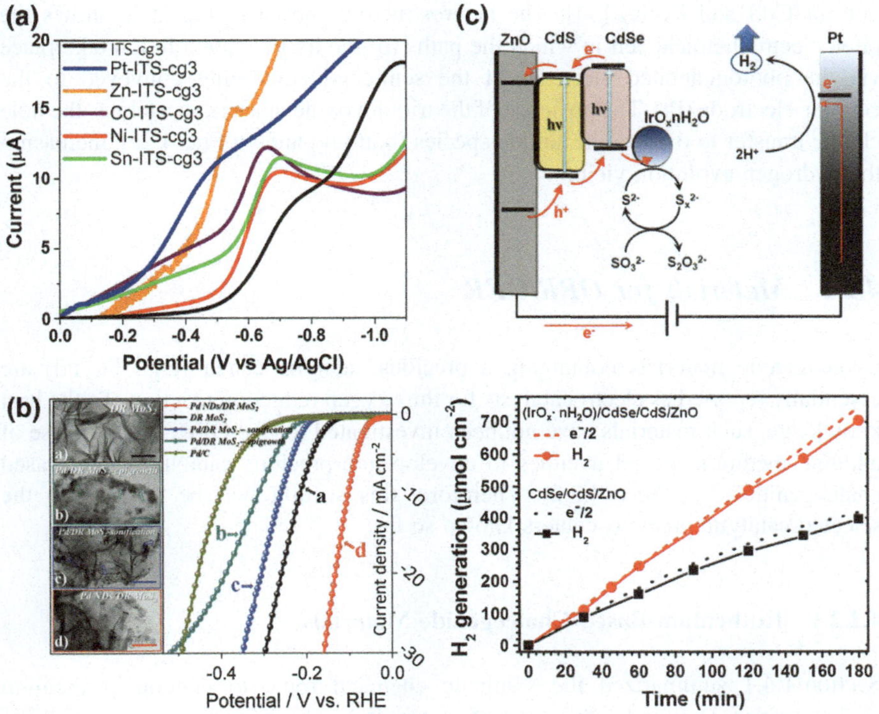

Fig. 4.2 **a** Hydrogen evolution current–potential characteristics at M-ITS-cg3 samples (M = Pt, Zn, Co, Ni, Sn). Adapted with permission from Ref. [55]. Copyright © 2014, American Chemical Society. **b** TEM images of composites: (a) DR-MoS_2; (b) Pd nanoparticle/DR-MoS_2 composite (epitaxial growth), (c) Pd nanoparticle/DR-MoS_2 (sonication), and (d) Pd ND/DR-MoS_2, in which arrows show Pd NPs aggregation and boundary sites of support (DR-MoS_2). The corresponding right panel shows the HER polarization curves of the Pd/DR-MoS_2 samples. Adapted with permission from Ref. [53]. Copyright © 2016, Royal Society of Chemistry. **c** Upper panel: photoelectrochemical hydrogen generation scheme; bottom panel: hydrogen produced at 0.6 V/RHE under an illumination of 1 sun (AM1.5G) on CdSe/CdS/ZnO nanowire electrode before, and after $IrO_x{\cdot}nH_2O$ modification. The 100% faradaic efficiency expected hydrogen evolution is shown in dashed lines. Adapted with permission from Ref. [54]. Copyright © 2013, American Chemical Society

(c) Pd NPs/DR-MoS_2 (sonication), and (d) Pd ND/DR-MoS_2 (epitaxial assembly of palladium nanodisks (PD NDs)). The HER on these systems clearly reflects the synergy of both catalytic entities in Pd ND/DR-MoS_2 as compared to their corresponding current–potential curves for Pd/C and DR-MoS_2 alone. The so-called synergy effect between Pd NDs and DR-MoS_2 shows an important chemical signature change in the emission lines for S *2p* and Mo *3d* (see Fig. 6, in Ref: [53]) assessing the fact of an electronic modification of the catalytic center via the interaction with the support. Furthermore, iridium was employed, in its oxidized form (IrO_x), since it is known that this material is a good oxygen evolution electrocatalysts (cf. Fig. 2.12), to form nanojunctions with chalcogenide semiconductors (quantum dots of CdS and CdSe) [54]. The heterostructure shown in Fig. 4.2c shows the photoelectrochemical cell in which the paths for the hydrogen evolution take place via the photogenerated electrons at the semiconductor being conveyed to the counter-electrode (Pt). The presence of the iridium oxide clusters accelerates the hole charge transfer to oxidize the sulfide species (bottom panel in Fig. 4.2c), increasing the hydrogen evolution yield.

4.2.2 Materials for ORR/OER

Chalcogenide materials containing a precious catalytic center (e.g., Pt, Pd) are essentially reported as electrocatalysts for the oxygen reduction reaction. To the best knowledge, such materials have not been investigated for the OER, since the use of alkaline medium opened avenues to develop non-precious catalytic center-based chalcogenides (see Sect. 4.3.2). Therefore, this section will be devoted to the selected catalytic precious centers known so far.

4.2.2.1 Ruthenium-Based Chalcogenide Materials

Section 1.4.1 summarized the synthetic chemical route to generate ruthenium chalcogenides. Definitely, this kind of materials has been investigated toward the oxygen reduction reaction in acid medium [45, 56–65]. The Chevrel-phase material, $Mo_4Ru_2Se_8$, was considered as a potential alternative to platinum-based electrocatalysts [66], and since then, ruthenium as catalytic center has been further developed [67].

Among the Chevrel phase, the pseudo-ternary material ($Mo_{4.2}Ru_{1.8}Se_8$), cf. Figs. 1.9, 1.10, showed the highest ORR activity with an overpotential of ca. 0.35 V, and the detected amount of hydrogen peroxide was around 2.8–3.9% in oxygen-saturated 0.5 M H_2SO_4 [68, 69] (see Table 4.1). The kinetic analysis of this kind of multi-electron charge transfer reaction suggested that the primary electron was the limiting rate of the reaction, and a first-order kinetics according to the Koutecky–Levich analysis. The slow kinetic may result from the limitation of favorable adsorption sites. Moreover, the cluster made of ($Mo_{6-2}Ru_2X_8$) can be considered as an electron reservoir responsible for the ORR activity, where the adsorbed molecule $O_{2,ad}$ can directly exchange electrons at the electrocatalyst's surface [70, 71]. This latter statement is supported by the in situ electrochemical EXAFS experiments and analysis of the material. The Ru-(Ru, Mo) distance was sensitive to the electrocatalytic oxygen reduction reaction, whereas the distance between Ru and Se remained insensitive [70]. This observation further assessed that the ORR took place on ruthenium centers.

Unlike the Chevrel phase, whose temperature synthesis in the solid state in quartz ampoule amounts to 1000 °C or higher and high pressure (0.4 GPa) [73], novel Ru-based chalcogenide catalysts were synthesized in mild conditions (<200 °C)

Table 4.1 Oxygen reduction reaction on sintered Chevrel-phase materials evaluated in 0.5 M H_2SO_4

Catalysts	Onset potential versus RHE/V	H_2O_2/%	Log $\|j_0\|$ mA/cm^{-2}	Refs.
$Mo_{4.2}Ru_{1.8}Se_8$	0.87	2.8–3.9	−0.85	[68]
$Mo_2Re_4Se_8$	0.77	<4	−1.7	[72]
Mo_6Se_6	0.60		−4.84	
$Mo_2Re_4S_8$	0.67		−3.5	
Mo_6S_8	0.30		−5.0	

in organic solvents [65, 74] and later on in aqueous conditions using commercial Ru black and Se powder as precursors without any carbonyl ligand involved in the preparation process [75, 76]. A summary of some relevant data obtained so far on ruthenium-based chalcogenide samples is abridged in Table 4.2. This table collects room-temperature (RT) data from the current–potential characteristics, generated in 0.5 M H_2SO_4 at a rotating speed of 1600 rpm, and a scan rate of 5 mV/s, namely the onset potential; the half-wave potential; the hydrogen peroxide production (at electrode potentials below 0.6 V vs. RHE); and the catalytic mass loading. Summarizing these data, generated in various laboratories, we learn that typical thermodynamic/kinetic parameters are as follows: (i) The onset potential is 0.85 ± 0.05 V/RHE, that is, ca. 0.1 V negative to that of Pt-based catalysts ca. 0.95 ± 0.05 V; (ii) the $E_{1/2}$ potential is 0.65 ± 0.05 V, again lower than that of Pt-based catalysts ca. 0.87 V; (iii) the maximum hydrogen peroxide production (ca. 4%) lies below 0.6 V/RHE, indicating that sufficient active sites are available to promote a four-electron pathway. It is important to recall that the ORR efficiency is dependent on the catalytic site density, whatever the nature of the catalytic center, thus the catalyst mass loading [77]; and (iv) the generated Ru_xSe_y in p-xylene (BP 140 °C) or 1,2-dichlorobenzene (BP 180 °C) showed higher catalytic activity toward the ORR as compared to the Chevrel phase (see Fig. 4.3). Indeed, for the same hydrodynamic conditions (i.e., 400 rpm), the Ru_xSe_y in powder form imbedded in a carbon paste assesses the above statement, as in Fig. 4.3b [78]. Aiming in producing the Chevrel phase, molybdenum atoms were added in the carbonyl synthesis route using non-aqueous solvents. It turned out that the resulting product was dubbed $(Ru_{1-x}Mo_x)_ySeO_z$ where the amount of incorporated Mo atoms was ca. 2–4% (mol/mol, related to Ru) [65].

The tiny amount of Mo metal (in doping quantities) showed a slightly higher ORR activity, and this phenomenon was ascribed to the fact that Mo atoms adsorb oxygen molecules which are reduced on Ru atoms, serving as ORR catalytic center. The success of Mo-modified Ru_xSe_y materials was reinforced by further showing the suppression of the catalytic center oxidation via the coordination with selenium atoms inducing ligand and strain effects, and in addition displaying tolerance to methanol. The material was also tolerant toward the phosphate poisoning [64]. The basis of the chemical stability toward oxidation in air of ruthenium nanoparticles to RuO_x lies in the fact that Se atoms are surface coordinated [87], reinforcing further

Table 4.2 Oxygen reduction reaction on ruthenium chalcogenides materials evaluated in 0.5 M H_2SO_4

Catalysts	Onset potential versus RHE/V	$E_{1/2}$ versus RHE/V	H_2O_2/%	Catalyst loading/μg cm^{-2}	Refs.
$(Ru_{1-x}Mo_x)_ySeO_z$	0.85	0.58	<4		[78]
$(Ru_{1-x}Mo_x)_ySeO_2$	0.90	0.72			[79]
Ru_xSe_y	0.83				[80]
Ru_xSe_y	0.87			57 ± 6	[81]
$Ru_{1.92}Mo_{0.08}SeO_4$	0.85	0.71	<1.5	40	[82]
$RuSe_{0.17}$	0.84	0.67	8.72	143	[83]
Ru	0.75	0.54		150	[28]
Ru/S	0.84	0.58			
Ru/Se	0.90	0.69			
Ru_xSe_y	0.90	0.71			
Pt/C	1.0	0.87			
$RuSe_x$/C	0.85	0.73			[18]
WO_3-$Ru_2Se_{0.1}$/C	0.90	0.59	0.67	156	[21]
$Ru_2Se_{0.1}$/C	0.90	0.50	1.9		
$RuSe_x$/C (Se, 6.3 wt%)	0.84	0.63		150	[84]
$RuSe_x$/C (Se, 11.8 wt%)	0.86	0.66			
$RuSe_x$/C (Se, 16.7 wt%)	0.89	0.68			
$RuSe_{0.15}$/C	0.81	0.70	2.06	28	[39]
$RuSe_{0.30}$/C	0.82	0.64	2.06		
$RuSe_{0.60}$/C	0.84	0.62	3.38		
RuSe/C	0.87	0.65	4.71		
$(Ru_xSe_y)_{xyl}$	0.89	0.68	10	56	[85]
$(RuFeSe)_{xyl}$		0.69	8		
$(RuFeSe)_{DCB}$		0.71	8		
Ru_xSe_y/C (20 wt%)	0.89	0.77		56	[86]

the positive effect of coordinated Se atoms toward the ORR process. The role of selenium atoms coverage on ORR was investigated, and it was found that on Ru_xSe_y a maximum ORR kinetic current density at 0.62 V/RHE occurs with 15 mol% Se [88], through a concomitant decrease of the hydrogen peroxide production with the selenium content [88, 89]. It was also found that the ORR enhancement increased with Se from 6.3 to 25 wt% [84]. The effect of selenium coverage can be directly detected when performing the electrochemical stripping of selenium. Figure 4.4a shows the different anodic potential profiles performed on 20 wt% Ru_xSe_y/C in 0.5 M H_2SO_4. The scan to 0.8 V (curve a) and 0.95 V (curve b) does not perturbate the material's ORR performance, as in Fig. 4.4b. However, scanning up to 1.05 V

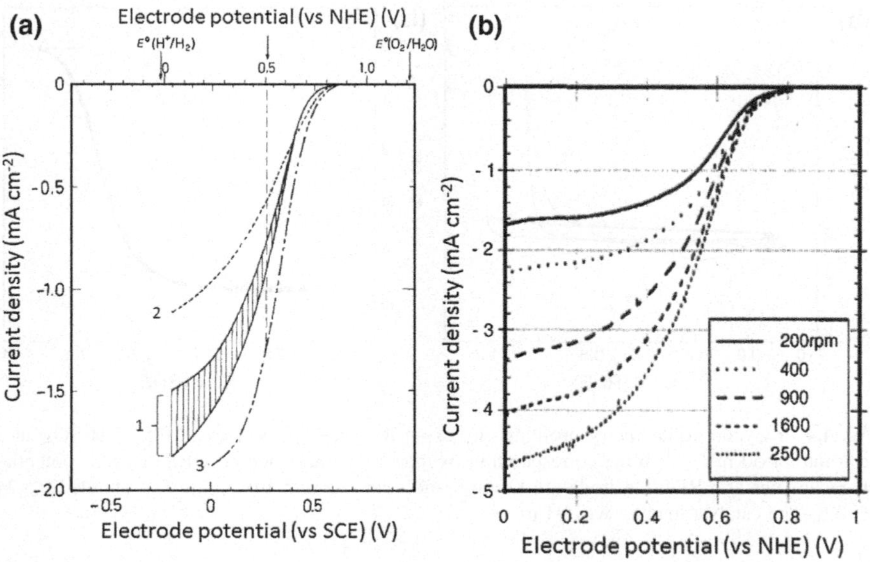

Fig. 4.3 **a** Oxygen reduction reaction in 0.5 M H_2SO_4 on sintered Chevrel phase (1) $Mo_{4.2}Ru_{1.8}Se_8$; (2) $Mo_{3.7}Ru_{2.3}Se_8$; (3) on massif Pt. Rotation rate 400 rpm; 5 mV/s. Adapted with permission from Ref. [66]. Copyright © 1987, Rights Managed by Nature Publishing Group. **b** Current–potential curves for the molecular oxygen reduction at RDE electrodes in 0.5 M H_2SO_4 of Ru_xSe_y powder embedded in carbon paste-electrodes. Adapted with permission from Ref. [78]. Copyright © 1995 Published by Elsevier Ltd.

(curve c), one can barely start to see the effect of the stripped selenium in the corresponding ORR (curve c). This phenomenon testifies that the ORR open-circuit potential of ruthenium selenide is limited to the surface-coordinated selenium oxidation. The selenium stripping can be accelerated when cycling up to 1.2 V, as demonstrated by the low ORR performance (curves s1 … s5) up to the formation of an oxide layer on ruthenium nanoparticles (curve s11). The nature of stabilizing chemically the ruthenium atoms' surface is due to a charge exchange between ruthenium and selenium. Selenium is a p-type semiconductor (E_g = 1.9 eV [90]) and via the coordination with ruthenium becomes metallic in Ru_xSe_y. This phenomenon was put in evidence using solid-state nuclear magnetic resonance (NMR) [91]. On the other hand, the local density of electronic states (LDOS) using the density functional theory (DFT) [92, 93], taking into consideration a one-third and a two-third monolayer coverage of selenium on Ru (0001), indicated no perceptible changes around the Fermi level of the Ru *d*-band center, as in Fig. 4.5. The coordinated selenium atoms tune the reactivity of the catalytic center since they become negatively charged, repelling negatively charged OH and O species, improving, as discussed above, the ORR activity.

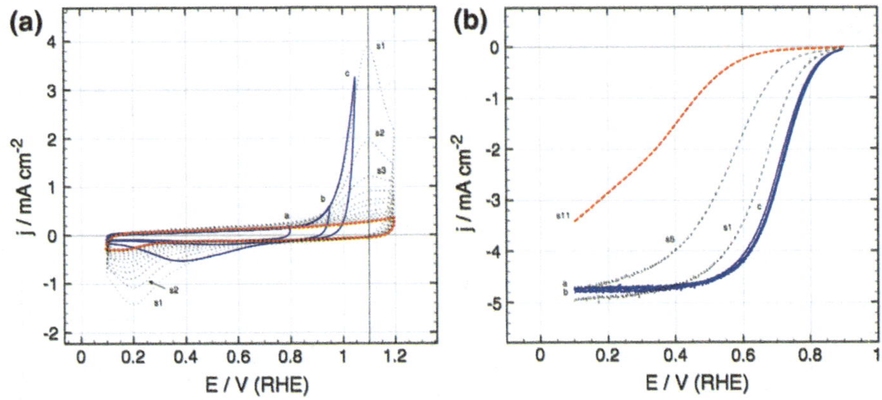

Fig. 4.4 **a** Cyclic voltammetry profiling curves on Ru_xSe_y/C in N_2-purged 0.5 M H_2SO_4, at a scan rate of 50 mV s^{-1}; **b** the corresponding oxygen reduction reaction (ORR) current–potential characteristics at a RDE electrode (1600 rpm), and scan rate of 5 mV s^{-1} in O_2-saturated 0.5 M H_2SO_4. The catalyst loading was 94 μg cm^{-2}

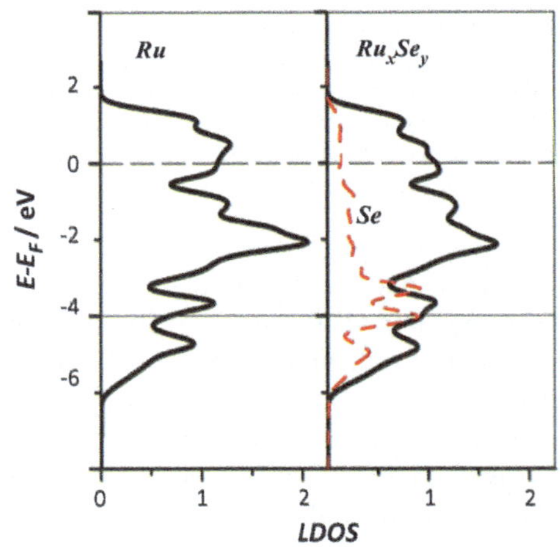

Fig. 4.5 Densities of the Ru *d*-electronic states (black solid lines) and Se *p*-states (red dashed line) calculated for Se (1/3)/Ru(0001) (upper panel) and for clean Ru(0001) (lower panel). Adapted with permission from Ref. [93]. Copyright © 2012, American Chemical Society

4.2.2.2 Platinum-, Rhodium-, Iridium-, Palladium-Chalcogenide-Based Materials

Since the decade of the 1980s, numerous articles have made reference to the properties of ruthenium chalcogenide, as a methanol-tolerant ORR electrocatalyst material. Attention has also been focused on other chalcogenide ORR cathodes containing other precious catalytic centers. In this respect, modified precious centers with chalcogenides have been reported so far, namely Pt_xS_y [94, 95], Pt_xSe_y [96, 97], Rh_xS_y [18, 28, 44, 98, 99], Rh_xSe_y [28], Ir_xS_y [48], Ir_xSe_y [47, 49], Pd_xS_y [41], and Pd_xSe_y [41, 42, 100].

Pt_xS_y One of the first synthesis, with regard to carbon-supported platinum sulfide, Pt_xS_y/C, was done reacting $(NH_4)_2PtCl_6$ with molecular sulfur [95]: $(NH_4)_2PtCl_6 + S \rightarrow (NH_4)_2Pt_xS_y + 3Cl_2$. The reaction for 24 h is done in the presence of carbon support. The electrocatalyst precursor ($(NH_4)_2Pt_xS_y$) is further heat-treated, in (75% v/v) N_2/H_2 atmosphere, at 350 °C to obtain Pt_xS_y/C, according to the pyrolytic chemical reaction: $(NH_4)_2Pt_xS_y/C \rightarrow Pt_xS_y/C + 2NH_3 + H_2S$. One key question in modifying via a chalcogen a catalytic center is to render the most active ORR catalytic center tolerant to small C_1 organic reactants, e.g., methanol. The oxidation of methanol in acid medium is a well-investigated process on platinum nanoparticles, e.g., [101, 102]. Thus, a key virtue of this reaction is to compare the activity of carbon-supported nanoparticulated materials Pt_xS_y vis-à-vis Pt, as shown in Fig. 4.6a. The marked different behavior in 0.5 M H_2SO_4 containing 0.5 M CH_3OH is clear. The catalytic center in the chalcogenide environment is active by 5% of the total oxidation current considering the forward scan. This property is further tested for the ORR process in the same medium. The Pt_xS_y/C kinetic evaluation compared to Pt/C presented a similar Tafel slope (80 mV dec^{-1}), Fig. 4.6b, in methanol free electrolyte, stressing the fact that the catalytic center is not perturbed by sulfur coordinated atoms. However, the presence of sulfur atoms increases the overpotential by 0.12 V with respect to the "naked" Pt NPs, at a current density of 0.2 mA/cm^2. The reason is rather similar to the phenomenon observed in Fig. 4.4 for Ru_xSe_y. The electrochemical oxidation of coordinated sulfur atoms with the ORR process is centered at around 0.12 V more negative than the oxidation of platinum (formation of a mixed potential). The oxidation of sulfur can be given by: $S^{2-} + 4H_2O \rightarrow SO_4^{2-} + 8H^+ + 8e^-$. Additionally, in agreement with the results depicted in Fig. 4.6a, the ORR in the presence of methanol is little affected (lower depolarization of ca. 0.03 V), when compared to Pt/C performance, for which the depolarization attains a negative shift of 0.37 V. The high tolerance of Pt_xS_y was assessed. Insights as to the chemical state of sulfur atoms was provided using in situ surface-enhanced Raman scattering (SERS) and DFT calculations [103]. The chemical state of sulfur on Pt NPs is indeed a S^{-2} (sulfide ion) interacting with a platinum atom detected by SERS at a wavenumber of 351 cm^{-1} (see insert in Fig. 4.6a). Further modification of Pt_xS_y with tungsten atoms was reported, $Pt_xW_yS_z/C$ [104]. In spite of the recorded good tolerance to methanol, and the relatively low ORR performance as compared to Pt_xS_y/C, there are no further insights as to the role played by W atoms in these materials.

Pt_xSe_y The role of selenium atoms was also established by modifying Pt NPs with selenium atoms. The initial material was prepared via a modified organic colloidal method in ethylene glycol, in which chloroplatinic acid and sodium citrate were dissolved. This solution was mixed in an organic solvent (tetrahydrofuran—THF) containing selenium and triphenyl phosphine, adding carbon as support while stirring, to finish the synthesis, after a pH adjustment of 12 in a Teflon-lined autoclave at 160 °C [105]. The recovered material was heat-treated under H_2/N_2 atmosphere at 200 °C. However, the formation of Pt_xSe_y can be performed by a

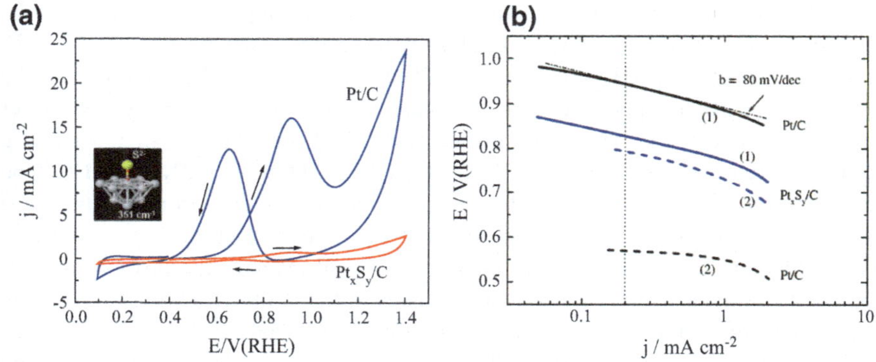

Fig. 4.6 **a** Methanol oxidation at Pt/C and Pt_xS_y/C showing that a 5% of the total methanol oxidation current is on Pt_xS_y/C. Insert: model describing the interaction between sulfide anions and platinum corresponding to a wavenumber of 351 cm^{-1}. **b** ORR corrected mass transfer Tafel plots for Pt/C and Pt_xS_y/C. Curves (1) in O_2-purged 0.5 M H_2SO_4 electrolyte, curves (2) in O_2− purged 0.5 M H_2SO_4 + 0.5 M CH_3OH electrolyte. The dashed-point line on curve (1) of Pt/C indicates the Tafel slope, and the dotted line at j = 0.2 mA cm^2 is an optical help to read the depolarization effect caused by the presence of methanol in the electrolyte. Adapted with permission from Ref. [95].

simple selenization in aqueous medium, a method which was used for ruthenium centers [106]. It turned out that this method was also effective for platinum centers departing either from commercial source or synthesized ones [97]. The electrocatalytic activity of aqueous generated platinum selenide (Pt_xSe_y) [96], see Fig. 4.7a, is rather similar to that recorded on Pt_xS_y (cf. Figure 4.6b). In this work, it was recognized that the increasing amount of surface coordinated chalcogenide decreases the ORR activity of platinum; therefore, a systematic investigation to find the optimum surface coordinated selenium was undertaken. One approach was using the cyclic voltammetry technique as shown in Fig. 4.7b. The Pt_xSe_y electrode material was submitted to anodic cycles opening the anodic branch from 0.9 to 1.4 V/RHE in N_2-saturated 0.5 M H_2SO_4 in the presence of 0.5 M CH_3OH at 25 °C. Similar to that exposed for Pt_xS_y, cf. Figure 4.6a, the methanol oxidation is enhanced as the majority of coordinated selenium atoms are stripped. The ORR measurement, Fig. 4.7c, after each cyclic potential profile scan, reveals that this process attains a maximum in the ORR activity, otherwise an optimum of surface coordinated selenium produces a catalytic material with the highest ORR activity and methanol tolerance. This is illustrated in Fig. 4.7d, after analysis of the half-wave potential data extracted from Fig. 4.7c. Each point in Fig. 4.7d shows the half-wave potential of maximum activity and depolarization effect caused by the methanol on selenium coordinated platinum centers. At the same time, each of these data is related to the effect of mixed-potential, which is triggered by the simultaneous oxidation of CH_3OH and the ORR process. On platinum, the chemisorption of methanol and thus its dissociation are attained when three adjacent Pt-sites on a crystallographic surface are present to form COH_{ads} species [107]. Therefore, on platinum chalcogenide materials this condition is not fulfilled in the presence of an

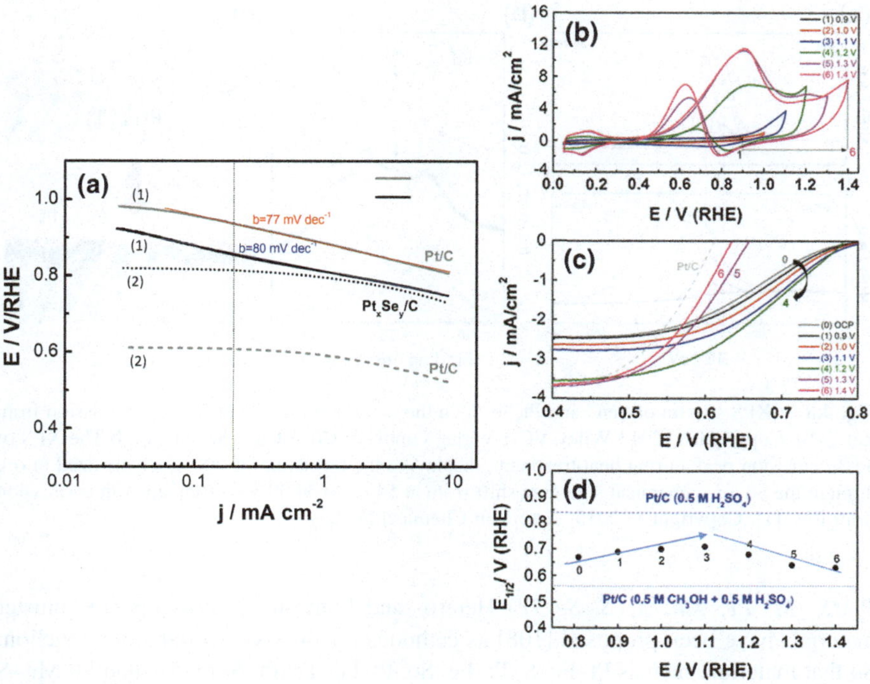

Fig. 4.7 **a** The ORR mass-transfer corrected Tafel plot for Pt/C and Pt_xSe_y/C electrodes: curves (1) recorded in 0.5 M H_2SO_4 electrolyte; curves (2) in 0.5 M CH_3OH + 0.5 M H_2SO_4 electrolyte. Adapted with permission from Ref. [96]. Copyright © 2014 Wiley-VCH Verlag GmbH & Co. KGaA, Weinheim. **b** Selenium stripping of the 20 wt% Pt_xSe_y/C via cyclic voltammetry (CVs) anodic profiling at 0.9 V to 1.4 V/RHE, at a scan rate of 50 mV/s; **c** ORR taken at 900 rpm on O_2-saturated electrolyte measured after CVs; and **d** the half-wave potential, $E_{1/2}$, collected from the ORR-RDE experiments. Measurement was done in N_2-saturated 0.5 M H_2SO_4 + 0,5 M CH_3OH at 25 °C. Adapted with permission from Ref. [97]. Copyright © 2013 WILEY-VCH Verlag GmbH & Co. KGaA, Weinheim

optimum sulfide or selenide coverage. It is important thus to recall that an equilibrium between these chalcogenides and free active sites has to be found. The chemical signature of platinum coordinated selenium in Pt_xSe_y is depicted in XPS Se 3d emission lines (see Fig. 4.8a). The main peak assigned to Se $3d_{5/2}$ is shifted by 0.6 eV toward lower binding energies [96]. A similar phenomenon was observed, since a shift of the Se $3d_{5/2}$ peak was around 0.4 eV, Fig. 4.8b, induced by the epitaxial formation of the layered semiconducting $PtSe_2$ at 270 °C, as shown in Fig. 4.8c. Herein, the phenomenon was attributed to the formation of the selenide ion (Se^{2-}) [1]. Although no evidence exists of the presence of a layered compound on Pt NPs, it becomes, however, clear that a phenomenon of charge transfer takes place, to explain the negative charge of coordinated selenium: Se^{δ} (where, $0 > \delta > -2$), in a similar way as to the phenomenon observed on Ru_xSe_y [91]. It is concluded that the charge transfer from platinum surface atoms to selenium drawn by XPS is at the origin of the improved methanol tolerance of Pt toward ORR.

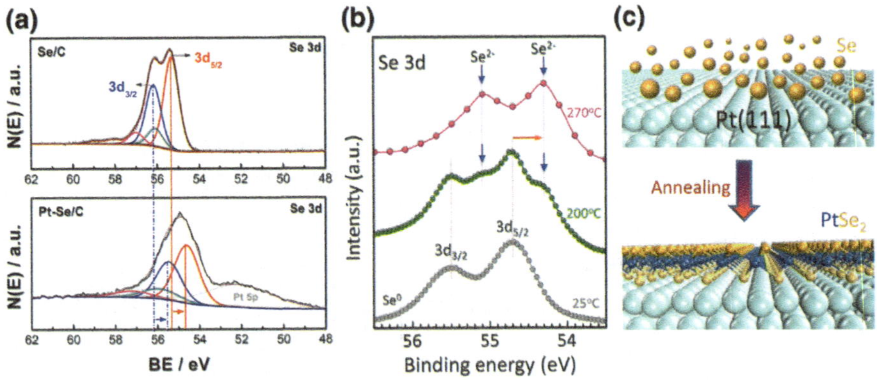

Fig. 4.8 **a** XPS spectra of Se/C and Pt_xSe_y/C in the Se 3*d* region. Adapted with permission from Ref. [96]. Copyright © 2014 Wiley-VCH Verlag GmbH & Co. KGaA, Weinheim. **b** The XPS of Se 3*d* emission peak during heat-treatment, displaying the formation of $PtSe_2$, schematized in (**c**), through the Se $3d_{5/2}$ chemical signature shift from at 54.80 to 54.39 eV. Adapted with permission from Ref. [1]. Copyright © 2015, American Chemical Society

PtMX (M: Fe, Mn; X: S, Se Te) Hetero- and homo-metal chalcogenide cluster materials have been proposed [108] as cathodes for the oxygen reduction reaction. So that materials such as Pt–Fe–S, Pt–Fe–Se, Pt–Fe–Te, Pt–S, Pt–Te, and Pt–Mn–S were dispersed on black carbon and used as cathodes for the ORR in methanol-oxygen fuel cells. Among those electrocatalysts, $PtMn_2S_2$ showed the highest activity in the half-cell measurements attaining an activity close to commercial Pt/C Etek material. Indeed, all the electrocatalysts were highly tolerant to methanol. This effect, as discussed above, comes from the presence of coordinated chalcogen atoms essentially. Since noble metal chalcogenides are the favored choice of electrocatalyst materials over pure metal and metal alloys, the following section shares alternatives to non-platinum chalcogenides discovered so far for the oxygen reduction activity in acid and the tolerance capacity in the presence of methanol.

M Sulfide (M: Rh, Ir, Pd) A typical synthetic chemical route for this kind of materials was described in Chap. 1 (Sect. 1.4.1). Herein, sulfurization or selenization can be done using alternative chemical routes; e.g., for Rh_xS_y or Rh/S, Rh black is reacted with S in suspension in xylene solvent previously purged with argon, and heat-treated for 24 h at 142 °C [28], whereas Ir_xS_y was obtained by precipitation in which heat-treated carbon black is put into a suspension containing sodium sulfite (Na_2SO_3) and chloroiridic acid (H_2IrCl_6) adding NaOH up to a pH > 12 leaving the reaction under stirring at 50 °C, 12 h [48]. The surface electrochemistry, in 0.5 M H_2SO_4, of Rh black, carbon-supported Ir centers, as well as their surface modified by sulfur, is compared in Fig. 4.9. The CV of Rh/S feature, as compared to Rh, is strongly modified, Fig. 4.9a. The oxidation of S starts at ca. 0.8 V, whereas the reduction starting at ca. 0.7 V extends over to 0.0 V. Due to sulfur surface modification of Rh, the ORR activity of Rh/S was lower than Rh

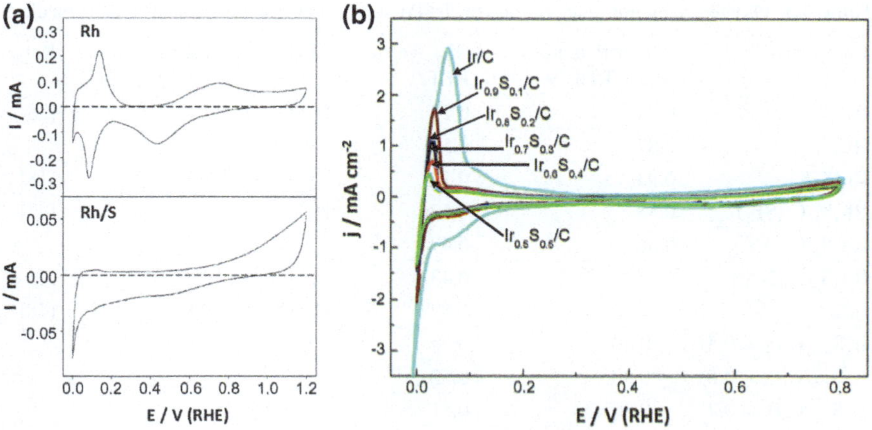

Fig. 4.9 **a** Cyclic voltammetry characteristics for Rh and Rh/S electrodes in 0.1 M H_2SO_4 at 20 mV/s. Reprinted with permission from Ref. [28]. Copyright 2006, The Electrochemical Society. **b** Cyclic voltammetry characteristics for Ir/C and Ir-S based materials as indicated in the figure in N_2-saturated 0.5 M H_2SO_4 at a scan rate of 20 mV/s. Adapted with permission from Ref. [48]. Copyright © 2011 Chinese Society of Particuology and Institute of Process Engineering, Chinese Academy of Sciences. Published by Elsevier B.V. All rights reserved

black [28]. Following the same pattern, the surface electrochemistry response of sulfur-modified carbon-supported iridium (Ir_xS_{1-x}/C) presents an analogous phenomenon; the hydrogen peaks depress with the sulfur content, Fig. 4.9b, compared to Ir/C. Moreover, an ORR enhancement (mass and surface specific activities) was reported with the catalyst: Ir_xS_{1-x}/C (x = 0.7) [48]. The common point from the sulfur surface-modification of Rh and/or Ir is their modified activity toward the ORR with a common strong methanol tolerance as compared to Pt/C. Similar to iridium, the palladium chalcogenide (Pd_xS_y) was prepared from palladium acetylacetonate dissolved in xylene. The addition of the chalcogen (sulfur) and carbon was done under nitrogen atmosphere and left reacting at 130 °C for 24 h [41]. The generated palladium-sulfur (Pd_xS_y) material showed an enhanced amorphization effect, as revealed by the XRD analysis, and an inactivity toward the ORR [41]. Although the reason for this inactivity was not further explored, the results led the authors to conclude that a formation of a new material phase was the reason of such a negative result. Some parameters such as the onset potential and half-wave potential are summarized in Table 4.3. Besides, the Tafel plot clearly displays the electrocatalytic differences of the chalcogenides compared to their metal centers in sulfuric and hydrochloric acid (Fig. 4.10).

First results on the implementation of a commercially available rhodium chalcogenide (30 wt% Rh_xS_y/C) in the mixed-reactant fuel cell technology were reported (see Chap. 6), because of its tolerance to methanol fuel [18]. The authors concluded that Rh_xS_y/C underachieved the performance of Ru_xSe_y/C. However, besides the ORR activity and stability of the chalcogenide Rh_xS_y/C, compared to Pt/C and Rh/

Table 4.3 Oxygen reduction reaction on Rh, Ir, Pd sulfide materials evaluated in acid media

Catalysts	[a]Onset Potential versus RHE/V	[a]$E_{1/2}$ versus RHE/V	Medium	Catalyst loading/μg cm^{-2}	Refs.
Rh	0.89	0.73	H_2SO_4	[b]150	[28]
Rh_xS_y	0.87	0.59			
Rh_xS_y/C	0.93	0.72		[b,d]63	[18]
Rh_xS_y/C (SF[e])	0.88	0.52	HCl	16	[45]
Rh_xS_y/C (AS[e])	0.86	0.47			
Rh_xS_y/C (NA[f])	0.75	0.43			
Ir/C	≤ 0.80	0.52	H_2SO_4	[c]200	[48]
Ir_xS_{1-x}/C (x = 0.5)	≤ 0.80	0.54			
Ir_xS_{1-x}/C (x = 0.6)	≤ 0.80	0.52			
Ir_xS_{1-x}/C (x = 0.7)	≤ 0.80	0.51			
Ir_xS_{1-x}/C (x = 0.8)	≤ 0.80	0.49			
Ir_xS_{1-x}/C (x = 0.9)	≤ 0.80	0.48			
Pd/C	0.72	0.42		[c]353	[41]
Pd_xS_x/C	N/A	N/A			

[a]Estimated data from current–potential curves
[b]At 1600 rpm
[c]At 2000 rpm
[d]Measured at 60 °C
[d]Sulfur ion-free synthesis
[e]Aqueous synthesis
[f]Non-aqueous synthesis

C, it showed resistance against depolarization in high concentrations of chloride ions (Cl^-) during the electrolysis of aqueous hydrochloric acid (HCl): $2HCl = 2H_2 + 2Cl_2$, Fig. 4.10b. This figure clearly depicts the differences in ORR activity obtained on rhodium-sulfide materials according to the chemical route of synthesis followed, on the one hand; and on the other hand, it is clearly shown that platinum is not an effective electrocatalyst in highly corrosive conditions since it is constantly poisoned by chloride ions. In this context, further fundamental results were generated on Rh_xS_y, since the synthesis of this material generates a multi-phase inorganic framework constituted of Rh_2S_3 and Rh_3S_4, present and active ORR phase in the so-called Rh_xS_y material. These data were derived from in situ-operando EXAFS measurements [44], as shown in Fig. 4.10d. This figure illustrates the Fourier Transforms between the absorbing atom (Rh) and its neighbors. The interaction of Rh with S atoms is centered at a lower coordination distance of 2.69 Å.

M Selenide (M: Rh, Ir, Pd)

Selenium is a popular element used to prepare transition metal selenides. Selenization in aqueous solvent can be used starting from commercial or self-synthesized carbon-supported metal centers, where the source of selenium is a selenium (IV) oxide. Herein, Rh black is reacted with Se in suspension in xylene solvent previously purged

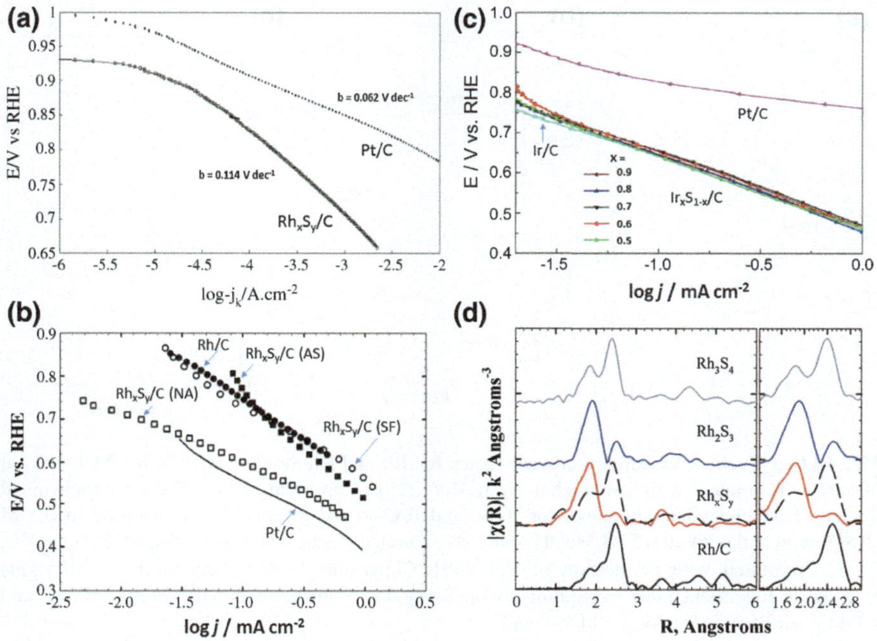

Fig. 4.10 **a** ORR Tafel plots on Rh_xS_y/C, and Pt/C in O_2-saturated 0.5 M H_2SO_4 at 60 °C. The scan rate used was 2 mV/s in the negative direction at 1600 rpm. Adapted with permission from Ref. [18]. Copyright © 2007 Elsevier Ltd. All rights reserved. **b** Tafel plots on Rh-based chalcogenides in O_2-saturated 0.5 M HCl. The catalysts loading was 30 wt% on carbon. Comparison was done with Rh/C and Pt/C electrode materials at 900 rpm and scan rate of 20 mV/s. See text for (NA, NS, and SF conditioning). Adapted with permission from Ref. [45]. Copyright © 2007 Published by Elsevier Ltd. **c** ORR Tafel plots on Ir_xS_{1-x} (x between 0.5 and 0.9) compared to Ir/C and Pt/C catalysts in O_2-saturated 0.5 M H_2SO_4 at 25 °C, scan rate of 5 mV/s, and 2000 rpm. Adapted with permission from Ref. [48]. Copyright © 2011 Chinese Society of Particuology and Institute of Process Engineering, Chinese Academy of Sciences. Published by Elsevier B.V. All rights reserved. **d** The nonphase-corrected Fourier Transform (FT) of Rh K-edge (23.22 keV) k^2-weighted $|\chi(R)|$ spectra in outgassed 1 M TFMSA. The spectrum of Rh_xS_y generated at 0.40 V, in outgassed 6 M TFMSA, is contrasted as dashed line on the 1 M TFMSA spectrum. Reprinted with permission of Ref. [44]. Copyright © 2009, American Chemical Society

with argon and heat-treated for 24 h at 142 °C [28]. The surface modification by selenium is visible in cyclic voltammetry experiments, at 20 mV/s in 0.1 M H_2SO_4, Fig. 4.11a, pointing out a similar phenomenon observed with sulfur-modified Rh surface, cf. Figure 4.9a, in which the common feature on Rh_xS_y and Rh_xSe_y is the suppression of the hydrogen and hydroxyl adsorption/reduction species present on Rh. Taking Iridium as the metal center, materials based on Ir–Se were synthesized using the carbonyl chemical route (see Chap. 1, Sect. 1.4.1) using as precursor dodecacarbonyl iridium: $Ir_4(CO)_{12}$, and selenium powder [49]. Similar to other metal centers, the surface electrochemistry of carbon-supported Ir_xSe_y/C is different to Ir/C. Selenourea (SeU)-assisted synthesis was also employed to generate Ir_xSe_y-(SeU). Again, the hydrogen adsorption/desorption region is suppressed, as well as the

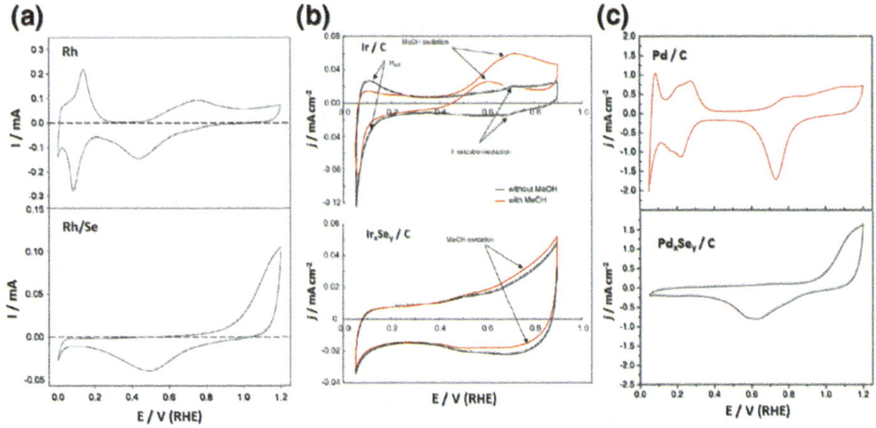

Fig. 4.11 **a** Cyclic voltammetry characteristics for Rh and Rh/Se electrodes in 0.1 M H_2SO_4 at 20 mV/s. Reprinted with permission from Ref. [28]. Copyright 2006, The Electrochemical Society. **b** Cyclic voltammograms for Ir–Se and Ir/C-coated glassy carbon electrode in 0.5 M H_2SO_4 with and without 0.5 M MeOH under N_2-saturated electrolyte. Potential scan rate: 5 mV/s, 25 °C. Reprinted with permission of Ref. [49]. Copyright © 2006 Elsevier B.V. All rights reserved. **c** Electrochemical measurements on Se-modified and unmodified Pd/C (1-PdSe/C and 1-Pd/C) surfaces in 0.1 M perchloric acid

hydroxide redox process, as in Fig. 4.11b. As similarly indicated in the figure, the Ir_xSe_y material possesses a good tolerance to methanol poisoning as compared to Ir/C or Pt/C [49, 109]. The effect of the ORR enhancement of Ir/C catalyst by covering the metal Ir centers with Se atoms is a positive effect, due to the strong methanol tolerance property, and the absence of a mixed-potential caused by the simultaneous reaction of ORR and methanol oxidation, as compared to Pt/C, is nil. Therefore, Ir_xSe_y is a tolerant catalyst for methanol oxidation and a promising catalyst to be considered as cathode for direct methanol membrane-less fuel cells (DMFCs). For the same ORR purpose, palladium-centered selenide (Pd_xSe_y/C) was developed for cathodes in DMFCs [172]. The selenium surface modification of palladium improved the ORR activity as compared to Pd/C. It seems that the Pd_xSe_y catalyst preserves the bulk structure of Pd nanoparticle, unlike S, as discussed above (cf. Table 4.3). As summarized in Table 4.4, and judging by the extracted half-wave potential data, the selenization of Rh, Ir, and Pd can favor the ORR kinetics in acid medium as compared to the non-modified catalytic center. But it can also show a negative potential shift, as in the case of rhodium.

An alternative method to produce Pd_xSe_y was reported [42]. However, the ORR activity on this hydrothermal generated material was ill-defined, and the conclusion of this work was that the electrocatalyst was based on PdSe, $PdSe_2$, and Pd_3Se (tetragonal and orthorhombic crystal structures). PdSe showed higher electrocatalytic activity than the Pd_3Se or $PdSe_2$ in 0.5 M H_2SO_4. Moreover, using a facile synthesis route, e.g., the selenization in aqueous solvent, our laboratory obtained the Pd_xSe_y material, presenting a similar surface electrochemistry feature in sulfuric acid, as well as in perchloric acid. Figure 4.11c displays the electrochemical

Table 4.4 Oxygen reduction reaction on Rh, Ir, Pd selenide materials evaluated in acid media

Catalysts	[a]Onset Potential versus RHE/V	[a]$E_{1/2}$ versus RHE/V	Medium	Catalyst loading/μg cm^{-2}	Refs.
Rh	0.89	0.73	H_2SO_4	[b]150	[28]
Rh_xSe_y	0.84	0.64			
Ir/C	0.90	0.60		[c]73	[49]
Ir_xSe_y/C	0.93	0.63			
Ir_xSe_y/C-(SeU)	0.97	0.70		[b]40	[109]
Pd/C	0.72	0.42		[d]353	[41]
Pd_xSe_x/C	ca. 0.70	0.45			
Pd/C	0.93	0.82	$HClO_4$	[e]342	This work
Pd_xSe_x/C	0.82	0.55			
Pd/C	0.97	0.82	KOH		
Pd_xSe_x/C	0.97	0.85			

[a] Estimated data from current–potential curves
[b]At 1600 rpm
[c]At 400 rpm
[d]At 2000 rpm
[e]At 900 rpm

behavior in 0.1 M $HClO_4$. The coordinated selenium atoms strongly modify the adsorption properties of Pd/C, since the hydrogen and oxide regions are suppressed. The oxidation wave that appears at 0.9 V corresponds to the coordinated selenium oxidation. Table 4.4 summarizes the ORR data obtained on this material in 0.1 M $HClO_4$ and 0.1 M KOH. It is worth to note that the ORR kinetics on the chalcogenide is strongly enhanced in alkaline medium, as compared to the acid one. The selenization of palladium using the non-aqueous medium strongly modifies the palladium core and/or surface, since the detection of phases such as Pd_7Se_4 and $Pd_{17}Se_{15}$ was revealed [110] by the XRD technique [111]. $Pd_{17}Se_{15}$ is a naturally occurring mineral known as Palladseite [112].

4.3 Non-precious Metal Center Chalcogenides

The major drawback of platinum group metal (PGM) electrocatalysts is the limited availability and elevated cost (cf. Fig. 4.1). The development of non-precious catalytic centers (or non-platinum group metal—NPGM) has been stimulated in the recent years with respect to the oxygen reduction reaction [67, 113–131], hydrogen evolution reaction [132–144], oxygen evolution reaction [120, 145, 146], and hydrogen oxidation reaction [147]. Metal oxides, carbides, nitrides, and carbonitrides constitute a good reliant among the NPGM electrocatalysts. However, the transition metal chalcogenide materials play a central role in this development (see below), since their incremental improvement regarding the activity and stability for some of the above mentioned electrochemical reactions has been proven.

4.3.1 Materials for HER/HOR

Non-precious electrocatalysts are generally not stable in acid medium. This is the reason why such materials (e.g., nickel-based) are preferentially integrated in alkaline electrolytes [148, 149]. However, the search of hydrogen evolution stable materials for the acid medium has led to some examples based on chalcogenides, e.g., MoS_2 [134, 136, 142–144], $MoSe_2$ [150–156], WSe_2 [157–161], $CoSe_2$ [133, 162–168], $NiSe_2$ [169–172], $FeSe_2$ [173], and other mixed-chalcogenides [135, 174, 175]. The survey of the recent literature sheds light on the fact that this family enriches the group of hydrogen-producing electrocatalysts for different uses in systems like low-temperature fuel cells, electrolyzers, and solar fuel devices.

4.3.1.1 Hydrogen Evolution Reaction (HER)

Non-precious transition metals, e.g., Fe, Ni, and Mo, form part of biological systems in which the generation of molecular hydrogen is catalyzed by metalloenzymes (hydrogenase) [176]. In this context, the concept of artificial photosynthesis assembling catalysts made of cobalt and phosphate ions to produce molecular oxygen from water [177] must be applied to another kind of electrocatalysts to produce hydrogen with emphasis on their engineering and electronic modification, as described below:

M Sulfide (M: Mo, W, Co) The molybdenum disulfide (MoS_2) is a layer-structured semiconductor (E_g = 1.75 eV) (cf. Figs. 1.4, 1.5, and 1.6). This material has been studied photoelectrochemically and outlined as a potential material for water splitting with visible light in the 1970s by Tributsch and Bennett [178]. It is worth to recall that those first experiments were performed on natural crystals (a few mm^2 to several cm^2), certainly with a poor bulk activity. Many years later, computational studies reported by Hinnemann et al. [144] predicted that nanostructured molybdenum disulfide possesses hydrogen evolution reaction (HER) activity. The analysis between MoS_2 metal-like (highly dispersed nanoparticles) and natural hydrogenases revealed that the Mo (1010) edge of MoS_2, Fig. 4.12a, has a hydrogen binding energy of ca. 0.08 eV, i.e., very close to the ideal value of 0 eV, Fig. 4.12b (cf. Figs. 2.6, 2.7). The HER performance of nanocrystals [179] of MoS_2 was correlated and corroborated with the number of its edge sites [143]. This finding was further developed to engineer an increased number of active edge sites on gold substrates [180], and on silica templates electrodepositing Mo followed by sulfurization with H_2S [181] to obtain large-area connecting thin films of MoS_2 with a tunable thickness showing outstanding HER electrocatalytic activity, with 0.15–0.2 V overpotential, and a Tafel slope of 50 mV/decade. Another synthetic approach was the variation of oxygen in MoS_2 to favor the formation of unsaturated sulfur sites, with the consequence of increasing the materials' conductivity [182]. As conferred in Chap. 1 (Sect. 1.2.2, cf. Fig. 1.6), there are two polytypes of MoS_2 (2H, and 1T). 2H-MoS_2 is thermodynamically (less electrical conducting) more stable but less

HER active than 1T-MoS_2. The exfoliation via Li intercalation of 2H-MoS_2 to obtain 1T-MoS_2 was done chemically. The obtained materials (1T) exhibited a dramatic HER enhancement [179], as indicated by the current density versus the electrode potential characteristics compared with platinum (Fig. 4.12c). The HER at $j = 10$ mA/cm^2 on the as grown 2H-MoS_2 was obtained at −0.320 V, whereas on the exfoliated material (1T-MoS_2) the same current density was obtained at −0.195 V. Additionally, amorphous molybdenum sulfide (MoS_x) films have been also considered good HER candidates. Experiments on such MoS_x (i.e., MoS_3 or MoS_2) films showed an excellent hydrogen evolution activity in acid medium achieving a current density of 15 mA/cm^2 at 0.20 V overpotential [183]. The reaction mechanism, apparently, is determined by the reduction and protonation of

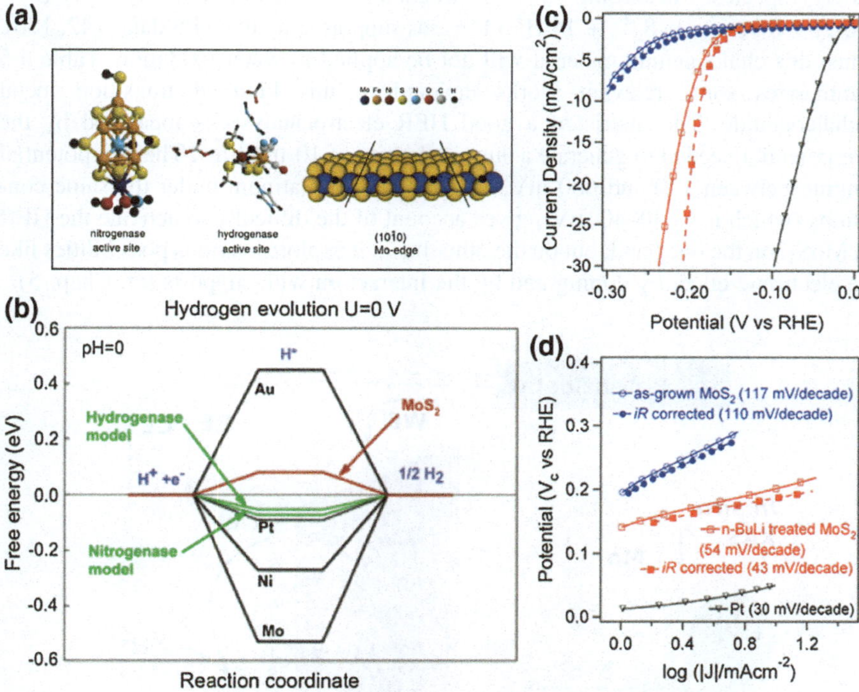

Fig. 4.12 **a** Schematics showing, at the left, the Nitrogenase FeMo cofactor (FeMoco) with three hydrogen atoms bound at the equatorial μ_2S sulfur atoms; at the center, the active site of the hydrogenase with one-bound hydrogen atom; at the right, the MoS_2 slab, where the unit cell in the x-direction is indicated (lines). The sulfur monomers are present at the Mo edge, in which hydrogen is bound every second sulfur atom (corresponding to a coverage of 50%). **b** The determined free energy diagram for hydrogen evolution, at a potential U = 0 relative to the standard hydrogen electrode at pH = 0 at standard conditions. The diagram includes data with other metals, e.g., Au, Pt, Ni, and Mo. The calculated data for MoS_2 corresponds to a hydrogen coverage from 25 to 50%. **a** and **b** Reprinted with permission from Ref. [144]. Copyright © 2005, American Chemical Society. **c** HER current–potential curves of chemically exfoliated MoS_2 nanosheets. **d** The corresponding Tafel plots, where filled symbols show iR-corrected data. **c** and **d** Reprinted with permission from Ref. [179]. Copyright © 2013, American Chemical Society

the terminal sulfide units. From the theoretical standpoint via the density functional theory (DFT) conducted by Li et al. [184], it is concluded that the terminal complex S_2^{2-} could be the true catalytically active site responsible for the low Tafel slope. Such terminal S_2^{2-} exists in MoS_3 and at the edge of crystalline MoS_2. However, data reported by Lassalle-Kaiser et al. [185], using in situ X-ray absorption spectroscopy (XAS) technique, support the idea that the rate determining HER on MoS_x is the protonation and reduction of interfacial Mo(III)(S_2) sites, involving directly a disulfide unit in molybdenum sulfide-based material, as schematized in Fig. 4.13. The deactivation of the HER active sites of molybdenum disulfide can take place if driven to electrode potentials higher than 0.98 V/NHE in acid electrolyte. The loss of MoS_2 HER activity was attributed to the irreversible surface oxidation. If eleven electrons are needed, this degradation is due to the generation of SO_4^{2-}, S_2^{2-}, and MoO_3 species, according to electrochemical reaction: $MoS_2 + 7H_2O \rightarrow MoO_3 + SO_4^{2-} + ½\, S_2^{2-} + 14\,H^+ + 11e^-$, as supported by the XPS data [142, 186]. Thus, this chalcogenide material will not be applied to water oxidation. Table 4.5 summarizes some relevant works devoted to this kind of transition metal dichalcogenide. The metric of a good HER electrocatalysts is measured by the overpotential needed to generate a current density of 10 mA/cm^2. The overpotential ranging between 120 and 680 mV, as compared to platinum under the same conditions (which is ca 30–40 mV), gives account of the difficulty to activate the HER on MoS_2, on the one hand, but on the other hand, it explores various possibilities like the electronic effect by doping and by the interaction with supports (cf. Chap. 5).

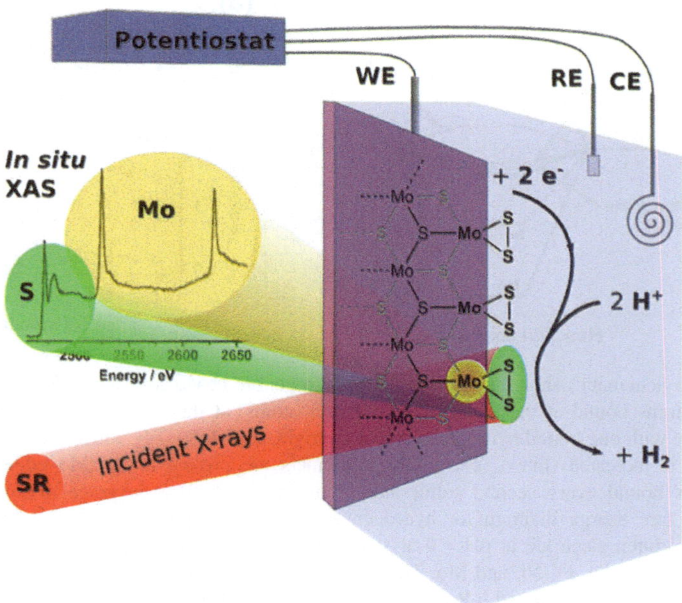

Fig. 4.13 Schematics of the synchrotron radiation incident beam (SR) to observe the sulfur K-edge, and Mo K-edge spectra and probe the HER catalytic centers of MoS_x films

Table 4.5 Hydrogen evolution reaction on Mo disulfides

Materials	Synthesis method	Morphology	Electrolyte	$-\eta$/V @ 10 mA/cm^2	Refs.
MoS_x	Electrodeposition	Amorphous films	1 M H_2SO_4	0.2	[183]
MoS_x				0.19	[191]
MoS_x/RGO/Ni	Thermolysis	3D MoS_x/graphene/Ni foam	0.5 M H_2SO_4	0.25	[192]
MoS_x/N-CNT	Chemical synthesis	MoS_x nanosheets		0.14	[193]
[a]MoS_2/RGO	Solvothermal	3D-like NPs		0.17	[194]
[a]MoS_2/RGO	Chemical synthesis	Granular, particles and nanosheets onto RGO		0.149	[195]
[a]MoS_2: Cu^{2+}/RGO				0.12	
MoS_2	Electrodeposition + sulfidization	Double-gyroid films		0.23	[181]
1T-MoS_2	Chemical exfoliation with Li	Nanosheets		0.20	[196]
MoS_2	Planetary mill	nanodots		0.28	[197]
MoS_2	CVD and H_2 annealing	Monolayers		~0.57	[198]
hH-MoS_2	Hydrothermal	Hollow structures μ-nanospheres		0.13	[199]
[b]mPF-MoS_2	Chemical synthesis	Mesoporous foam		0.21	[200]
[b]mPF-Co-MoS_2				0.156	
Co-MoS_2	Chemical synthesis	Nanoparticles		0.135	[201]
Ni-MoS_2	Magnetron sputtering deposition of Ni	Nanoclusters	2 mM $HClO_4$/0.1 M $NaClO_4$	0.68	[202]

[a]RGO—reduced graphene—hybrid material
[b]mPF—mesoporous foam

However, the gathered data on molybdenum disulfide motivated the application of the isostructural tungsten disulfide (WS_2; E_g = 2.1 eV) for the HER deposited onto carbon supports [142], exfoliated 2H-WS_2 (semiconductor) to 1T-WS_2 (metallic nanosheets) [187, 188], sulfidization of nanoarrays of WS_2 to modulate the increase of the HER activity via metal sulfide composition [189], through the growth of WS_2 onto oxidized carbon fiber (oCF) via the hydrothermal process [190] building 3D nanostructures probed to be more effective than the same material onto carbon fiber (CF) (see Fig. 4.14).

A further achievement to enhance the electrocatalytic activity was done by a monolayer dendritic growth of ternary chalcogenide: $WS_{2(1-x)}Se_{2x}$. The abundant edge sites generated were evidenced by sulfurization and selenization of tungsten trioxide powders [174].

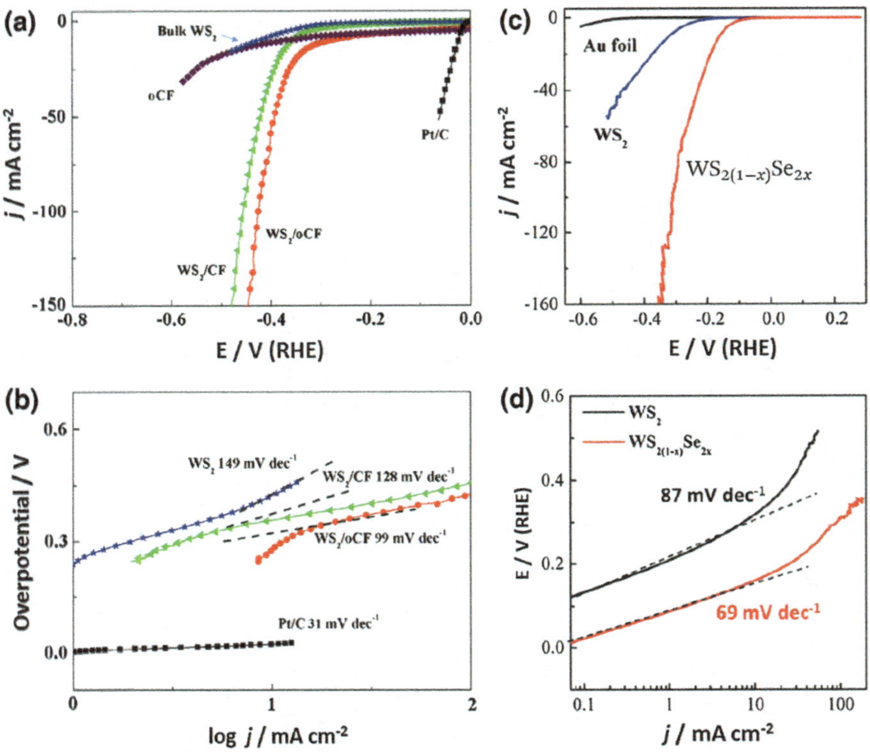

Fig. 4.14 **a** HER current–potential characteristics on WS_2 massif materials and supported onto various carbon-based materials: oCF (oxidized carbon fiber), CF (carbon fiber), and compared to Pt/C. Measurements were performed in N_2-saturated 0.5 M H_2SO_4. **b** The corresponding Tafel plots. **a** and **b** Adapted with permission from Ref. [190]. Copyright © 2017 Elsevier B.V. All rights reserved. **c** HER current–potential characteristics on monolayer WS_2, and monolayer dendritic $WS_{(1-x)}Se_{2x}$ flakes in N_2-saturated 0.5 M H_2SO_4 electrolyte, at room temperature. Gold foil is shown for comparison and used as support. **d** The corresponding Tafel plots of WS_2, and $WS_{(1-x)}Se_{2x}$ on Au foils. **c** and **d** adapted with permission from Ref. [174]. Copyright © 2017, Royal Society of Chemistry

Unlike the TMD-layered materials, cf. Fig. 1.4, cobalt disulfide (CoS_2) is a narrow bandgap ferromagnetic material, while the isostructural pyrite (FeS_2; E_g = 0.95 eV) is a diamagnetic semiconductor, long considered as a solar energy material [203], and the low-band-gap semiconductor nickel disulfide (NiS_2, E_g = 0.3 eV [204]). The pyrite structure is characterized by the presence of anions pairs X_2^{2-} (X: S, Se, Te), as a consequence of the increased binding energy of the *d*-orbitals of the transition metal, and considered as better HER materials than bulk layered materials [205]. In this context, by a specific morphology control, Faber et al. [206] demonstrated that the HER can be tailored on metallic cobalt disulfide with different morphologies, such as films, microwire, and nanowire (Fig. 4.15a). This figure shows the current density, referred to the geometric surface area, and contrasted with the morphology (F, MW, and NW), as a function of the applied electrode potential, together with the Tafel analysis (Fig. 4.15b). By thermal

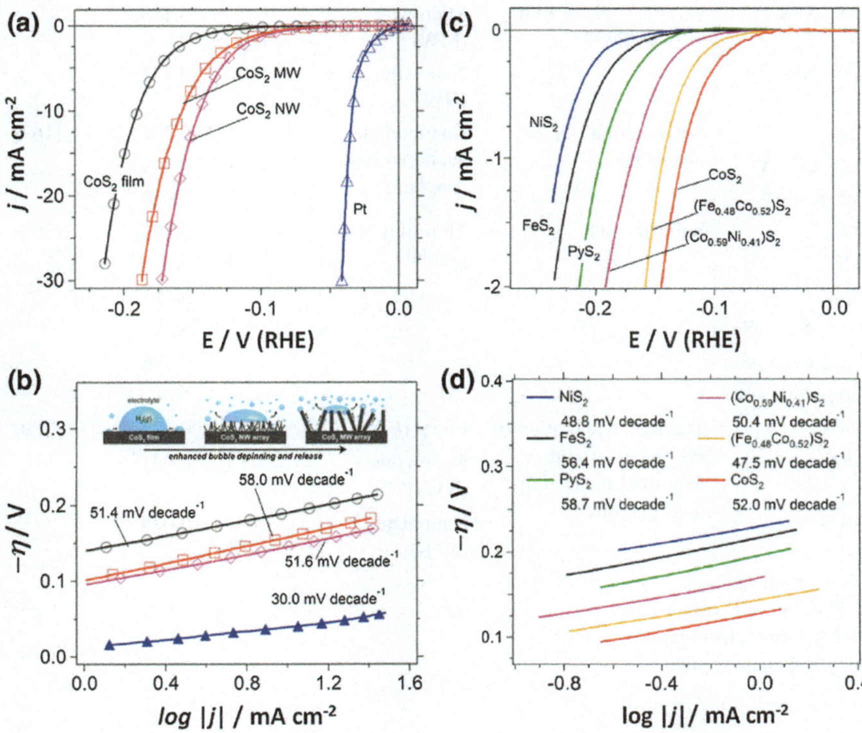

Fig. 4.15 **a** Current–potential characteristics on CoS_2 films, MW array, and NW array electrodes for the HER electrocatalysis. Pt is also shown for comparison in H_2-purged 0.5 M H_2SO_4 electrolyte at RT. **b** Tafel plot of data presented in (**a**). The electrode morphology is shown in the inset of (**b**). Figures **a** and **b** were adapted with permission from Ref. [206]. Copyright © 2014, American Chemical Society. **c** The HER iR-corrected current–potential characteristics of FeS_2, CoS_2, NiS_2, PyS, $(Fe_{0.48}Co_{0.52})S_2$, and $(Co_{0.59}Ni_{0.41})S_2$ thin-film materials on graphite supports. Measurements done in H_2-purged 0.5 M H_2SO_4 electrolyte at RT. **d** Tafel analysis of the data shown in (**c**). Figures **c** and **d** adapted with permission from Ref. [50]. Copyright © 2014, American Chemical Society

Table 4.6 Hydrogen evolution reaction on W-, Co-disulfides

Materials	Synthesis method	Morphology	Electrolyte	$-\eta$/V @ 10 mA/cm^2	Refs.
WS_2	CVD on W foil, and Li-exfoliation	NPs as grown	0.5 M H_2SO_4	0.33	[187]
1T-WS_2		Oven exfoliated		0.18	
1T-WS_2		μ-wave exfoliated		0.15	
WS_2	Chemical synthesis and autoclave	Vertical grown arrays		0.47	[190]
WS_2/CF[a]				0.37	
WS_2/oCF[b]				0.33	
WS_2	CVD-sulfurization, selenization	Monolayer		0.310	[174]
$WS_{2(1-x)}Se_{2x}$		Nanoflakes		0.156	
CoS_2	Thermal sulfidation on deposited metal films onto graphite	Film (F)		0.190	[206]
CoS_2-MW		Microwire (MW)		0.158	
CoS_2-NW		Nanowire (NW)		0.145	
CoS_2	Solvothermal on nanobelts	Nanoparticles anchored onto nanobelts		0.206	[167]
CoS_2/DETA[c]				0.149	
CoS_2/$CoSe_2$[d]				0.080	
CoS_2	Sulfurization	Thin film onto graphite		0.192	[50]
FeS_2				N/A	
NiS_2				N/A	
$(Fe_{0.07}Ni_{0.93})S_2$				N/A	
$(Fe_{0.48}Co_{0.52})S_2$				0.196	
$(Co_{0.59}Ni_{0.41})S_2$				N/A	
CoPS	Thermal evaporation of PS precursors on deposited metal films onto graphite	Films (Fs)		0.128	[209]
CoPS		Nanowires (NWs)		0.061	
CoPS		Nanoplates (NPls)		0.048	

[a]CF—carbon fibers
[b]oCF—oxidized carbon fibers
[c]DETA—diethylenetriamine
[d]A hybrid chalcogenide

sulfurization of electron-beam evaporated metal thin films previously deposited onto graphite substrates, Faber et al. prepared CoS_2 and $M_xCo_{1-x}S_2$ (M: Fe, Ni) pyrite structure compounds [50]. The HER activity of all these pyrite-phase thin films is displayed in Fig. 4.15c. The results offer a clear HER performance difference within the same pyrite structure. Cobalt disulfide achieves the cathodic current of 10 mA cm^{-2} at an overpotential of −0.192 V, and additional data are summarized in Table 4.6. The result on CoS_2 is comparable to the best MoS_2 (T-MoS2) reported so far (see Table 4.5). The Tafel plot in Fig. 4.15c discloses a similar Tafel slope, around 50−60 mV/decade. One further interesting aspect in tailoring activities among the pyrite structure is the electronic modification by introducing (alloying) CoS_2 with Fe and Ni, since the HER performance of NiS_2 and FeS_2 is lower. Certainly, the intrinsic conductivity of the materials plays an important role since, as mentioned above, CoS_2 possesses a metal-like behavior. It is also worth to mention, besides the electrocatalytic application, that the properties of making $Co_{1-x}Fe_xS_2$ alloys are to raise the conductivity. Due to the alloy, low-temperature conduction electrons, tunable with x, provide fundamental investigation for the development of spintronic devices [207], and solar applications [208]. The HER electrocatalytic improvement on tungsten- and cobalt disulfide, as summarized in Table 4.6, is due to the various modifications of the catalytic center, via synthesis route, nanoalloying, and the interaction with the support.

M Selenide (M: Mo, W, Co) Computational studies proposed that the Gibbs free energy for the H adsorption on $MoSe_2$ (E_g = 1.64 eV [210]) edge is much closer to the thermoneutral state, and $MoSe_2$ possesses a higher H coverage as compared to MoS_2 [156, 211]. Edge-terminated MoX_2 (X: S and Se) films were prepared in a single-zone furnace (550 °C) in the presence of sulfur or selenium powder on Mo films, previously prepared by electron-beam evaporation, followed by a thermal annealing [212]. This process produced vertically aligned layers, so that the layers were perpendicular to the substrate favoring the exposition of Van der Waals gap for reaction. The HER evaluation (current–potential curves and Tafel plot) of such films grown on glassy carbon substrates is displayed in Fig. 4.16a, b, respectively [211]. The data in this figure demonstrate that active sites lying on edges of $MoSe_2$ is rather similar to that of MoS_2. Moreover, the results generated on MoS_2, cf. Table 4.5, set as HER benchmark electrocatalysts the MoS_2/RGO [194], due to the lowest onset potential and Tafel slope. In this context, Tang et al. [156] reported the synthesis of hybrids $MoSe_2$/RGO synthesized via the hydrothermal reaction of Na_2MoO_4 and hydrazine hydrate Se in water at 200 °C. Rich folded edges parallel to RGO support revealed basal planes of the chalcogenide onto RGO maintaining the 2H-$MoSe_2$ structure. Indeed, in sharp contrast to $MoSe_2$, $MoSe_2$/RGO showed lower onset potential with a Tafel slope closing at 69 mV/decade (see Fig. 4.16c, d). It is ostensible that the Gibbs free energy of hydrogen adsorption (higher H coverage) on $MoSe_2$ is closer to the thermoneutral value than that of MoS_2, see Fig. 4.12b, and consistent with computational results [211]. This hierarchy is also reported when comparing different liquid exfoliated dichalcogenides [213]. $MoSe_2$ outperformed the HER of MoS_2 (Table 4.7). The effect on $MoSe_2$ was reinforced in hybrid material composites: $MoSe_2$/SWNT (single wall nanotube). Enhanced HER activity was also reported on $MoSe_2$ [152] using the colloidal chemical synthesis. Through a chemical synthetic way using an organic solvent and $Mo(CO)_6$, and selenium powders, as reported some time ago [214], $MoSe_2$ nanostructures possessing a network-like, and flower-like morphologies were obtained. The key to obtain such morphologies was the variation in the synthesis of the ratio of

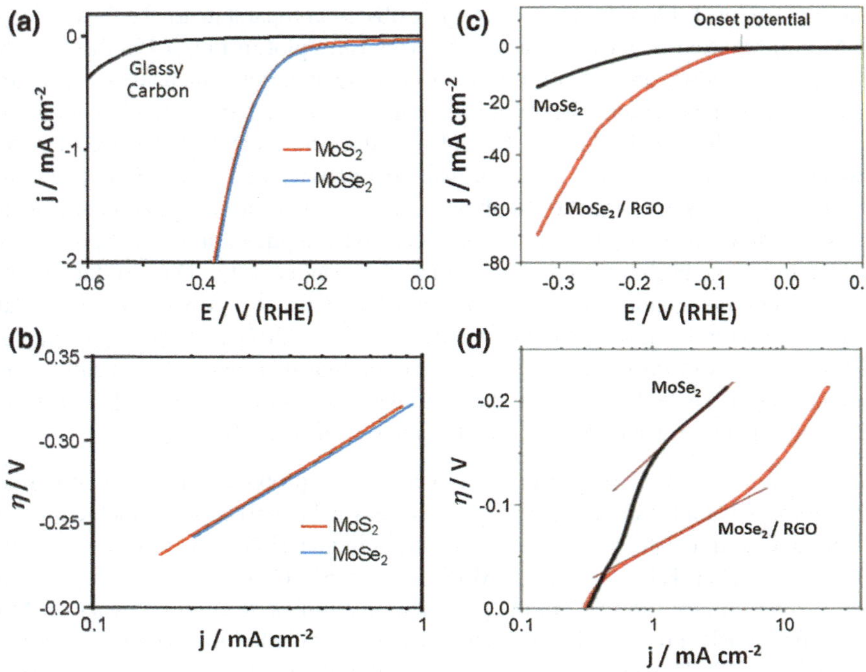

Fig. 4.16 **a** HER current–potential characteristics, in N_2-saturated 0.5 M H_2SO_4 solution, of edge-terminated MoS_2, and $MoSe_2$ films. Glassy carbon is shown as a blank. **b** The corresponding Tafel plot of data in (**a**). Figures **a** and **b** adapted with permission from Ref. [212]. Copyright © 2013, American Chemical Society. **c** HER current–potential characteristic of pure $MoSe_2$ and $MoSe_2$/RGO in acidic 0.5 M H_2SO_4 at a potential scan rate of 10 mV/s. **d** The Tafel plots of data in (**c**) recorded on glassy carbon electrodes with a catalyst loading of 160 μg/cm^2 at a scan rate of 5 mV/s. Figures **c** and **d** adapted with permission from Ref. [156]. Copyright © 2013, Royal Society of Chemistry

oleylamine (OAm) to oleic acid (OA). Thus, OAm: OA = 1:0 produced a network and OAm: OA = 1:1 produced a restructured network, whereas OAm: OA = 0:1 produced nanoflowers (see Fig. 4.17). Although the growth mechanism is not established yet, this complex chemical interplay shows versatility and an opportunity to create important exposure of active sites.

Interlayer engineering allowed to produce 1T-$MoSe_2$ nanosheets via solvothermal synthesis of flower-like assemblies [155]. The obtained nanosheets showed expanded interlayer spacing of ca. 1.17 nm, compared to 0.646 nm of 2H-$MoSe_2$. The conversion of 1T to 2H was obtained by a simple temperature heat treatment at 300 °C. The recorded HER activity on sample 1T-$MoSe_2$ outperformed 2H-$MoSe_2$, see Table 4.7, with a combined Volmer–Heyrovsky reaction mechanism. Another alternative way to enhance the HER on $MoSe_2$ was its growth on nitrogen-doped carbon nanotubes (NCNTs) to obtain hybrids: $MoSe_2$/NCNTs [154]. The strategy was based on the preparation of MoO_3/PANI (PANI: polyaniline) nanohybrids.

Table 4.7 Hydrogen evolution reaction on Mo diselenide

Materials	Synthesis method	Morphology	Electrolyte	−η/V @ 10 mA/cm^2	Refs.
2H-$MoSe_2$	Hydrothermal process and annealing	Nanoflowers, ultrathin sheets	0.5MH_2SO_4	~0.26	[156]
2H-$MoSe_2$/RGO		Folded edgenanosheets		0.15	
$MoSe_2$	Selenization transport in a single-zone furnace	Perpendicular layers grown onto substrate		[a]0.45	[212]
[b]$MoSe_2$/SWNT	Probe sonication of powders in sodium cholate hydrate	~100 nm avg. exfoliated nanosheets		0.22	[213]
[b]MoS_2/SWNT				0.28	
[c]$MoSe_2$	Colloidal chemical synthesis	Nanonetwork		0.30	[152]
[d]$MoSe_2$		Corrugated		0.23	
[e]$MoSe_2$		Nanoflowers		0.23	
1T-$MoSe_2$	Solvothermal synthesis of $MoCl_5$ with SeO_2 in octylamine at 180 °C	Flower-like assemblies		0.17	[155]
2H-$MoSe_2$		Annealing treatment 300 ° C		0.56	
$MoSe_2$/NCNTs	Hydrothermal	Nanosheets on NCNTs		0.102	[154]
$MoSe_2$/MoO_3	Hydrothermal	(h1) Thin rod agglomerated flowers		0.23	[157]
$MoSe_2$/MoO_3		(h2) Marigold-like flowers		0.30	
Cu_2MoS_4/$MoSe_2$	Top down method, sonication, solvothermal, autoclave	Nanodots on Cu_2MoS_4 basal planes		0.17	[215]
$MoSe_2$/RGO-M(f)	Gas phase reaction	Crumpled and sphere-like morphology		0.21	[216]
$MoSe_2$-1-180	Hydrothermal	[g]100%-2H		0.355	[150]
$MoSe_2$-1-200		52%-2H;48% 1T		0.163	

[a]Data extrapolated; [b]Composites with 10 wt%-SWNT-mass loading of 1.5 mg/cm^2; [c]oleylamine (OAm) to oleic acid (OA) = 1:0; [d]oleylamine (OAm) to oleic acid (OA) = 1:1; [e]oleylamine (OAm) to oleic acid (OA) = 0:1; [f]M means a mass loading of 20 wt%; [g]Flower-like; [h1]at 200 °C; [h1]at 150 °C

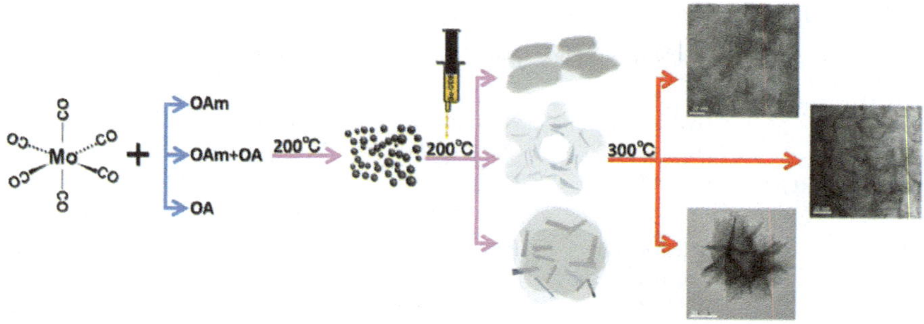

Fig. 4.17 Scheme showing the chemical route of synthesis of $MoSe_2$ leading to various morphologies, as shown on the high-magnification TEM images: Nanonetwork (upper, middle); Nanoflower (bottom). Adapted with permission from Ref. [152]. Copyright © 2017 Elsevier Ltd. All rights reserved

Thereafter, a thermal treatment in hydrazine hydrate solution converted by an anion exchange reaction, the molybdenum oxides in $MoSe_2$/PANI, heat-treated further at 600 °C. Small-size nanosheets (some Se–Mo–Se atomic layers) and the expanded (002) planes of the NCNTs (obtained by conversion of PANI nanorods during the heat-treatment) support were produced. This hybrid, indeed, produced a current density of 10 mA/cm^2 at an overpotential of 0.102 V. Due to the potential versatility, the use of the hydrothermal method for some dichalcogenide materials' morphology control seems to be very common, as recently summarized [151]. In this framework, working with reaction temperature and residence time, hybrids of $MoSe_2$/MoO_3 with marigold-like flowers (150 °C), and thin rod agglomerated flowers (200 °C) morphologies were produced and tested for the HER in acid medium [157]. The rod-like petals showed higher performance in terms of overpotential than marigold-like flower morphology (see Table 4.7). However, the data reported on this type of hybrid materials are far from other data summarized in the table. Since basal planes of layered materials are rather inactive for the HER, hybrid structures are being explored aiming at using 0D (zero-dimensional) $MoSe_2$ materials. In this sense, nanostructures of Cu_2MoS_4/$MoSe_2$ have been explored [215]. The authors report that an overpotential of 0.17 V at 10 mA/cm^2 was obtained, with excellent stability. An optimum $MoSe_2$ nanodot (ND) size was found to be 30 nm. Furthermore, using the spray-drying process composites of $MoSe_2$/RGO with a high content of RGO were produced [216]. An optimum mass loading of RGO was 20 wt%. However, the high HER activity of such a composite material was attributed to the morphology generated during the synthesis, namely the RGO crumple structure and empty nanovoids with ultrafine nanocrystal layers of molybdenum diselenide. This fact assesses the catalytic center-support interaction (see Chap. 5), as the synergetic effect. The role of the material's phase (2H vs. 1T) toward the HER was investigated, in which a new strategy synthetic route was devised [150]. This technique consisted in a control of the reductant $NaBH_4$ to $NaMoO_4.2H_2O$ and selenium, with the temperature. The molar variation of $NaBH_4$ to $NaMoO_4.2H_2O > 1$ allowed the synthesis of phase 1T-$MoSe_2$. So that the

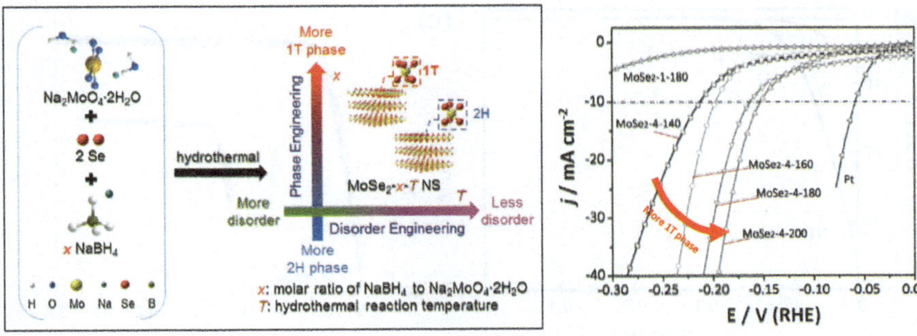

Fig. 4.18 Induced phase- and disorder-controlled synthesis of $MoSe_2$ nanosheets (NSs) via the hydrothermal technique by tuning $NaBH_4$ relative to $Na_2MoO_4{\cdot}2H_2O$ (x) and the reaction temperature (T). Right panel: iR-corrected HER current–potential curves of various $MoSe_2$, showing the contribution of the increasing 1T phase. Pt wire is depicted for comparison. Adapted with permission from Ref. [150]. Copyright © 2017 Wiley-VCH Verlag GmbH & Co. KGaA, Weinheim

synthesis engineering of this method is shown in Fig. 4.18a, showing besides a synergy regulation of both crystal phase and disorder.

The growing interest in transition metal dichalcogenides (TMDs) has also been extended to tungsten and cobalt center materials coordinated to selenium atoms. For optimum HER conditions, the computational studies revealed that molybdenum and tungsten edges and selenium edges in $MoSe_2$ ($E_g = 1.8$ eV) and WSe_2 ($E_g = 1.6$ eV) should be the major sites for the electrocatalytic activity [211]. Therefore, the synthesis of materials should be focused on edges design in the nanocatalyst. In this sense, 2D tungsten diselenide nanosheets were produced and supported onto carbon nanotubes via the solvothermal method [159], similar to the process used for $MoSe_2$ (cf. Fig. 4.18). Herein, additionally, acid-treated CNTs in DMF were added. The HER on bare and CNT-supported WSe_2 materials were measured, Fig. 4.19a, observing that WSe_2/CNT showed a superior activity (Table 4.8). The difference in the performance between the bare and CNT-supported electrocatalysts was attributed to an efficient charge transport through the CNT support of the multiple exposed catalytic sites in intimate contact either on CNTs [159] or graphene nanosheets (RGO) [158, 160]. Other hybrid composites WSe_2/WO_3 prepared via the solvothermal technique, controlling the reaction temperature and residence time, showed different morphologies tested for the HER in acid medium [157] following the same strategy as that mentioned above (cf. Table 4.7). Although a certain progress is reported, these data remain far from those obtained by supporting WSe_2 onto RGO and/or CNTs, Table 4.8.

Cobalt diselenide ($CoSe_2$) is a semi-metal that exists in two types of structure: orthorhombic (Marcasite) and cubic (Pyrite), cf. Fig. 1.4. In conjunction with small-gap semiconductors, e.g., Si, it was studied as co-catalyst in (p-Si/$CoSe_2$ (Marcasite)) for the photoelectrochemical evolution of hydrogen in acid medium (cf. Chap. 3) with photocurrents of up to 9 mA/cm^2 at 0 V/RHE was reported by Basu et al. [217], and to further improve the stability and efficiency, $CoSe_2$ was embedded in carbon nitride (C_3N_4) to form the heterostructure: (p-Si/C_3N_4-$CoSe_2$

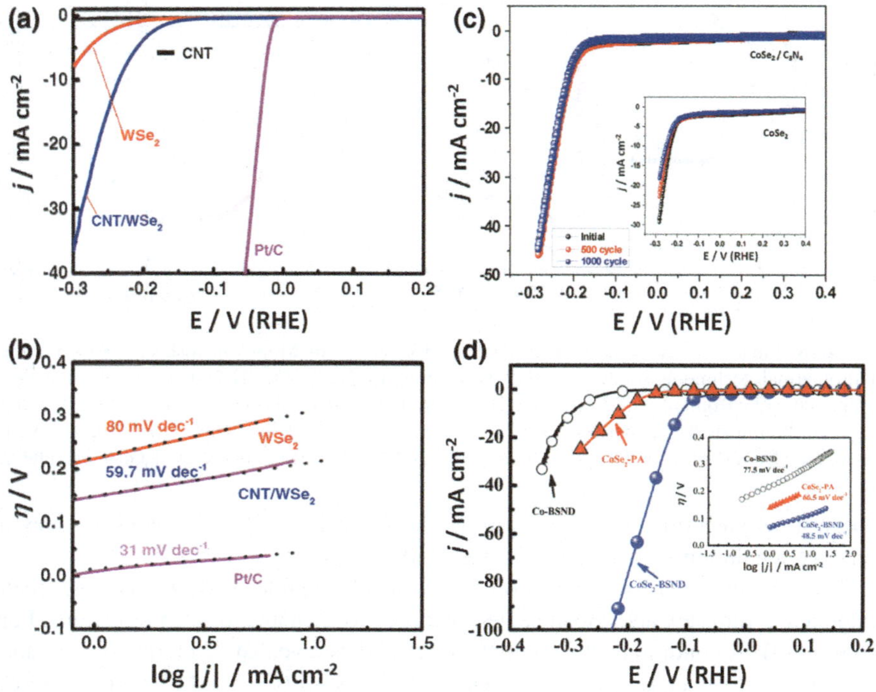

Fig. 4.19 **a** HER-current-potential for bare WSe_2, and CNT/WSe_2 in 0.5 M H_2SO_4. Commercial Pt/C is shown for comparison, as well as the support CNT. **b** The corresponding Tafel plots of data of figure (**a**). Figures **a** and **b** adapted with permission from Ref. [159]. Copyright © 2016 Elsevier B.V. All rights reserved. (**c**) The HER on $CoSe_2$ material. The stability test (0–1000 cycles) is contrasted on the C_3N_4-supported and C_3N_4-non-supported chalcogenide (inset). Figures adapted with permission from Ref. [218]. Copyright © 2016, American Chemical Society. **d** The HER performance, in Ar-purged 0.5 M H_2SO_4, of pyrite-type beaded stream-like cobalt diselenide nanoneedles ($CoSe_2$-BSND) formed onto titanium foils. The insert shows the corresponding Tafel plot. Adapted with permission from Ref. [168]. Copyright © 2016, Royal Society of Chemistry

(Marcasite)) as photocathode [164] to additionally demonstrate the solar hydrogen production in a photoelectrochemical cell (PEC), delivering a photocurrent of −4.89 mA/cm^2 at 0 V/RHE [218]. Interestingly, the HER on C_3N_4-$CoSe_2$, in acid medium, was verified, demonstrating the positive effect in performance and stability with the carbon nitride (Table 4.8, Fig. 4.19c). In contrast, it seems that the $CoSe_2$ (Pyrite) can be conditioned to a specific morphology as to display a higher HER performance. In this direction, Lee et al. [168] reported the synthesis of this chalcogenide in form of nanoneedles on flexible titanium foils via the Co_3O_4 nanoneedle array template with selenium vapor. The enhanced electrochemical activity of this system, see Fig. 4.19c, is apparently facilitated by 3D structure. Lamellar structured cobalt diselenide nanosheets onto titanium foils were synthesized via the hydrothermal method by Xiao et al. [162]. Different morphologies of the materials were obtained via the catalyst mass loading variation, volume, and reaction time. The HER-optimized material was found to be 0.83% of $CoSe_2$ onto titanium plate, (Fig. 4.19d, Table 4.8). Again, the enhanced activity is due to the direct growth of the cobalt diselenide onto titanium plate. The use of metal organic

Table 4.8 Hydrogen evolution reaction on W, Co diselenide

Materials	Synthesis method	Morphology	Electrolyte	-η/V @ 10 mA/cm^2	Refs.
WSe_2	Solvothermal	Graphene-like structure	0.5MH_2SO_4	0.31	[159]
WSe_2/CNTs		Edge-rich hybridized nanosheets		0.21	
WSe_2/RGO		Nanoflowers anchored onto RGO		0.23	[158, 160]
WSe_2/WO_3	Hydrothermal	Rhombic dodecahedron		0.48	[157]
WSe_2/WO_3		Nanorods		0.37	
$CoSe_2$	Hydrothermal	Nanorods		0.23	[218]
$CoSe_2/C_3N_4$		Nanorods grafted on C_3N_4		0.21	
[c]$CoSe_2$-PA	Hydrothermal and pyrolysis	Particles on Ti foil		0.20	[168]
[d]$CoSe_2$-BSND		Nanoneedle arrays		0.12	
$CoSe_2$	Hydrothermal: absence or presence of Ti foils	Nanobelts		0.338	[162]
$CoSe_2$/Ti		Perpendicular nanosheets		0.152	
$CoSe_2$@DC	Pyrolysis of Co-MOF, oxidation and selenization	NPs embedded in carbon nanotubes		0.132	[164]
$CoSe_2$/DC	Chemical synthesis with DC	NPs onto defective carbon		0.23	
MOF-$CoSe_2$	Pyrolysis of Co-MOF, and selenization	$CoSe_2$ nanoparticles anchored in N-doped MOF		0.23	[165]
$CoSe_2$/CFN	Co-CFN olid-state reaction and selenization at 500 °C	3D nanonetlike carbon fibers with highly dispersed NPs		0.133	[133]
$CoSe_2$ NS@CP	Hydrothermal in the presence of CP	Dense and perpendicularly packed nanosheets		0.162	[163]
$CoSe_2$/CNT	Spray pyrolysis for CoO-CNT and selenization	Microsphere composite		0.174	[166]

[a]Data extrapolated; [b]Composites with 10 wt%-SWNT-mass loading of 1.5 mg/cm^2; [c]PA-Particles on Ti foil; [d]BSND—Beaded Stream-like Nanoneedle

framework (MOF) [219], as a chemical precursor, is an interesting strategy. Cobalt-based MOFs (ZIF-67) was employed to generate cobalt nanoparticles embedded in carbon (Co@C), after pyrolysis in Ar atmosphere at 800 °C, and washing with acid overnight. The heat treatment of Co@C in air at 350 °C formed Co_3O_4@C, the chemical precursor for cobalt chalcogenide, after selenization at 450 °C to obtain the so-called chalcogenide nanoparticles embedded defective carbon: $CoSe_2$@DC [164]. The HER data generated on this material and the comparison with DC supported one are summarized in Table 4.8. However, although using similar chemical precursors, Lin et al. [165] reported the synthesis of cobalt diselenide using cobalt-based MOF (ZIF-67) and thereafter selenization to obtain MOF-$CoSe_2$. The overpotential of HER of this material at j = 10 mA/cm^2 was 0.23 V. This performance is lower than the former $CoSe_2$@DC issued from similar chemical precursors. Thus, apparently the oxidation chemical step after pyrolysis [164] is a key factor. The morphology and conductivity of the support are essential ingredients to enhance active sites for the HER performance of cobalt dichalcogenide. In this connection, highly dispersed $CoSe_2$ NPs were deposited on 3D nanonet-like carbon fibers [133], by the in situ growth of interlayer expanded lamellar cobalt diselenide nanosheets on carbon paper substrate ($CoSe_2$NS@CP) [163], or by producing macroporous $CoSe_2$/CNT composites microspheres [166]. All these approaches led to an enhancement of cobalt dichalcogenide toward the HER, as noted in Table 4.8.

Another metal center dichalcogenide based on Ni has been reported as an efficient 3D electrode for the HER [169]. The solid-state selenization reaction at 450–600 °C allowed the formation of $NiSe_2$ on Ni foam. The morphology obtained depended on the selenization temperature. More active sites were, apparently, created with the selenization temperature, so that the activity toward the HER results more effective on the sample generated at 600 °C, as noted at a current density of 10 mA/cm^2. The overpotential data were: 0.19 V (450 °C), 0.16 V (500 °C), 0.15 V (550 °C), and 0.14 V (600 °C), respectively. Again, for the TMDs discussed so far, the electrocatalytic activity (active sites) of the HER is dominated by the morphology and the electrical conductivity of the material.

4.3.1.2 Hydrogen Oxidation Reaction (HOR)

Non-precious chalcogenide materials are essentially reported as electrocatalysts for the oxygen reduction reaction (ORR), oxygen evolution reaction (OER), see Sect. 4.3.2, and as discussed above, for the hydrogen evolution reaction (HER). To the best knowledge, such materials have not been investigated for the hydrogen oxidation reaction (HOR), neither in acid nor in alkaline medium due to lack of activity both in acid and in alkaline media. However, the commencement of alkaline membrane fuel cells development, the HOR in alkaline conditions ($2H_2 + 4OH^- \rightarrow 4H_2O + 4e^-$) has attracted an increasing interest since then. Even, if chalcogenides have not been developed yet, it is interesting to mention that non-precious catalytic centers have been investigated for HOR in alkaline polymer

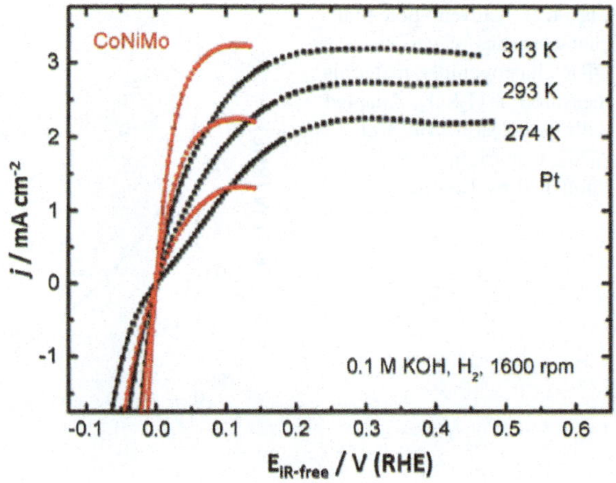

Fig. 4.20 HOR current–potential characteristics on electroplated CoNiMo recorded in H_2-saturated 0.1 M KOH electrolyte at the temperature indicated, 1600 rpm, and compared to Pt/C. Reprinted with permission from Ref. [147] Copyright © 2014, Royal Society of Chemistry

electrolyte fuel cell (APEFCs), as reported by Lu et al. [220] with chromium surface-modified Ni NPs and by Hu et al. [221] with tungsten surface-modified Ni NPs. The system still has some difficulties, first, since the availability of the alkaline membrane is limited, and second, a non-precious anode is not straightforward. In this connection, a ternary Ni-based metallic thin film (NiMoCo) electrodeposited onto gold substrate was reported [147]. As shown in Fig. 4.20, the non-precious trimetallic material outperforms the HOR under the same condition to platinum. The activity was sustained at potentials lower than 0.1 V to avoid oxidation of the material, leading to a passivation of nickel. Likewise, the effects are recorded on Ni–Cr and Ni–W toward the HOR; Ni–Mo also offers an enhanced activity, compared to Ni. Thus, in spite of the existence of tiny amounts of Co, it builds hydroxide as well as nickel. The enhanced activities related to the hydrogen bonding energy must be favorable in the ternary NiMoCo.

4.3.2 Materials for ORR/OER

Massive chalcogenide materials (sulfides, selenides, tellurides) with non-precious metal centers, e.g., Co, Ni, W, Mo, Ti, Ta, have been reported as electrocatalysts for the oxygen reduction reaction (ORR) in acid medium by Baresel et al. in 1974 [222]. The authors concluded that cobalt sulfide and cobalt-nickel sulfides showed the best ORR activity in 2 M H_2SO_4, whereas chalcogenides with the same metal in selenide or telluride environment were less active. The reason for such a difference was not established, but attributed to a correlation between the ORR activity and the decrease in the energy difference between the O_2 2*p*-orbital and the highest occupied *d*-orbital of the transition metal coordinated to sulfide; nonetheless, the materials were not sufficiently stable to sustain the reaction in a fuel cell system. In the same line, electrocatalysts containing non-precious metal centers and

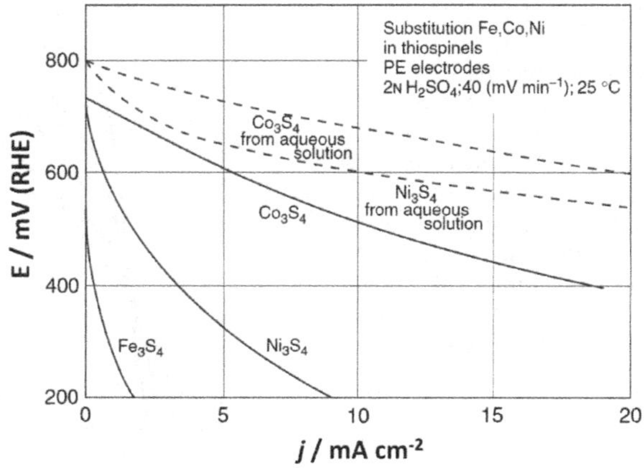

Fig. 4.21 Current–potential characteristics of various ORR chalcogenides materials measured in H_2SO_4. Adapted with permission from Ref. [223]. Copyright © 1975 Published by Elsevier Ltd

chalcogenide in the spinel structure (AB_2X_4) were also reported in 1975 by Behret et al. [223]. These materials (A_3X_4: A = Co, Ni, Fe; X = S, or A = Mn, Fe, Co, Ni, Cu, Zn; B = Ti, V, Cr, Fe, Co, Ni; X = S, Se, Te, O), synthesized at 300 °C (solid-state reaction), demonstrated a non-negligible ORR activity in 1 M H_2SO_4 medium. The Co_3S_4 and Ni_3S_4 present an open-circuit potential of 0.8 V/RHE, as shown in Fig. 4.21. The figure clearly shows, within the spinel structure, the electrocatalytic effect of the metals: Co > Ni > Fe. In 0.1 M KOH solution. Kitayama et al. [224] reported that thiospinel Co_3S_4 is an effective catalyst for the ORR, assessing that the multi-electron charge transfer (four electrons) proceeded on cobalt sulfide electrode. The contributions of Baresel et al. [222] and Behret et al. [223] is the hallmark into the use of "exotic" materials based on chalcogenides for the ORR. On the other hand, the OER process with a more important overpotential due to the complex oxidation process (cf. Chap. 2, Sect. 2.2.2) [225] has been restricted to the use of Ir and Ru oxide-based electrocatalyst (cf. Fig. 2.12). The important issues for application is the stability of the material and the support (see Chap. 5). In working conditions, the dissolution, in acid as well as in alkaline, of RuO_2 is higher than IrO_2 [226]. In this context, Co and Ni oxide-based materials have shown comparable OER activities to IrO_2 [227]. From these facts, Co and Ni metal centers for the OER process are essential ingredients in the development of novel electrocatalyst, and cobalt sulfide has been recently reported for the OER [228–230].

4.3.2.1 Oxygen Reduction Reaction (ORR)

M-Sulfide (M: Co, W, Mo, Fe). Following the research on massive chalcogenide materials, cobalt sulfide was taken as a model to understand the activity reported in the 1970s. Quantum computational calculation, using as a model Co_9S_8, determined that the (202) surface was partially OH-covered, at an onset potential of 0.74 V/RHE, thus active for ORR [231], and further predicting that oxygen bonding also takes place on a non-metallic site, i.e., sulfur. The synthesis of nanoparticulated (50–80 nm) cobalt sulfide was first performed using an in situ free-surfactant approach using cobalt carbonyl ($Co_2(CO)_8$) in organic solvents mixed with carbon, and oleic acid and tri-octylphosphine oxide (TOPO) as surfactants to prevent the particle aggregation, followed by sulfurization (and/or selenization) [225]. Thus, carbon-supported Co_3S_4 was generated. The ORR electrocatalytic activity of 20 w % Co_3S_4/C in acid medium is displayed in Fig. 4.22a. The effect of blocking sites by surfactant is also displayed on the sample Co_3S_4-01, whereas the surfactant-free samples (Co_3S_4-02; Co_3S_4-03) show higher activity. Herein, the predicted ORR onset potential read on the current–potential characteristic approaches the predicted value of 0.74 V. Using physical methods, chalcogenide thin films were prepared by magnetron sputtering [232]. The electrocatalytic nature, of pyrite generated materials, namely CoS_2, and (Co-Ni)S_2 thin films deposited onto glassy-carbon substrates, was examined in acid medium and compared to NiS_2 (Fig. 4.22b). The Tafel plot (inset) displays these data against platinum thin-film electrode; among the chalcogenides, (Co-Ni)S_2 is better, but still 1.5–2 orders of magnitude lower than Pt. Moreover, it seems that the ORR activity differences among these chalcogenide materials is determined by the synergy effect of Co and Ni in the pyrite structure, as evidenced by Raman and XDR techniques. The hydrothermal chemical route was employed to generate CoS_2 and investigated toward the ORR in alkaline medium [124]. It is worth to note that the authors obtained CoS_2 NPs of ca. 31 nm that showed tolerance to small organics, like methanol, and that the ORR activity depends on the catalytic site density (mass loading) similar to the phenomenon reported for platinum, cf. Fig. 2.19, and other non-platinum catalytic centers, e.g., $CoSe_2$ [233]. Nanocrystalline CoS_2 was generated by hydrothermal process, by Jirkovský et al., [234]. By optimizing the weight ratio CoS_2/C (physical mixture),

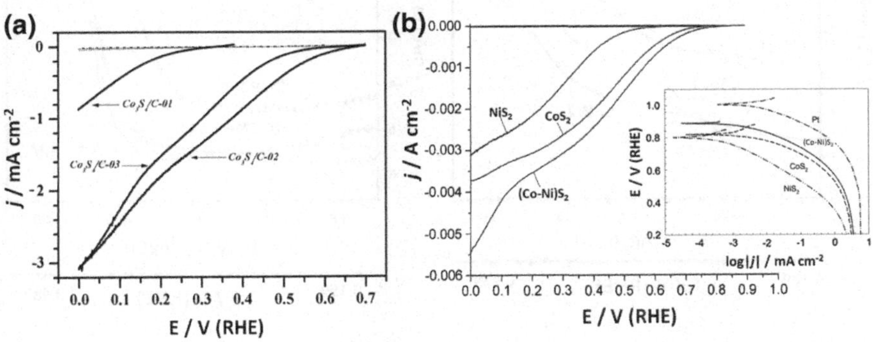

Fig. 4.22 **a** Linear voltammetry, in O_2-saturated 0.5 M H_2SO_4 at room temperature, and 1600 rpm, for 20 wt% Co_3S_4/C-01, −02 and −03. Scan rate of 5 mV/s. The catalyst loading was 11 μg cm^{-2}. Dashed line recorded under nitrogen-saturated electrolyte. Reprinted with permission from Ref. [225]. Copyright © 2008, American Chemical Society. **b** ORR performance of NiS_2, CoS_2 and (Co,Ni)S_2 films in 0.1 M $HClO_4$ solutions at 2000 rpm and scan rate of 5 mV/s. The insert shows the corresponding Tafel plots for the same materials compared to Pt. Adapted with permission from Ref. [232]. Copyright © 2008 Elsevier Inc. All rights reserved

the authors determined experimentally that the highest ORR activity in acid medium was obtained with a ratio of CoS_2/C of 1. The high hydrogen peroxide production on this system was suppressed by increasing the catalyst dispersion, attaining a performance similar to Ru_xSe_y.

Some chalcogenides have been tested as bifunctional electrocatalyst for ORR and OER (cf. Fig. 2.2; see next Sect. 4.3.2.2). Hybrid electrocatalyst can provide this possibility. WS_2 sheets on carbon nanotubes (CNTs) connected via tungsten carbide (WC) that provide low-resistance for charge transfer have been engineered, as an example [235]. The interconnected structure and growth of WS_2 sheets on CNTs via WC were confirmed by XRD and XPS. The ORR electrocatalytic performance in 0.1 M KOH at 1600 rpm is displayed in Fig. 4.23. The number of WS_2 sheets (ca. 4–5 layers) in the sample noted CSW2 showed the best performance, as compared to CSW1 (2–3 layers) and CSW3 (10 layers), clearly demonstrating an optimum in the number of catalytic sites with the conductivity of the support of the complex hybrid structure schematized in the inset of Fig. 4.23a and Table 4.9. The hybrid structure was also tested for the OER and reported in Sect. 4.3.2.2. Since the WS_2 and MoS_2 are semiconducting materials, it was thought that doping these materials could be beneficial to catalyze the ORR (HER). However, the effect of Nb and Ta as p-type dopants on WS_2, and MoS_2, produces, on the one hand, a high percentage of 1T-phase in WS_2 and, on the other hand, a negative electrocatalytic effect on the ORR, as well as on the HER compared to their undoped counterparts [236]. This work shows that various challenges of fundamental interest in the TMDs materials science need to be pursued.

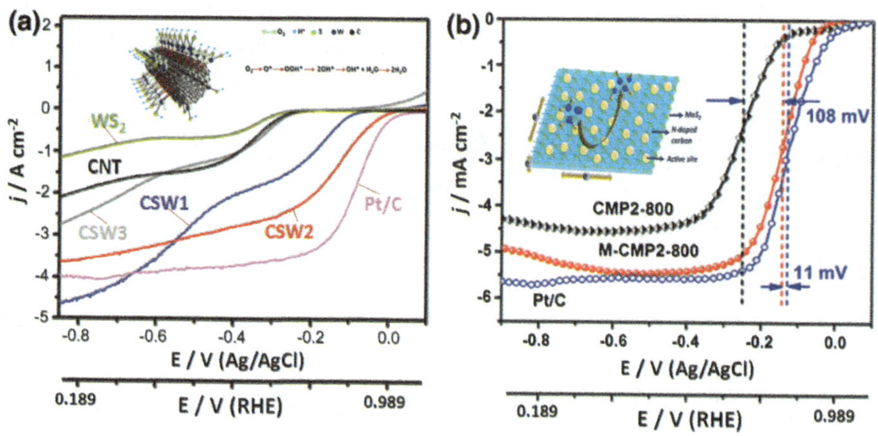

Fig. 4.23 **a** ORR-performance in O_2-saturated 0.1 M KOH electrolyte hybrid structured WS_2-CNT materials noted as CSW1 to CSW3. WS_2, CNT, and Pt/C serve as comparison. The inset shows the morphology of the hybrid materials and the possible ORR mechanism. Adapted with permission from Ref. [235]. Copyright © 2017 Wiley-VCH Verlag GmbH & Co. KGaA, Weinheim. **b** ORR performance of MoS_2-based hybrid materials (M-CMP2-800) in O_2-saturated 0.1 M KOH. Measured at 1600 rpm and 5 mV/s. The inset schematizes the surface process. Adapted with permission from Ref. [241]. Copyright © 2016 WILEY-VCH Verlag GmbH & Co. KGaA, Weinheim. The RHE scale for both figures was calculated: $E = E_{AgCl} + 0.222 + 0.059\ pH$

Table 4.9 Oxygen reduction reaction on Co, W Mo, Fe sulfide materials

Catalysts	Onset potential versus RHE/V	$E_{1/2}$ versus RHE/V	Medium	Catalyst loading/μg cm^{-2}	Refs.
Co_3S_4-02	0.68	NA	0.5 M H_2SO_4	11	[225]
CoS_2	0.82	NA	0.1 M $HClO_4$	[e]TF	[232]
(Co-Ni)S_2	0.89	NA			
CoS_2/C (ratio = 1)	≈ 0.80	≈ 0.72		1.5 μm thick layer	[234]
CoS_2	0.94	0.71	0.1 M KOH	100	[124]
[a]CSW1	0.90	0.55		200	[235]
[b]CSW2	0.98	0.82			
[c]CSW3	0.75	0.55			
MoS_2	0.80	0.64		283	[238]
P-MoS_2-0.2	0.96	0.80			
MoS_2	0.76	0.64		283	[239]
O-MoS_2-87	0.94	0.80			
MoS_2/N-RGO	0.89	0.77		35	[240]
[d]CMP2-800	NA	0.751		600	[241]
[e]M-CMP2-800	1.01	0.859			
MoS_2/G-500 (0.5:1)	0.84	0.75		600	[242]
MoS_2/G-500 (1:1)	0.91	0.80			
MoS_2/G-500 (1.5:1)	0.85	0.72			
MoS_2/Pd	0.97	0.81		10	[244]
FeS_2	0.77	NA	0.1 M $HClO_4$	[e]TF	[246]
(Fe-Co)S_2	0.80	NA			

[a]0.06 mmol WS_2/WC/CNT; [b]0.125 mmol WS_2/WC/CNT; [c]0.25 mmol WS_2/WC/CNT; [d]Micro-porous carbon generated from pyrazine-containing polymer; [e]MoS_2 in Micro-porous carbon generated from pyrazine-containing polymer; [e]TF: thin film

Prior to any doping, comparable phenomenon (exfoliation) on MoS_2 toward the HER (cf. Table 4.5), the Mo edges of MoS_2 NPs were tested to be responsible for the ORR activity [237]. Instead, this work also reports on the effect of NPs size. These exfoliated nanoobjects were obtained via a combination of sonication and centrifugation (1000–12,000 rpm), from MoS_2, dispersed in dimethyl formamide (DMF). The material obtained at 12,000 rpm, corresponding to 2 nm MoS_2, showed the largest limiting current at 0 V/RHE (−1 V/SCE). Coming back into the context, the doping with low-electronegative phosphorous (via annealing at 600 °C sulfourea, red phosphorous and MoO_3) [238] and oxygen (through hydrogen peroxide incorporation in the synthesis after pyrolysis of ammonium molybdate, thiourea, and layered g-C_3N_4 template) [239] into the plane of MoS_2, as confirmed by XPS, induced positive effect toward the ORR in alkaline medium, as assessed by the half-wave potential shift into the positive values; see samples P-MoS_2-0.2 and O-MoS_2-87 (higher oxygen content) in Table 4.9. The combination of highly conductive support like nitrogen-doped graphene (N-RGO) with molybdenum sulfide can compensate for the semiconducting property of the chalcogenide. By a simple physical mixing of commercial MoS_2 and NRGO, Zhao et al. [240] reported ORR improvement of the system. The authors claim that the MoS_2/N-RGO hybrid obtained by loading MoS_2 sheets onto N-RGO through ultrasonication improved the electrocatalytic ORR activity via the exposed active edges as well as the synergistic effect and reduced contact resistance at the interface of MoS_2/N-RGO, but the reported performance is still poor (Table 4.9). In this regard, two-dimensionally core-shelled system, 2D MoS_2/N-CMP-T (nitrogen-doped porous carbon) hybrid, was synthesized by pyrolysis of 2D porous polymer/MoS_2 sandwiches. In the structure of the hybrid, both sides of ultrathin MoS_2 nanosheets are electrically contacted with the nitrogen-doped porous carbon treated at 800 °C (M-CMP2-800) [241] (see inset in Fig. 4.23b). This material showed an excellent ORR in 0.1 M KOH. Herein, contrary to non-supported materials, the strong interaction between nitrogen-doped porous carbon and the chalcogenide via covalent functionalization is favorable to boost the ORR process (Fig. 4.23b, Table 4.9). Using the hydrothermal synthesis chemical route, Arunchander et al. [242] developed flower-like MoS_2 incorporated onto graphene. The hybrid material MoS_2/G-500 obtained, via a-two step method, i.e., MoS_2, by hydrothermal, then mixing with graphene and pyrolyzing at various temperatures (400, 500, 600 °C), was evaluated in alkaline medium. Interesting results were obtained that allowed the authors to pursue the test, as cathode, in an alkaline electrolyte membrane fuel cell (AEMFC), a system giving a peak power density of 29 mW/cm^2.

Previous to this work of Arunchander et al. [242], Zhou et al. [243] reported hierarchical MoS_2-reduced graphene oxide (MoS_2-rGO), prepared by the hydrothermal procedure, in which hierarchical MoS_2 nanosheets grew onto rGO sheets. This material delivered a modest ORR activity in alkaline medium with an onset potential of 0.8 V/RHE, a bit positive compared to MoS_2 (0.7 V/RHE). This latter is in agreement with other data on similar systems reported so far (cf. Table 4.9). Paths toward a facile synthesis route of hybrids based on TMDs materials have been devised to promote synergistic effects, as documented in Table 4.9. In addition, other composites having electrocatalytic NPs onto the chalcogenide have been reported [244]. Using the sonochemical route, composites of MoS_2/Pd were generated. Based on the idea that catalytic metal NPs can strongly hybridize on graphene layers, such catalytic NPs could interact similarly onto exfoliated layers of MoS_2 under a high intensity ultrasonic irradiation. As a result, the combination of the ultrathin nanosheets of MoS_2 as support and a highly dispersed Pd NPs is favorable for the ORR (see Table 4.9). As an important class of 2D materials (like graphene) MoS_2 has been very widely studied in the last years, because of its intriguing properties, and computational studies appear as a nice tool to design and further develop the material. In this connection, it is predicted that heteroatoms doping can be beneficial for the ORR process, since the energy barrier and overpotential can largely decrease if a MoS_2 monolayer is doped with Nitrogen [245]. Out of these 2D TMDs, the pyrite compound was also reported as an ORR electrocatalyst material. Similar to CoS_2 and $(Co\text{-}Ni)S_2$ (see Table 4.9), FeS_2 and $(Fe\text{-}Co)S_2$ thin films deposited onto glassy carbon substrates were produced by the magnetron sputtering method and examined in acid medium [246]. The difference in these materials was the binding energy of S $2p_{3/2}$ signal in $(Fe\text{-}Co)S_2$, which was shifted to higher binding energies compared to that from FeS_2 films. Again, the electronic structure of pyrite changes upon addition of cobalt atoms. This phenomenon impinges a positive effect toward the ORR process, although rather poor against platinum thin films, since the magnitude of current, at an electrode potential of 0.6 V, is three orders or magnitude lower.

M Selenide (M: Co, W, Mo, Fe) Soft chemistry based on the pyrolysis of transition metal carbonyl complexes and selenium in powder form in the refluxing conditions of an organic solvent has been applied to the synthesis of metal selenides based, e.g., on Co, Fe, Mn metal centers [247]. A method which was widely used in the synthesis of Ru-based selenides (Chap. 1, Sect. 1.4). Since this method presents some drawbacks (e.g., unreacted selenium or toxicity of some metal carbonyls), its modification avoiding organic solvents led to the use of green solvents like water and/or using a free-surfactant method. This approach was employed to generate Co_3S_4 compounds described in Fig. 4.22a.

Taking advantage of the dispersing property of black carbon, carbon-supported (20 wt%) $CoSe_2$ NPs were obtained [225]. Figure 4.24a displays the effect of heating during the synthesis on two samples $CoSe_2$/C-01 (conventional heating) and $CoSe_2$/C-02 (microwave heating). These first data opened the path to further improve materials synthesis aiming at enhancing the ORR activity. Like the pyrite-type (cubic structure) $3d$ transition cobalt disulfide, CoS_2, cobalt diselenide, $CoSe_2$, is a metallic compound, whose valence bands are composed essentially of strongly hybridized Co $3d$ and Se $4p$ states [248], which are certainly key to favor electrocatalytic reactions (cf. Fig. 1.10). From a structural point of view, cobalt diselenide has two structural phases: orthorhombic and cubic. It was found that the cubic phase is more active than the orthorhombic one for the ORR in acid medium [249, 250]. The in situ surfactant-free method allowed to produce, using conventional heating, such nanoobjects in mild conditions. The as-prepared sample developed the orthorhombic phase under a heat treatment under nitrogen atmosphere between 250 and 300 °C (avg. crystallite size of 14 nm, o-$CoSe_2$-300C), whereas the cubic phase was formed between 400 and 430 °C (avg. crystallite size of 26 nm, c-$CoSe_2$-430C). The various ORR activities on the samples are contrasted in Fig. 4.24a. With respect to the structural aspects of the material, a clear improvement was obtained, without considering any influence of the support, with a 3.5 e^-/ molecular oxygen charge transfer promotion. An aspect of paramount importance is the material's tolerance to methanol during the ORR. Figure 4.24d shows this phenomenon, compared to platinum in acid medium, with the carbon-supported sample (o-$CoSe_2$-300/C). One remarks a practically complete tolerance to CH_3OH [250], since there is no mixed-potential, which is present in Pt/C. Electronic

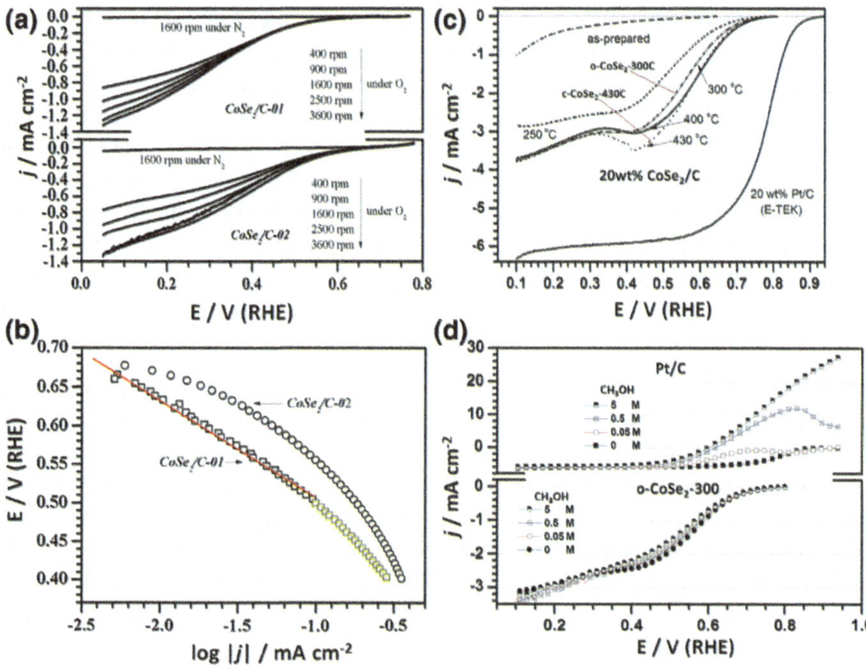

Fig. 4.24 **a** ORR current–potential RDE curves of $CoSe_2$/C-01, synthesized by the conventional heating, and $CoSe_2$/C-02 synthesized by the microwave heating, in O_2-saturated 0.5 M H_2SO_4 at room temperature. **b** The corresponding Tafel plot, after the mass-transfer correction of data of figure (**a**). Figures **a** and **b** adapted with permission from Ref. [130]. Copyright © 2008 Wiley-VCH Verlag GmbH & Co. KGaA, Weinheim. **c** ORR current–potential RDE curves of $CoSe_2$/C NPs as prepared, and annealed at the indicated temperatures in O_2-saturated 0.5 M H_2SO_4 at 2500 rpm, and 25 °C, scan rate of 5 mV/s. **d** ORR characteristics in the presence of methanol of $CoSe_2$/C (heat-treated at 300 °C) compared to Pt/C, at 2500 rpm, and 5 mV/s. Figures **c** and **d** Reprinted with permission from Ref. [250]. Copyright © 2010 Wiley-VCH Verlag GmbH & Co. KGaA, Weinheim

modification of the catalytic center is a key question. An attempt was done to produce Co–W–Se material [251], synthesized by the carbonyl chemical route and selenium powder in xylene organic solvent (cf. Fig. 1.13). XPS results specified positive binding energies shift of Co $2p_{3/2}$, and W $4f_{7/2}$, in Co–W–Se materials, with respect to the element in their binary environment ($CoSe_2$, WSe_2), thus strongly indicative of a charge transfer of the metals (Co, W) to selenium. This apparent electronic modification did not improve the multi-electron charge transfer, since the limiting current obtained was essentially a two-electron process (Table 4.10). This process depends on the number of the catalytic site density and is independent of the nature of the catalytic center [77] (cf. Fig. 2.19). To illustrate this effect on a non-precious catalytic center, Fig. 4.25 shows the production of hydrogen peroxide with mass loading of $CoSe_2$. [233] measured by the rotating ring-disk electrode (RRDE) technique. Assuming a similar thickness on each sample, one can correlate

Table 4.10 Oxygen reduction reaction on Co, W, Mo selenide materials

Catalysts	Onset potential versus RHE/V	$E_{1/2}$ versus RHE/V	Medium	Catalyst loading/μg cm^{-2}	Refs.
[a]$CoSe_2$/C-01	0.72	NA	0.5 M H_2SO_4	11	[130, 225]
[b]$CoSe_2$/C-02	0.72	NA			
o-$CoSe_2$-300C/C	0.78	[c]≈0.56		22	[250]
c-$CoSe_2$-430C/C	0.81	[c]≈0.59			
Co-W-Se/C	0.76	[c]≈0.60		270	[251]
CoSe/C	0.82	[c]≈0.70		450	[253]
$CoSe_{2.0}$/C	0.637	NA		100	[254]
$CoSe_{3.0}$/C	0.705	NA			
$CoSe_2$/C	0.81	[c]≈0.68		609	[255]
Pt/$CoSe_2$	1.01	NA		50	[256]
[d]CoW_0Se/C	0.79	NA		250	[263]
[d]$CoW_{1.49}$Se/C	0.81	NA			
[e]$CoSe_2$/N-C (2)	0.79	0.609		450	[257]
[e]$CoSe_2$/N-C (5)		0.663			
[e]$CoSe_2$/N-C (10)		0.711			
$CoSe_2$/C	0.85	0.71	0.1 M KOH	44	[127]
[f]$CoSe_2$/GCP	0.93	0.81		51	[258]
$CoSe_2$/NCNH	0.9	0.81		214	[259]
CoSe/C	0.79	0.70		285	[260]
CoSe/NG25(1000)	0.92	0.86			
WSe_2	0.88	[c]≈0.77		100	[262]
r-WS_2	0.97	[c]≈0.83			
Pt-WSe_2/C-800	[c]≈0.82	[c]≈0.66	0.5 M H_2SO_4		[264]
$MoSe_2$	0.88	[c]≈0.77	0.1 M KOH	100	[262]
r- $MoSe_2$	0.97	[c]≈0.83			

[a]Via conventional heating; [b]Via microwave heating; [c]Estimated data from plots; [d]Orthorhombic; [e]The number in parenthesis means the ratio (Se/Co) used in the synthesis; [f]Material obtained a 750 °C

the catalytic site density. The figure shows that typical catalyst mass loadings can be limited to a percentage of hydrogen peroxide production. For non-precious catalytic metal centers, the low mass loading affects the thickness and therefore the ORR kinetics, and an increased mass transport loss at high current densities can take place [252]. Microwave-assisted polyol process for cobalt selenide has been pursued on nanoporous carbon using Co(II) acetate and sodium selenite as precursors. The material generated (CoSe) corresponded to a hexagonal phase (Freboldite). The ORR catalytic activity of CoSe/C was examined in acid medium, showing also a very high tolerance in the presence of small organics, namely methanol, ethanol, 2-propanol, ethylene glycol, and formic acid in a concentration span of 2–10 M

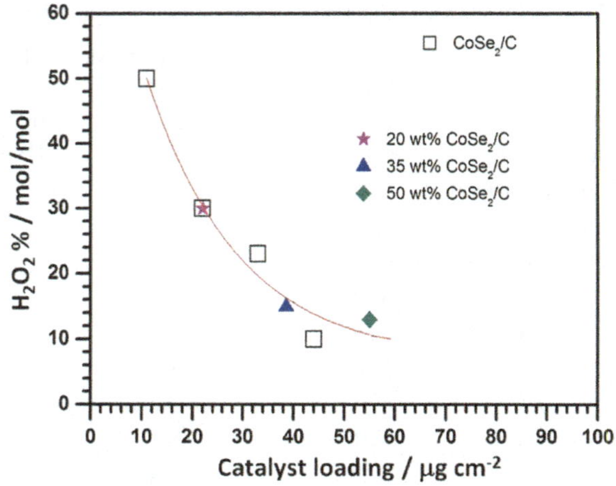

Fig. 4.25 Hydrogen peroxide production as a function of the $CoSe_2$ catalyst mass loading per unit surface. The full symbols represent three different catalyst/carbon ratios. Adapted with permission from Ref. [233]. Copyright © 2009 Elsevier Ltd. All rights reserved

[253]. This property of CoSe/C is interesting in mixed-reactant systems (see Chap. 6). Moreover, employing the similar microwave technique, Li et al. [254] reported the synthesis of carbon-supported $CoSe_2$/C to understand the role of how selenium affects the ORR activity. Cobalt acetate and selenium dioxide were used as precursors with different Se/Co mole ratios, and samples heat-treated at 400 °C to obtain the pyrite (cubic) structure. For Se/Co from 2 to 4, the crystallite size decreased, and the ORR activity shifted positive with Se/Co ratio, leveling at Se/Co = 3. This material ($CoSe_{3.0}$/C) showed the best activity toward ORR and stability in acid solution. Furthermore, the $CoSe_2$ (pyrite structure), unlike the microwave technique, was synthesized by a wet-chemical route using cobalt salts ($CoSO_4.H_2O$), selenium dioxide (SeO_2), and a reducing agent $NaBH_4$, followed by a heat treatment at 600 °C [255]. The carbon-supported material gave similar results, assessing that $CoSe_2$ activity, in acid medium, is higher than that on thin films of CoSe hexagonal phase. In addition, the evaluation of the hydrogen peroxide production was done with the mass loading. One way to boost cobalt diselenide was the synthesis of carbon-free Pt/$CoSe_2$ nanobelt as ORR methanol-tolerant catalysts. Following the solvothermal route to produce $CoSe_2$/DETA (diethylenetriamine) nanobelts, Gao et al. [256] reported that Pt NPs (avg. size of 8.3 nm) can be simply deposited by reaction exchange onto $CoSe_2$/DETA nanobelts. This material presented an onset potential similar to Pt/C catalyst. However, the ORR current–potential shape exhibited a broad mixed kinetic diffusion region (absence of a diffusion-limiting current). Remarkably, the nanohybrid material was insensitive to high concentrations of methanol (5 M) whereas, as is well-known, the ORR activity on Pt/C catalyst is significantly reduced even at low concentration (0.05 M). The effect of synergy of tungsten on cobalt selenide was undertaken by pyrolysis of corresponding carbonyl compounds in 1,6-hexanediol solvent. The so-called W-doped Co–Se catalysts showed a high catalytic activity for the ORR in acid solution. The generated catalysts presented structural characteristics of orthorhombic $CoSe_2$ with presence of tungsten oxide (WO_3). The presence of this oxide was put in evidence by

XPS. Indeed, the binding energy (E_B) of W $4f_{7/2}$ in the Co–W–Se compound is 3.1 eV higher than that of WSe_2 (W $4f_{7/2}$ = 32.2 eV) and similar to E_B of WO_3 (W $4f_{7/2}$ = 35.4 eV). Thus, it is thought that the oxide in Co–W–Se modifies the electron density of selenium. Since, non-precious metal centers lack of sufficient activity and stability, one devised way to boost; e.g., the stability, in acid medium, of $CoSe_2$, is using other kinds of supports. Through a detailed synthesis, and characterization of a variety of N-doped carbons, Chao et al. [257] reported the synthesis of $CoSe_2$ (pyrite structure) using the best N-Carbon (generated at 800 °C) containing 6.1 wt% and Co/Fe 4 wt% (issued from the chemical precursors used in the preparation). The chalcogenide was obtained by the carbonyl route, blending $Co_2(CO)_8$, N-Carbon, in xylene solvent, and thereafter addition of selenium in xylene submitted to the solvent reflux conditions. Thereafter, it was heat-treated at 400 °C. Herein, to assess the material's phase structure, the authors varied the ratio of Se/Co (= 2, 5, 10, and 25). Thus, with a ratio of Se/Co = 2, the material's structure was dominated by the orthorhombic phase; when Se/Co > 5, the cubic phase emerged; otherwise, sharper XRD lines of the cubic phase were present at Se/Co = 25. The impact of the material's synthesis is reflected in the ORR process (Table 4.10). However, Fig. 4.26 contrasts the stability effect of the Vulcan–Carbon vis-à-vis the N-Carbon supports. It is clearly shown that the activity and stability of $CoSe_2$ pyrite (half-wave potential shift) are better on the N-Carbon. No insights were further given for this effect. So far, the data shortly summarized in Table 4.10 report on the ORR of cobalt selenide in acid medium. The alkaline medium is a medium where non-precious metal centers show outstanding ORR activities. One of the first works to evaluate the carbon-supported pyrite $CoSe_2$, in alkaline medium, was reported by Feng and Alonso-Vante [127]. The results, as compared to acid medium [233], cf. Fig. 4.24c, for the same material confirmed a more facile ORR kinetics, stability, Fig. 4.26c, and tolerance to methanol tested up to 5 M concentration (Fig. 4.26d). In the alkaline medium, again, the HO_2^- release depends on the catalytic site density. The metal organic framework (Co-MOF: ZIF-67) was proposed as support that provides during the thermal carbonization, N-doped graphitic carbon. The product, $CoSe_2$/GCP, was obtained by ZIF-67 combustion in the presence of selenium, in argon atmosphere, at various defined temperatures (450–900 °C). The material generated at 750 °C showed the highest ORR performance [258]. Although the current–potential characteristic shape was somewhat dominated by kinetic factors, the authors claim that $CoSe_2$/GCP outperformed the activity of commercial platinum catalysts. Other types of supports based on nitrogen-doped carbon nanohorns [259], and nitrogen-doped RGO has been reported [260]. In the former one, as depicted in Fig. 4.27a, the nanohorns were nitrogen doped with urea and heat-treated at 800 °C, serving as a support to cobalt diselenide, made from cobalt chloride and selenium oxide reduced by $NaBH_4$, to obtain $CoSe_2$/NCNH. In the latter one, nitrogen-doped reduced graphite oxide was prepared by high-temperature pyrolysis from a polypyrrole/GO composite, serving also as support to cobalt diselenide, made from cobalt sulfate and selenium oxide reduced by $NaBH_4$, followed by a heat treatment of 600 °C, to obtain $CoSe_2$/NCNH (see Fig. 4.27b). The pyrite ($CoSe_2$) and the hexagonal (CoSe) phases perform better on N-doped carbons, as compared to their

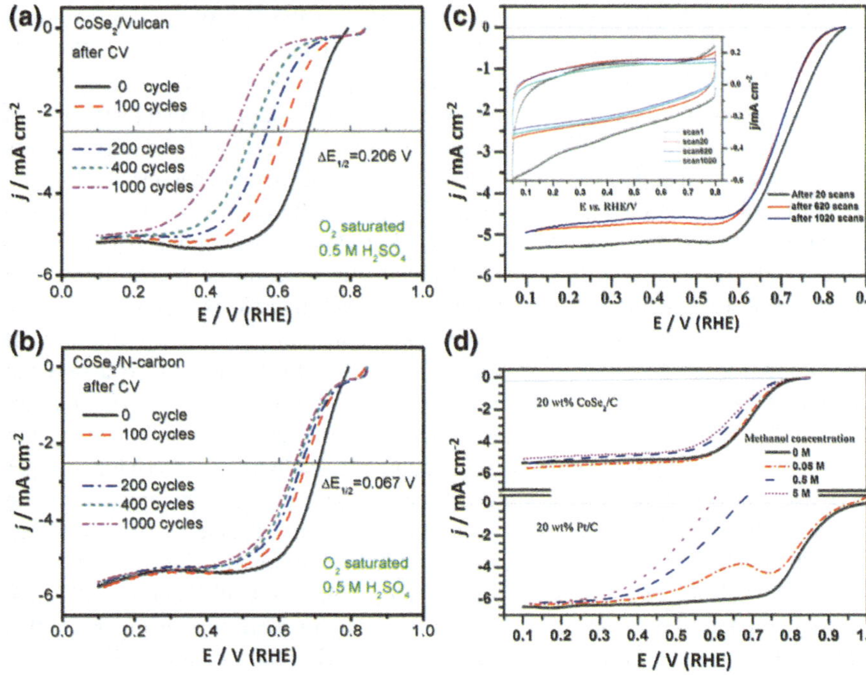

Fig. 4.26 Stability tests performed in O_2-saturated H_2SO_4, 1600 rpm, and 10 mV/s. for **a** Carbon–Vulcan-supported $CoSe_2$; **b** N-doped carbon-supported $CoSe_2$. The total activity decays after 1000 CV cycles is noted by the change of the half-wave potential ($\Delta E_{1/2}$). Figures **a** and **b** reprinted with permission from Ref. [257]. Crown copyright © 2013 Published by Elsevier Ltd. All rights reserved. **c** Stability test in O_2-saturated 0.1 M KOH at 2500 rpm of carbon-supported cubic $CoSe_2$ (20 wt% mass loading) after cyclic voltammetry scans in N_2-saturated 0.1 M KOH at 25 °C (see inset). **d** Linear sweep scans (5 mV/s) at 2500 rpm in 0, 0.05, 0.5 and 5 M methanol electrolyte containing 0.1 M KOH saturated with O_2 at 25 °C. The lower panel compares the Pt/C performance under the same conditions. The gray line was recorded in N_2-saturated 0.1 M KOH containing 5 M methanol. Figures **c** and **d** reprinted with permission from Ref. [127]. Copyright © 2012 Elsevier Ltd. All rights reserved

counterparts on carbon (e.g., Vulcan XC-72) toward the ORR. The important ingredient in N-doped carbons, as confirmed by XPS, is the concentration of pyridinic-specific bonding on carbon. Furthermore, the morphology plays an important role, since these N-doped carbons (single-wall carbon nanohorns-SWCNH; Graphite) develop during the pyrolysis process more active sites and increase pore size and volume, affecting favorably the ORR process, as confirmed by the current–potential curves in Fig. 4.27c, d.

The proliferation of Pyrite ($CoSe_2$) toward the ORR leads to the examination of other metal center selenides, such as layered compounds: WSe_2 and $MoSe_2$ for the same finality. Through their exfoliation [261], they offer an interesting surface chemistry flexibility to engage electrocatalytic processes. In this background, the layered TMDs (2D materials), considered as a graphene analogue, combined with

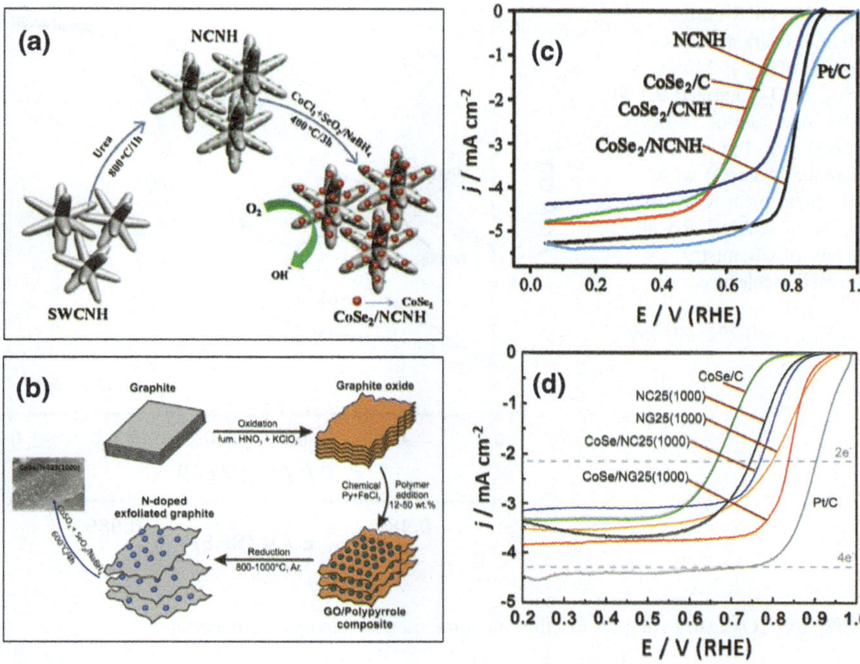

Fig. 4.27 **a** Schematic synthesis of NCNH (nitrogen-doped carbon nanohorns)-supported $CoSe_2$. **b** Schematic synthesis of GO and N-doping process thereof onto which hexagonal-CoSe is chemically deposited. **c** ORR performance of various C-, CNH-, NCNH-supported $CoSe_2$ electrocatalysts in oxygen-saturated 0.1 M KOH, 1600 rpm, and a scan rate of 5 mV/s. **d** ORR performance of various carbon-supported hexagonal CoSe. In both cases Pt/C served as reference. Figures **a** and **c** adapted with permission from Ref. [259]. Copyright © 2015 Wiley-VCH Verlag GmbH & Co. KGaA, Weinheim. Figures **b** and **d** adapted from Ref. [260]

graphene (reduced graphene oxide—RGO) generate nanocomposite materials (g-WSe_2 and g-$MoSe_2$) that take advantage of the conductivity of RGO, and the catalytic centers of the semiconducting materials. Nanocomposites of g-WSe_2 and g-$MoSe_2$ were synthesized [262] chemically using exfoliated GO, WCl_6 (or $MoCl_5$) with selenium powder in oleylamine, and refluxed at 280 °C in nitrogen atmosphere, further centrifugated and finally heating the product at 450 °C. The interaction of RGO with the TMDs was assessed by Raman spectroscopy. The ORR performance, compared with their unsupported counterparts, is contrasted in Fig. 4.28. The composite electrodes show a favorable ORR performance, confirming the significant contribution of the single-layer graphene (Table 4.10). Further modification of the WSe_2 has been sought using porous carbon and as a co-catalyst Pt NPs.

Pan et al. [264] reported Pt–WSe_2/C-800 as the best ORR electrocatalyst in acid medium, exhibiting higher tolerance to methanol. However, judging the ORR characteristics of this material (cf. Fig. 6a in Ref. [264]) under the same hydrodynamic conditions (although not in the same electrolyte medium), the performance

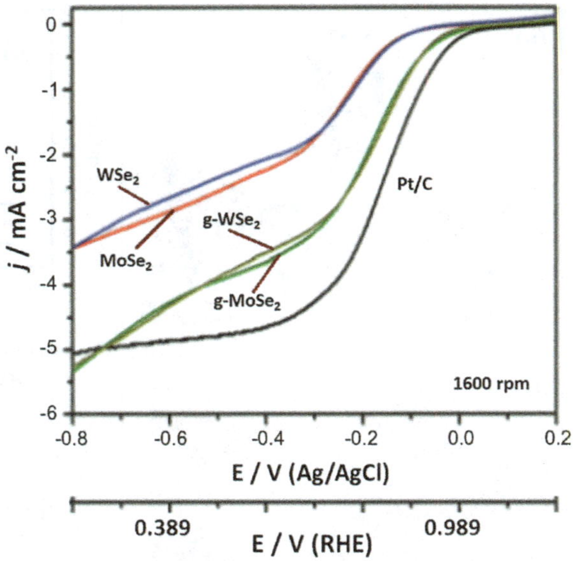

Fig. 4.28 ORR-current potential characteristics of $MoSe_2$, WSe_2, g–$MoSe_2$, g–WSe_2, in O_2-saturated 0.1 M KOH, scan rate of 10 mV/s, at 1600 rpm. Pt/C is used as reference. Adapted with permission from Ref. [262]. Copyright © 2015, Royal Society of Chemistry. The RHE scale was calculated: $E = E_{AgCl} + 0.222 + 0.059$ pH

Table 4.11 Oxygen evolution reaction on some metal chalcogenide materials

Catalysts	[a]Onset Potential versus RHE/V	η_{OER}@ 10 mA cm^{-2}/V	Medium	Catalyst loading/μg cm^{-2}	Refs.
$CoSe_2$	1.60	0.52	0.1 M KOH	200	[284]
$Mn_3O_4/CoSe_2$	≈1.47	0.45			
NG- $CoSe_2$	1.52	0.366			[285]
$CoTe_2$	[a]≈1.52	0.357		250	[286]
CoTe	[a]≈1.5	0.365			
$CeO_2/CoSe_2$	≈1.39	0.288		200	[287]
$Ni_xFe_{1-x}Se_2$-DO	≈ 1.4	0.19	1 M KOH	–	[288]
$Zn_{0.76}Co_{0.24}S/CoS_2$	1.55	0.43		1000	[290]

of Pt–WSe_2/C-800 remains at the level of WSe_2 (Fig. 4.28; Table 4.10). Additional materials' designs using the hydrothermal approach have been proposed, e.g., multi-functional WSe_2/CdS, $MoSe_2$/CdS [265] and WSe_2/MoO_3, $MoSe_2/WO_3$ [157]. Moreover, the ORR current–potential characteristics of these nanohybrid materials, prepared at 200 °C, with peculiar morphologies, are kinetically controlled, although the trend for the onset potential was: $MoSe_2/MoO_3$ (0.84 V/RHE) > WSe_2/WO_3 (0.78 V/RHE), whereas on the nanohybrid dots (NHDs) the trend was $MoSe_2$:CdS (−0.46 V/RHE) < WSe_2:CdS (−0.26 V/RHE), obtained by linear sweep voltammetry. The bifunctionality of these hybrids materials was also investigated.

4.3.2.2 Oxygen Evolution Reaction (OER)

The oxygen evolution reaction (OER) is a fundamental process in the electrocatalytic water splitting (cf. Fig. 2.3). Together with the oxygen reduction (ORR), both processes are also fundamental for rechargeable metal–air batteries. The best electrocatalysts for the OER are precious metal oxides-based materials, specifically, RuO_2 [266–269], IrO_2 [267, 268], recently revised [226], and theoretically rationalized [270]. As a matter of fact, the comparison of OER electrocatalysts is made at a current density of 10 mA cm^{-2} considered as a metric relevant to an integrated solar water splitting device [271] and a figure of merit to compare electrocatalytic activity [268]. The subject matter of this section is devoted to discuss recent progress done so far on metal chalcogenide materials. Due to the versatile physical–chemical properties, pyrite materials are merging as potential materials for the OER.

Cobalt-based nanomaterials have become popular to tailor OER interesting electrocatalysts, including cobalt oxide [272, 273], mixed oxides [274, 275], hydro (oxy)oxides [276], phosphates [277, 278], perovskites [279–282], and chalcogenides [235, 283]. Gao et al. [284] reported that $CoSe_2$ can oxidize water delivering a current density of 10 mA cm^{-2} at an overpotential of 0.45 V in alkaline solution (0.1 M KOH), with an enhanced stability. The material was based on a hybrid between cobalt diselenide and manganese oxide: ($Mn_3O_4/CoSe_2$). This material was synthesized by the polyol method where Mn_3O_4 NPs (average particle size 15.7 nm) were anchored, via a heterogeneous nucleation process, on $CoSe_2$/DETA (precursor used to produce Pt/$CoSe_2$, cf. Table 4.11). The OER performance on $Mn_3O_4/CoSe_2$ is contrasted with those of single counterparts, Fig. 4.29. The XPS analysis on this material revealed that the binding energy of Co $2p_{3/2}$ decreases in the presence of Mn_3O_4 indicating a charge transfer from the oxide sites to $CoSe_2$. As a consequence, this phenomenon creates a synergy effect between the generated Lewis (oxide) and acid (pyrite) sites improving the OER activity. With the same finality, a cobalt composite electrocatalyst was designed using the graphene to generate nitrogen-doped cobalt-based composite. The synthesis was performed via the hydrothermal reduction approach using graphene oxide (GO) sheets onto which $CoSe_2$ nanobelts were nucleated at 180 °C, in a medium containing DETA/H_2O, to obtain NG–$CoSe_2$–nanobelt composite [285]. The ORR performance of this composite outperforms pure $CoSe_2$ nanobelts, and Pt/C, and that of RuO_2 at current densities higher than 10 mA cm^{-2}. The synergy phenomenon, attested by XPS, and Raman spectroscopies, is into play between the nitrogen moieties (pyridinic, pyrrolic, graphitic) of the support (rGO) and the chalcogenide. Moreover, a slightly more OER efficient pyrite material was found with $CoTe_2$ [286]. The synthesis of the material was done by a chemical transformation process of ultrathin Te nanowires (1D), as templates, of hierarchical $CoTe_2$ and CoTe nanofleeces via the polyol reduction chemical route, and heat-treated at 200 °C, 220 °C, to generate, by changing the concentration of Co(acac) from simple to double, $CoTe_2$ (orthorhombic) and CoTe (hexagonal), respectively, as revealed by XRD measurements. The electrocatalytic OER activities of both materials showed that $CoTe_2$

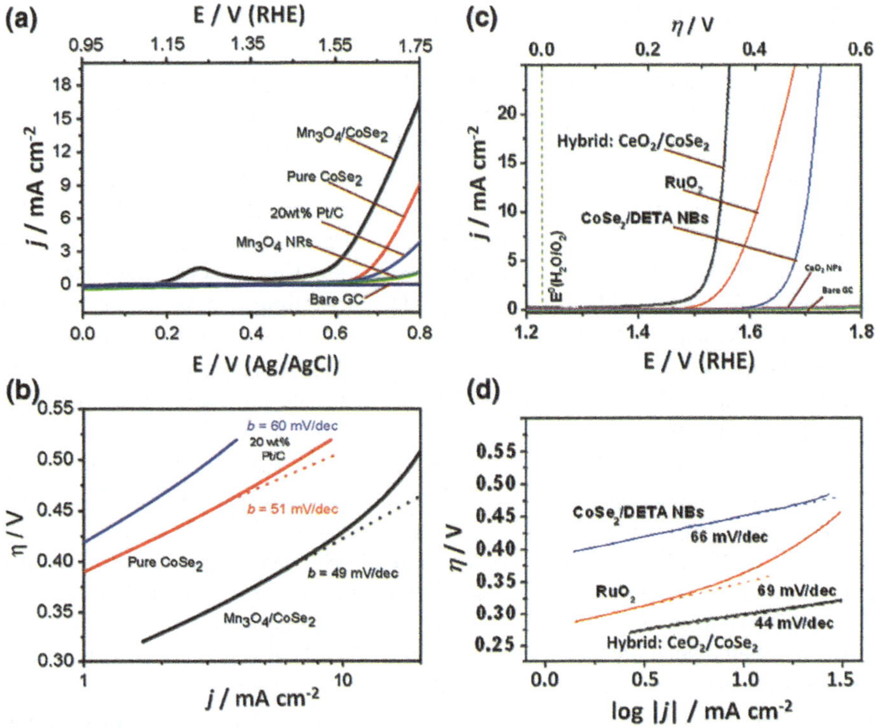

Fig. 4.29 **a** OER current potential characteristics on glassy carbon (GC), Mn_3O_4 NRs, pure $CoSe_2$/DETA NBs (DETA = diethylenetriamine; NBs = nanobelts), and hybrid $Mn_3O_4/CoSe_2$, in O_2-saturated 0.1 M KOH (pH ~ 13) using a mass loading of 200 μg/cm^2. **b** The corresponding Tafel plot of data in figure (**a**). 20 wt% Pt/C was used as a reference material. Figures **a** and **b** adapted with permission from Ref. [284]. Copyright © 2012, American Chemical Society. **c** iR-corrected OER current–potential characteristics, at 5 mV/s, in O_2-purged 0.1 M KOH, on glassy carbon (GC) and GC modified with materials, such as RuO_2 (as a OER reference electrocatalysts), $CoSe_2$/DETA NBs, and $CeO_2/CoSe_2$ hybrid (catalyst loadings of 200 μg/cm^2. (**d**) The corresponding Tafel plot of data in figure (**c**). Figures **c** and **d** adapted with permission from Ref. [287]. Copyright © 2014 Wiley-VCH Verlag GmbH & Co. KGaA, Weinheim

was more active than CoTe, Pt/C, and RuO_2 in 0.1 M KOH (see Table 4.11). It appears that the superior performance of $CoTe_2$ versus CoTe can be due to the high surface area, and the electrical conductivity, as assessed by BET and capacitance measurements. From a mechanistic point of view, the facets (120—orthorhombic) and (101—hexagonal) of these two-structural phases dominate and determine the difference in the overpotential, as rationalized by DFT calculations. Moreover, by coupling chemical and electrical properties in functional materials a superior OER activity could be developed. The grafting of $CoSe_2$ to oxygen-vacancies rich oxide (CeO_2) represents an example [287]. The oxygen vacancies present in CeO_2 NPs can favor the binding of adsorbates as compared to other oxides. The anchoring of CeO_2 NPs onto ultrathin $CoSe_2$/DETA (diethylenetriamine), by the polyol reduction method at 278 °C, was favored by the heterogeneous nucleation onto the amino groups present on the surface of $CoSe_2$/DETA. The oxide coverage was modulated by the amount of the chemical precursor (Ce(acac)). The OER activity of the hybrid $CeO_2/CoSe_2$ casted onto GC, depicted in Fig. 4.29c, can afford a current density of 10 mA cm^{-2} at an overpotential of ca. 0.28 V, a value smaller than that of RuO_2, and that of $Mn_3O_4/CoSe_2$, Fig. 4.29a. Here, the synergy between the oxide and the chalcogenide is reinforced and favorable to the OER kinetics. As confirmed by XPS, the binding energy of Co $2p_{3/2}$ in $CeO_2/CoSe_2$ is shifted to negative values as related to the same signal in pure $CoSe_2$, assessing again, for this system, a charge transfer from the oxide to the chalcogenide, similar phenomenon to that reported on $Mn_3O_4/CoSe_2$. Moreover, particular attention must be paid on the in situ electrochemical oxidation of metal sulfides and/or selenides for OER, since the metal oxides derived thereof are responsible for OER, as in the case of Ni–Fe–Se, a selenide-derived oxide (DO) [288]. Indeed, the combinatorial screening of metal oxides revealed an important number of potential OER trimetallic oxide materials, where Ni and Co are essential non-precious metal elements for water oxidation in alkaline medium [289]. Moreover, $Zn_{0.76}Co_{0.24}S/CoS_2$ that derived from an oxide materials ($ZnCo_2O_4$) was reported to be an excellent OER electrocatalysts in alkaline medium [290].

4.3.2.3 Bifunctional Chalcogenides (ORR/OER)

Bifunctional electrocatalysts for the oxygen reduction reaction (ORR) and the oxygen evolution reaction (OER) play a critical role in energy converting systems, e.g., metal—air battery, in low-temperature fuel cells (FCs) (cf. Figs. 2.2, 2.3, and Sect. 2.2.2. Mesoporous non-precious catalysts based on cobalt, molybdenum, and nitrogen (Co_3Mo_3N) have been successfully developed as bifunctional electrocatalyst for the ORR/OER in alkaline medium for Li–O_2 battery [291]. In this context, ultrathin cobalt-based oxide nanostructures have been developed for such a purpose in zinc–air batteries [273]. As discussed in the sections above, transition metal chalcogenides are nowadays potential electrocatalysts for ORR and OER. Figure 4.30 shows as an example the ORR/OER bifunctionality performance of two electrocatalytic materials (Pt/C and IrO_2/C or RuO_2/C) considered as reference ones, either for the ORR or for the OER processes (Table 4.12). On these precious metal centers, in alkaline medium, we can appreciate that there is no OER/ORR symmetry reaction due to then reaction kinetics differences discussed in Chap. 2 (cf. Figs. 2.10, and 2.12). Moreover, we can appreciate that IrO_x/C [292] shows a better bifunctional performance compared to RuO_2 [293] and Pt. The metrics used to evaluate novel materials is the difference in energy defined as $\Delta E = E_{OER@j=10} - E_{ORR@j=-3}$. $E_{OER@j=10}$ is the OER potential at a current density of 10 mA cm^{-2}, and $E_{ORR@j=-3}$ is the ORR potential at a current density −3 mA cm^{-2} (see Fig. 4.30). Thus, the smaller the ΔE is, the closer the system is to an ideal reversible oxygen electrode. Table 4.12 summarizes some significant data. Few chalcogenide electrocatalysts have been developed as bifunctional materials. Moreover, Liu et al. [294] reported that hybrids of $NiCo_2S_4$ nanoparticles grown on

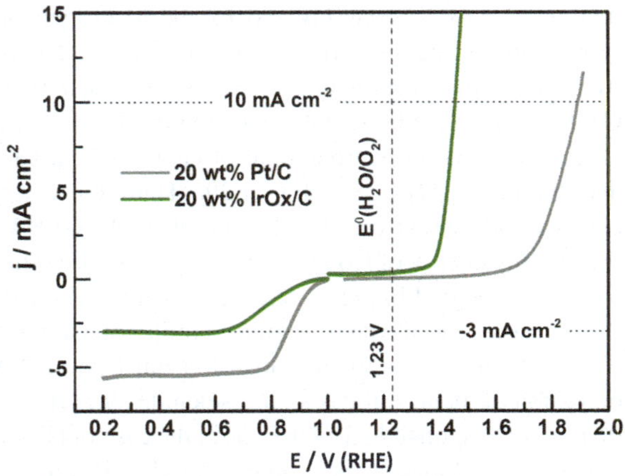

Fig. 4.30 Data in Table 4.12 can be compared to the oxygen electrode activities for 20 wt% Pt/C and 20 wt% IrO_x/C catalysts within ORR and OER potential window. Data obtained in O_2-saturated 0.1 M KOH. Scan rate 5 mV/s; rotation rate 1600 rpm, at RT. Figure modified and adapted with permission from Ref. [292]. Copyright © 2017 Elsevier Ltd. All rights reserved

Table 4.12 Bifunctional ORR/OER materials

Catalysts	E_{ORR}/V at j = −3 mA cm^{-2}	E_{OER}/V at j = 10 mA cm^{-2}	Oxygen electrode ΔE/V = E_{OER}- E_{ORR}	Refs.
20 wt% Pt/C	0.86	1.90	1.04	[292]
20 wt% IrO_x/C	0.61	1.45	0.84	
RuO_2/C	0.09	1.60	1.49	[293]
$NiCo_2S_4$@N/S-rGO	0.76	1.70	0.94	[294]
$Co_{0.5}Fe_{0.5}S$@N-MC	0.78	1.64	0.86	[295]
$NiCo_2O_4$	0.75	1.72	0.97	[296]
$NiCo_2O_4$-A1	0.78	1.62	0.84	[297]

Metal chalcogenide compared with Ni–Co–oxides form and precious centered reference electrodes

graphene (S- and N-doped rGO) can be considered as a novel bifunctional oxygen electrocatalyst. The low activity of a physical mixture of the same components assessed the importance of the interaction between the support and the catalytic center (cf. Chap. 5). The hybrid material showed a ΔE of 0.94 V. In the same line, Shen et al. [295] proposed a nanostructured materials based on cobalt–iron double sulfides covalently entrapped in nitrogen-doped mesoporous graphitic carbon: $Co_{1-x}Fe_xS$@N-MC, where the best ORR and OER performances were obtained with $Co_{0.5}Fe_{0.5}S$@N-MC, cf Table 4.12. However, in these two works, the chemical signature of the materials after OER process was not reported. This fact allows to think that surface electrocatalytic precursor must be a derived oxide. In this sense, it is reasonable to compare also the ORR/OER bifunctionality performance of similar metal centers in oxide form. This is the case of spinel nanowires of $NiCo_2O_4$ [296], and this oxide is prepared with varying weight ratio of polymer, to tune the effect of the morphology, and as chemical precursors, nickel acetylacetonate ($C_{10}H_{14}NiO_4$) and cobalt acetylacetonate ($C_{10}H_{14}CoO_4$) to obtain a porous rod-like structures of 1D $NiCo_2O_4$-A1 [297].

Chalcogenide-based materials can be a concept to develop bifunctional ORR/OER materials for the alkaline medium. Such centers must be provided with a chemical interaction with the support, so that the surface electrochemical precursor could be stabilized forming oxide layers for the OER.

References

1. Wang Y, Li L, Yao W, Song S, Sun JT, Pan J, Ren X, Li C, Okunishi E, Wang Y-Q, Wang E, Shao Y, Zhang YY, H-t Yang, Schwier EF, Iwasawa H, Shimada K, Taniguchi M, Cheng Z, Zhou S, Du S, Pennycook SJ, Pantelides ST, Gao H-J (2015) Monolayer $PtSe_2$, a new semiconducting transition-metal-dichalcogenide, epitaxially grown by direct selenization of Pt. Nano Lett 15(6):4013–4018. https://doi.org/10.1021/acs.nanolett.5b00964
2. Huang Z, Zhang W, Zhang W (2016) Computational search for two-dimensional MX2 semiconductors with possible high electron mobility at room temperature. Materials 9(9):716
3. He T, Kreidler E, Xiong L, Ding E (2007) Combinatorial screening and nano-synthesis of platinum binary alloys for oxygen electroreduction. J Power Sources 165(1):87–91
4. Kulesza PJ, Miecznikowski K, Baranowska B, Skunik M, Zlotorowicz A, Chojak M, Lewera A, Fiechter S, Bogdanoff P, Dorbandt I Enhancement of the electrocatalytic reduction of oxygen at tungsten oxide modified carbon-supported $RuSe_x$ nanoparticles and Co-porphyrin centers. In: ECS Transactions, 2007. pp 11–17
5. Colmenares L, Jusys Z, Behm RJ (2006) Electrochemical surface characterization and O_2 reduction kinetics of Se surface-modified Ru nanoparticle-based RuSey/C catalysts. Langmuir 22(25):10437–10445
6. Stassen WN, Heyding RD (1968) Crystal structures of $RuSe_2$, $OsSe_2$, $PtAs_2$, and α-$NiAs_2$. Can J Chem 46(12):2159–2163. https://doi.org/10.1139/v68-351
7. Sheu J-S, Shih Y-S, Lin S-S, Huang Y-S (1991) Growth and characterization of $RuSe_2$ single crystals. Mater Res Bull 26(1):11–17. https://doi.org/10.1016/0025-5408(91)90033-I
8. Park H-Y, Yoo SJ, Kim SJ, Lee S-Y, Ham HC, Sung Y-E, Kim S-K, Hwang SJ, Kim H-J, Cho E, Henkensmeier D, Nam SW, Lim T-H, Jang JH (2013) Effect of Se modification on RuSey/C electrocatalyst for oxygen reduction with phosphoric acid. Electrochem Commun 27:46–49. https://doi.org/10.1016/j.elecom.2012.10.032
9. Kühne HM, Jaegermann W, Tributsch H (1984) The electronic band character of Ru dichalcogenides and its significance for the photoelectrolysis of water. Chem Phys Lett 112(2):160–164. https://doi.org/10.1016/0009-2614(84)85013-7
10. Shukla AK, Raman RK (2003) Methanol-resistant oxygen-reduction catalysts for direct methanol fuel cells. Annu Rev Mater Res 33(1):155–168. https://doi.org/10.1146/annurev.matsci.33.072302.093511
11. Bernal-Alvarado J, Vargas-Luna M, Solorza-Feria O, Mondragon R, Alonso-Vante N (2000) Photoacoustic characterization of n-$RuSe_2$ semiconductor pellets. J Appl Phys 88(6): 3771–3772
12. Jiang X, Mayers B, Wang Y, Cattle B, Xia Y (2004) Template-engaged synthesis of $RuSe_2$ and $Pd_{17}Se_{15}$ nanotubes by reacting precursor salts with selenium nanowires. Chem Phys Lett 385(5–6):472–476. https://doi.org/10.1016/j.cplett.2004.01.033
13. Üzengi Aktürk O, Tomak M (2015) Adsorption of RuSex (x = 1–5) cluster on Se-doped graphene: First principle calculations. Appl Surf Sci 347:808–815. https://doi.org/10.1016/j.apsusc.2015.04.124

14. Dembinska B, Kiliszek M, Elzanowska H, Pisarek M, Kulesza PJ (2013) Enhancement of activity of RuSex electrocatalyst by modification with nanostructured iridium towards more efficient reduction of oxygen. J Power Sources 243:225–232. https://doi.org/10.1016/j.jpowsour.2013.05.120
15. Lewera A, Miecznikowski K, Hunger R, Kolary-Zurowska A, Wieckowski A, Kulesza PJ (2010) Electronic-level interactions of tungsten oxide with unsupported Se/Ru electrocatalytic nanoparticles. Electrochim Acta 55(26):7603–7609. https://doi.org/10.1016/j.electacta.2010.03.038
16. Chiao S-P, Tsai D-S, Wilkinson DP, Chen Y-M, Huang Y-S (2010) Carbon supported $Ru_{1-x}Fe_xSe_y$ electrocatalysts of pyrite structure for oxygen reduction reaction. Int J Hydrogen Energy 35(13):6508–6517. https://doi.org/10.1016/j.ijhydene.2010.04.032
17. Papageorgopoulos DC, Liu F, Conrad O (2007) Reprint of "A study of Rh_xS_y/C and $RuSe_x$/C as methanol-tolerant oxygen reduction catalysts for mixed-reactant fuel cell applications". Electrochim Acta 53(2):1037–1041
18. Papageorgopoulos DC, Liu F, Conrad O (2007) A study of Rh_xS_y/C and $RuSe_x$/C as methanol-tolerant oxygen reduction catalysts for mixed-reactant fuel cell applications. Electrochim Acta 52(15):4982–4986
19. Kolary-Zurowska A, Zieleniak A, Miecznikowski K, Baranowska B, Lewera A, Fiechter S, Bogdanoff P, Dorbandt I, Marassi R, Kulesza PJ (2007) Activation of methanol-tolerant carbon-supported $RuSe_x$ electrocatalytic nanoparticles towards more efficient oxygen reduction. J Solid State Electrochem 11(7):915–921
20. Kulesza PJ, Miecznikowski K, Baranowska B, Skunik M, Fiechter S, Bogdanoff P, Dorbandt I (2006) Tungsten oxide as active matrix for dispersed carbon-supported $RuSe_x$ nanoparticles: enhancement of the electrocatalytic oxygen reduction. Electrochem Commun 8(5):904–908
21. Kulesza PJ, Miecznikowski K, Baranowska B, Skunik M, Kolary-Zurowska A, Lewera A, Karnicka K, Chojak M, Rutkowska I, Fiechter S, Bogdanoff P, Dorbandt I, Zehl G, Hiesgen R, Dirk E, Nagabhushana KS, Boennemann H (2007) Electroreduction of oxygen at tungsten oxide modified carbon-supported $RuSe_x$ nanoparticles. J Appl Electrochem 37(12):1439–1446
22. Vogel W, Kaghazchi P, Jacob T, Alonso-Vante N (2007) Genesis of Ru_xSe_y Nanoparticles by Pyrolysis of $Ru_4Se_2(CO)_{11}$: A Combined X-ray in Situ and DFT Study. J Phys Chem C 111:3908–3913
23. Haas S, Zehl G, Dorbandt I, Manke I, Bogdanoff P, Fiechter S, Hoell A (2010) Direct Accessing the Nanostructure of Carbon Supported Ru—Se Based Catalysts by ASAXS. J Phys Chem C 114(51):22375–22384. https://doi.org/10.1021/jp103340t
24. Shen M-Y, Chiao S-P, Tsai D-S, Wilkinson DP, Jiang J-C (2009) Preparation and oxygen reduction activity of stable RuSex/C catalyst with pyrite structure. Electrochim Acta 54(18):4297–4304. https://doi.org/10.1016/j.electacta.2009.02.081
25. Malkhandi S, Yangang Y, Rao V, Bund A, Stimming U (2011) Synthesis and Electrochemical Study of Antimony-Doped Tin Oxide Supported RuSe Catalysts for Oxygen Reduction Reaction. Electrocatalysis 2(1):20–23. https://doi.org/10.1007/s12678-010-0033-y
26. Kozlova EA, Safatov AS, Kiselev SA, Marchenko VY, Sergeev AA, Skarnovich MO, Emelyanova EK, Smetannikova MA, Buryak GA, Vorontsov AV (2010) Inactivation and Mineralization of Aerosol Deposited Model Pathogenic Microorganisms over TiO_2 and Pt/TiO_2. Environ Sci Technol 44(13):5121–5126. https://doi.org/10.1021/es100156p
27. Montiel M, García-Rodríguez S, Hernández-Fernández P, Díaz R, Rojas S, Fierro JLG, Fatás E, Ocón P (2010) Relevance of the synthesis route of Se-modified Ru/C as methanol tolerant electrocatalysts for the oxygen reduction reaction. J Power Sources 195(9):2478–2487
28. Cao D, Wieckowski A, Inukai J, Alonso-Vante N (2006) Oxygen reduction reaction on ruthenium and rhodium nanoparticles modified with selenium and sulfur. J Electrochem Soc 153(5):A869–A874

29. Cheng H, Yuan W, Scott K (2007) Influence of thermal treatment on RuSe cathode materials for direct methanol fuel cells. Fuel Cells 7(1):16–20
30. Scott K, Shukla AK, Jackson CL, Meuleman WRA (2004) A mixed-reactants solid-polymer-electrolyte direct methanol fuel cell. J Power Sources 126(1–2):67–75
31. Cheng H, Yuan W, Scott K, Browning DJ, Lakeman JB (2007) The catalytic activity and methanol tolerance of transition metal modified-ruthenium-selenium catalysts. Appl Catal B: Environ 75(3–4):221–228
32. Cheng H, Yuan W, Scott K (2006) The influence of a new fabrication procedure on the catalytic activity of ruthenium-selenium catalysts. Electrochim Acta 52(2):466–473
33. Serov AA, Min M, Chai G, Han S, Kang S, Kwak C (2008) Preparation, characterization, and high performance of RuSe/C for direct methanol fuel cells. J Power Sources 175(1): 175–182
34. Zehl G, Schmithals G, Hoell A, Haas S, Hartnig C, Dorbandt I, Bogdanoff P, Fiechter S (2007) On the structure of carbon-supported selenium-modified ruthenium nanoparticles as electrocatalysts for oxygen reduction in fuel cells. Angew Chem Int Ed 46(38):7311–7314
35. Zehl G, Bogdanoff P, Dorbandt I, Fiechter S, Wippermann K, Hartnig C (2007) Carbon supported Ru-Se as methanol tolerant catalysts for DMFC cathodes. Part I: preparation and characterization of catalysts. J Appl Electrochem 37(12):1475–1484
36. Wippermann K, Richter B, Klafki K, Mergel J, Zehl G, Dorbandt I, Bogdanoff P, Fiechter S, Kaytakoglu S (2007) Carbon supported Ru-Se as methanol tolerant catalysts for DMFC cathodes. Part II: preparation and characterization of MEAs. J Appl Electrochem 37(12): 1399–1411
37. Racz A, Bele P, Cremers C, Stimming U (2007) Ruthenium selenide catalysts for cathodic oxygen reduction in direct methanol fuel cells. J Appl Electrochem 37(12):1455–1462
38. Nagabhushana KS, Dinjus E, Bönnemann H, Zaikovskii V, Hartnig C, Zehl G, Dorbandt I, Fiechter S, Bogdanoff P (2007) Reductive annealing for generating Se doped 20 wt% Ru/C cathode catalysts for the oxygen reduction reaction. J Appl Electrochem 37(12):1515–1522
39. Colmenares L, Jusys Z, Behm RJ (2007) Activity, selectivity, and methanol tolerance of Se-modified Ru/C cathode catalysts. J Phys Chem C 111(3):1273–1283
40. Ohtani T, Ikeda K, Hayashi Y, Fukui Y (2007) Mechanochemical preparation of palladium chalcogenides. Mater Res Bull 42(11):1930–1934. https://doi.org/10.1016/j.materresbull.2006.12.012
41. Serov AA, Cho S-Y, Han S, Min M, Chai G, Nam KH, Kwak C (2007) Modification of palladium-based catalysts by chalcogenes for direct methanol fuel cells. Electrochem Commun 9(8):2041–2044
42. Madhu Singh RN (2011) Palladium selenides as active methanol tolerant cathode materials for direct methanol fuel cell. Int J Hydrogen Energy 36(16):10006–10012. https://doi.org/10.1016/j.ijhydene.2011.05.069
43. Sarkar S, Sampath S (2014) Equiatomic ternary chalcogenide: PdPS and its reduced graphene oxide composite for efficient electrocatalytic hydrogen evolution. Chem Commun 50(55):7359–7362. https://doi.org/10.1039/C4CC02364A
44. Ziegelbauer JM, Gatewood D, Gullá AF, Guinel MJF, Ernst F, Ramaker DE, Mukerjee S (2009) Fundamental investigation of oxygen reduction reaction on rhodium sulfide-based chalcogenides. J Phys Chem C 113(17):6955–6968
45. Ziegelbauer JM, Gulla AF, O'Laoire C, Urgeghe C, Allen RJ, Mukerjee S (2007) Chalcogenide electrocatalysts for oxygen-depolarized aqueous hydrochloric acid electrolysis. Electrochim Acta 52(21):6282–6294
46. Ziegelbauer JM, Murthi VS, O'Laoire C, Gullá AF, Mukerjee S (2008) Electrochemical kinetics and X-ray absorption spectroscopy investigations of select chalcogenide electrocatalysts for oxygen reduction reaction applications. Electrochim Acta 53(17):5587–5596
47. Liu G, Zhang H (2008) Facile synthesis of carbon-supported Ir_xSe_y chalcogenide nanoparticles and their electrocatalytic activity for the oxygen reduction reaction. J Phys Chem C 112(6):2058–2065

48. Ma J, Ai D, Xie X, Guo J (2011) Novel methanol-tolerant Ir–S/C chalcogenide electrocatalysts for oxygen reduction in DMFC fuel cell. Particuology 9(2):155–160. https://doi.org/10.1016/j.partic.2010.05.015
49. Lee K, Zhang L, Zhang JJ (2007) A novel methanol-tolerant Ir-Se chalcogenide electrocatalyst for oyxgen reduction. J Power Sources 165(1):108–113
50. Faber MS, Lukowski MA, Ding Q, Kaiser NS, Jin S (2014) Earth-abundant metal pyrites (FeS_2, CoS_2, NiS_2, and their alloys) for highly efficient hydrogen evolution and polysulfide reduction electrocatalysis. J Phys Chem C 118(37):21347–21356. https://doi.org/10.1021/jp506288w
51. Roger I, Shipman MA, Symes MD (2017) Earth-abundant catalysts for electrochemical and photoelectrochemical water splitting. Nat Rev Chem 1:0003. https://doi.org/10.1038/s41570-016-0003
52. Shim Y, Yuhas BD, Dyar SM, Smeigh AL, Douvalis AP, Wasielewski MR, Kanatzidis MG (2013) Tunable Biomimetic Chalcogels with Fe_4S_4 Cores and $[Sn_nS_{2n+2}]^{4-}$(n = 1, 2, 4) Building Blocks for Solar Fuel Catalysis. J Am Chem Soc 135(6):2330–2337. https://doi.org/10.1021/ja311310k
53. Qi K, Yu S, Wang Q, Zhang W, Fan J, Zheng W, Cui X (2016) Decoration of the inert basal plane of defect-rich MoS_2 with Pd atoms for achieving Pt-similar HER activity. J Mater Chem A 4(11):4025–4031. https://doi.org/10.1039/C5TA10337A
54. Seol M, Jang J-W, Cho S, Lee JS, Yong K (2013) Highly efficient and stable cadmium chalcogenide quantum dot/ZnO nanowires for photoelectrochemical hydrogen generation. Chem Mater 25(2):184–189. https://doi.org/10.1021/cm303206s
55. Shim Y, Young RM, Douvalis AP, Dyar SM, Yuhas BD, Bakas T, Wasielewski MR, Kanatzidis MG (2014) Enhanced photochemical hydrogen evolution from Fe4S4-based biomimetic chalcogels containing M2 + (M = Pt, Zn Co, Ni, Sn) centers. J Am Chem Soc 136(38):13371–13380. https://doi.org/10.1021/ja507297p
56. Lee JW, Popov BN (2007) Ruthenium-based electrocatalysts for oxygen reduction reaction-a review. J Solid State Electrochem 11(10):1355–1364
57. Suarez-Alcantara K, Rodriguez-Castellanos A, Duron-Torres S, Solorza-Feria O (2007) $Ru_xCr_ySe_z$ electrocatalyst loading and stability effects on the electrochemical performance in a PEMFC. J Power Sources 171(2):381–387
58. Alonso-Vante N (2008) Tailoring of metal cluster-like materials for the molecular oxygen reduction reaction. Pure Appl Chem 80(10):2103–2114
59. Hara Y, Minami N, Itagaki H (2008) Electrocatalytic properties of ruthenium modified with Te metal for the oxygen reduction reaction. Appl Catal A-Gen 340(1):59–66
60. Hara Y, Minami N, Itagaki H (2008) Electrocatalytic properties of ruthenium modified with Te metal for the oxygen reduction reaction. Appl Catal A 340(1):59–66. https://doi.org/10.1016/j.apcata.2008.01.036
61. Garsuch A, Michaud X, Wagner G, Klepel O, Dahn JR (2009) Templated Ru/Se/C electrocatalysts for oxygen reduction. Electrochim Acta 54(4):1350–1354
62. Delacôte C, Lewera A, Pisarek M, Kulesza PJ, Zelenay P, Alonso-Vante N (2010) The effect of diluting ruthenium by iron in Ru_xSe_y catalyst for oxygen reduction. Electrochim Acta 55(26):7575–7580
63. Gago AS, Morales-Acosta D, Arriaga LG, Alonso-Vante N (2011) Carbon supported ruthenium chalcogenide as cathode catalyst in a microfluidic formic acid fuel cell. J Power Sources 196(3):1324–1328
64. Ghoshal S, Jia Q, Li J, Campos F, Chisholm CRI, Mukerjee S (2017) Electrochemical and In Situ Spectroscopic Evidences toward Empowering Ruthenium-Based Chalcogenides as Solid Acid Fuel Cell Cathodes. ACS Catal 7(1):581–591. https://doi.org/10.1021/acscatal.6b02417
65. Alonso-Vante N, Giersig M, Tributsch H (1991) Thin layer semiconducting cluster electrocatalysts for oxygen reduction. J Electrochem Soc 138(2):639–640
66. Alonso-Vante N, Tributsch H (1986) Energy conversion catalysis using semiconducting transition metal cluster compounds. Nature 323(6087):431–432

67. Lee K, Alonso-Vante N, Zhang J (2014) Transition metal chalcogenides for oxygen reduction electrocatalysts in PEM fuel cells. In: Chen Z, Dodelet J-P, Zhang J (eds) Non-noble metal fuel cell catalysts, 1st edn. Wiley-VCH Verlag GmbH & Co, KGaA, pp 157–182
68. Alonso-Vante N, Jaegermann W, Tributsch H, Hönle W, Yvon K (1987) Electrocatalysis of oxygen reduction by chalcogenides containing mixed transition metal clusters. J Am Chem Soc 109(11):3251–3257
69. Alonso-Vante N (2003) Chevrel phase and cluster-like chalcogenide materials. In: Vielstich W, Lamm A, Gasteiger H (eds) Handbook of fuel cells, vol 2. Wiley, Chichester, pp 534–543
70. Alonso-Vante N, Fieber-Erdmann M, Rossner H, Holub-Krappe E, Giorgetti C, Tadjeddine A, Dartyge E, Fontaine A, Frahm R (1997) The catalytic centre of transition metal chalcogenides vis-à-vis the oxygen reduction reaction: an in situ electrochemical EXAFS study. J Phys IV France 7 (2 Part 2)
71. Alonso-Vante N (1996) Electrocatalyse par l'intermédiaire des centres métalliques de composés de métaux de transition. Réduction de l'oxygène moléculaire. J Chim Phys Phys-Chim Biol 93(4):702–710
72. Alonso-Vante N, Schubert B, Tributsch H (1989) Transition metal cluster materials for multi-electron transfer catalysis. Mater Chem Phys 22(3–4):281–307
73. Sergent M, Chevrel R (1973) Sur de nouvelles phases séléniées ternaires du molybdène. J Solid State Chem 6(3):433–437. https://doi.org/10.1016/0022-4596(73)90235-1
74. Alonso-Vante N (1998) Inert for selective oxygen reduction of oxygen and method for the production thereof. Germany Patent WO1997DE02453 19971016; DE19961044628 19961017
75. Campbell SA (2004) Non-noble metal catalysts for the oxygen reduction reaction. United States Patent 7125820
76. Zaikovskii VI, Nagabhushana KS, Kriventsov VV, Loponov KN, Cherepanova SV, Kvon RI, Bonnemann H, Kochubey DI, Savinova ER (2006) Synthesis and Structural Characterization of Se-Modified Carbon-Supported Ru Nanoparticles for the Oxygen Reduction Reaction. J Phys Chem B 110(13):6881–6890
77. Alonso-Vante N (2010) Platinum and Non-Platinum Nanomaterials for the Molecular Oxygen Reduction Reaction. ChemPhysChem 11(13):2732–2744
78. Alonso-Vante N, Tributsch H, Solorza-Feria O (1995) Kinetics studies of oxygen reduction in acid medium on novel semiconducting transition metal chalcogenides. Electrochim Acta 40(5):567–576
79. Alonso-Vante N, Bogdanoff P, Tributsch H (2000) On the origin of the selectivity of oxygen reduction of ruthenium-containing electrocatalysts in methanol-containing electrolyte. J Catal 190(2):240–246
80. Alonso-Vante N, Borthen P, Fieber-Erdmann M, Strehblow HH, Holub-Krappe E (2000) In situ grazing incidence X-ray absorption study of ultra thin Ru_xSe_y cluster-like electrocatalyst layers. Electrochim Acta 45(25–26):4227–4236
81. Le Rhun V, Garnier E, Pronier S, Alonso-Vante N (2000) Electrocatalysis on nanoscale ruthenium-based material manufactured by carbonyl decomposition. Electrochem Commun 2(7):475–479
82. Schmidt TJ, Paulus UA, Gasteiger HA, Alonso-Vante N, Behm RJ (2000) Oxygen reduction on $Ru_{1.92}Mo_{0.08}SeO_4$, Ru/carbon, and Pt/carbon in pure and methanol-containing electrolytes. J Electrochem Soc 147(7):2620–2624
83. Bron M, Bogdanoff P, Fiechter S, Hilgendorff M, Radnik J, Dorbandt I, Schulenburg H, Tributsch H (2001) Carbon supported catalysts for oxygen reduction in acidic media prepared by thermolysis of $Ru_3(CO)_{12}$. J Electroanal Chem 517(1–2):85–94
84. Inukai J, Cao D, Wieckowski A, Chang K-C, Menzel A, Komanicky V, You H (2007) In situ synchrotron X-ray spectroscopy of ruthenium nanoparticles modified with selenium for an oxygen reduction reaction. J Phys Chem C 111(45):16889–16894

85. Solorza-Feria O, Ramírez-Raya S, Rivera-Noriega R, Ordoñez-Regil E, Fernández-Valverde SM (1997) Kinetic studies of molecular oxygen reduction on $W_{0.013}Ru_{1.27}Se$ thin films chemically synthesized. Thin Solid Films 311(1–2):164–170. doi:http://dx.doi.org/10.1016/S0040-6090(97)00685-8
86. Delacote C, Bonakdarpour A, Johnston CM, Zelenay P, Wieckowski A (2009) Aqueous-based synthesis of ruthenium-selenium catalyst for oxygen reduction reaction. Faraday Discuss 140:269–281
87. Dassenoy F, Vogel W, Alonso-Vante N (2002) Structural studies and stability of cluster-like Ru_xSe_y electrocatalysts. J Phys Chem B 106(47):12152–12157
88. Bron M, Bogdanoff P, Fiechter S, Dorbandt I, Hilgendorff M, Schulenburg H, Tributsch H (2001) Influence of selenium on the catalytic properties of ruthenium-based cluster catalysts for oxygen reduction. J Electroanal Chem 500(1–2):510–517
89. Bogolowski N, Nagel T, Lanova B, Ernst S, Baltruschat H, Nagabhushana KS, Boennemann H (2007) Activity of selenium modified ruthenium-electrodes and determination of the real surface area. J Appl Electrochem 37(12):1485–1494
90. Bhatnagar AK, Reddy KV, Srivastava V (1985) Optical energy gap of amorphous selenium: effect of annealing. J Phys D Appl Phys 18(9):L149
91. Babu PK, Lewera A, Jong HC, Hunger R, Jaegermann W, Alonso-Vante N, Wieckowski A, Oldfield E (2007) Selenium becomes metallic in Ru-Se fuel cell catalysts: An EC-NMR and XPS investigation. J Am Chem Soc 129(49):15140–15141
92. Tritsaris GA, Nørskov JK, Rossmeisl J (2011) Trends in oxygen reduction and methanol activation on transition metal chalcogenides. Electrochim Acta 56(27):9783–9788
93. Stolbov S (2012) Nature of the Selenium Submonolayer Effect on the Oxygen Electroreduction Reaction Activity of Ru(0001). J Phys Chem C 116(12):7173–7179. https://doi.org/10.1021/jp2072952
94. Gago AS, Gochi-Ponce Y, Feng Y-J, Esquivel JP, Sabaté N, Santander J, Alonso-Vante N (2012) Tolerant chalcogenide cathodes of membraneless micro fuel cells. Chemsuschem 5 (8):1488–1494. https://doi.org/10.1002/cssc.201200009
95. Gochi-Ponce Y, Alonso-Nunez G, Alonso-Vante N (2006) Synthesis and electrochemical characterization of a novel platinum chalcogenide electrocatalyst with an enhanced tolerance to methanol in the oxygen reduction reaction. Electrochem Commun 8(9):1487–1491
96. Ma J, Canaff C, Alonso N (2013) The effect of tuning and origin of tolerance to organics of platinum catalytic centers modified by selenium. Phys Status Solidi (a) 211(9):2030–2034. https://doi.org/10.1002/pssa.201330148
97. Ma J, Gago AS, Vogel W, Alonso-Vante N (2013) Tailoring and Tuning the Tolerance of a Pt Chalcogenide Cathode Electrocatalyst to Methanol. ChemCatChem 5(3):701–705. https://doi.org/10.1002/cctc.201200777
98. Ziegelbauer JM, Gatewood D, Gulla AF, Ramaker DE, Mukerjee S (2006) X-ray absorption spectroscopy studies of water activation on an Rh_xS_y electrocatalyst for oxygen reduction reaction applications. Electrochem. Solid State Lett 9(9):A430–A434
99. Gullá AF, Gancs L, Allen RJ, Mukerjee S (2007) Carbon-supported low-loading rhodium sulfide electrocatalysts for oxygen depolarized cathode applications. Appl Catal A-Gen 326(2):227–235
100. Kukunuri S, Naik K, Sampath S (2017) Effects of composition and nanostructuring of palladium selenide phases, Pd_4Se, Pd_7Se_4 and $Pd_{17}Se_{15}$, on ORR activity and their use in Mg-air batteries. J Mater Chem A 5(9):4660–4670. https://doi.org/10.1039/C7TA00253J
101. Park S, Xie Y, Weaver MJ (2002) Electrocatalytic pathways on carbon-supported platinum nanoparticles: comparison of particle-size-dependent rates of methanol, formic acid, and formaldehyde electrooxidation. Langmuir 18(15):5792–5798
102. Solla-Gullon J, Vidal-Iglesias FJ, Feliu JM (2011) Shape dependent electrocatalysis. Annu Rep Sect "C" (Phys Chem) 107:263–297
103. Park I-S, Xu B, Atienza DO, Hofstead-Duffy AM, Allison TC, Tong YJ (2011) Chemical state of adsorbed sulfur on Pt nanoparticles. ChemPhysChem 12(4):747–752. https://doi.org/10.1002/cphc.201000998

104. Alonso G, Gochi Y, Barbosa R, Arriaga LG, Alonso N (2006) Oxygen reduction reaction and PEM fuel cell performance of a chalcogenide platinum material. ECS Trans 3(1):189–197. https://doi.org/10.1149/1.2356137
105. Wang R-F, Liao S-J, Liu H-Y, Meng H (2007) Synthesis and characterization of Pt-Se/C electrocatalyst for oxygen reduction and its tolerance to methanol. J Power Sources 171(2):471–476
106. Timperman L, Gago AS, Alonso-Vante N (2011) Oxygen reduction reaction increased tolerance and fuel cell performance of Pt and Ru_xSe_y onto oxide-carbon composites. J Power Sources 196(9):4290–4297
107. Wasmus S, Küver A (1999) Methanol oxidation and direct methanol fuel cells: a selective review. J Electroanal Chem 461(1–2):14–31
108. Grinberg VA, Pasynskii AA, Kulova TL, Maiorova NA, Skundin AM, Khazova OA, Law CG (2008) Tolerant-to-methanol cathodic electrocatalysts based on organometallic clusters. Russ J Electrochem 44(2):187–197. https://doi.org/10.1134/s1023193508020055
109. Dembinska B, Dobrzeniecka A, Pisarek M, Kulesza PJ (2015) Selenourea-assisted synthesis of selenium-modified iridium catalysts: evaluation of their activity toward reduction of oxygen. Electrochim Acta 185:162–171. https://doi.org/10.1016/j.electacta.2015.10.105
110. Kukunuri S, Austeria PM, Sampath S (2016) Electrically conducting palladium selenide (Pd_4Se, $Pd_{17}Se_{15}$, Pd_7Se_4) phases: synthesis and activity towards hydrogen evolution reaction. Chem Commun 52(1):206–209. https://doi.org/10.1039/C5CC06730H
111. Mora-Hernández JM, Estudillo Wong LA, Alonso Vante N (to be published) The chemical surface modification of nanoparticulated palladium with selenium and its covalent interaction toward the ORR enhancement in Alkaline Medium
112. Anthony JW, Bideaux RA, Bladh KW, Nichols MC, Eds. (2003) Handbook of Mineralogy. Mineralogical Society of America, Chantilly, VA 20151–1110, USA., http://www.handbookofmineralogy.org/
113. Charreteur F, Jaouen F, Ruggeri S, Dodelet J-P (2008) Fe/N/C non-precious catalysts for PEM fuel cells: Influence of the structural parameters of pristine commercial carbon blacks on their activity for oxygen reduction. Electrochim Acta 53(6):2925–2938
114. Bashyam R, Zelenay P (2006) A class of non-precious metal composite catalysts for fuel cells. Nature 443(7107):63–66
115. Shao M, Chang Q, Dodelet J-P, Chenitz R (2016) Recent advances in electrocatalysts for oxygen reduction reaction. Chem Rev 116(6):3594–3657. https://doi.org/10.1021/acs.chemrev.5b00462
116. Jaouen F, Proietti E, Lefevre M, Chenitz R, Dodelet J-P, Wu G, Chung HT, Johnston CM, Zelenay P (2011) Recent advances in non-precious metal catalysis for oxygen-reduction reaction in polymer electrolyte fuel cells. Energy Environ Sci 4(1):114–130
117. Tang Y, Chen T, Guo W, Chen S, Li Y, Song J, Chang L, Mu S, Zhao Y, Gao F (2017) Reduced graphene oxide supported MnS nanotubes hybrid as a novel non-precious metal electrocatalyst for oxygen reduction reaction with high performance. J Power Sources 362:1–9. https://doi.org/10.1016/j.jpowsour.2017.07.019
118. Fu S, Zhu C, Song J, Du D, Lin Y (2017) Metal-organic framework-derived non-precious metal nanocatalysts for oxygen reduction reaction. Adv Energy Mater:1700363-n/a. https://doi.org/10.1002/aenm.201700363
119. Mora-Hernández J, Luo Y, Alonso-Vante N (2016) What can we learn in electrocatalysis, from nanoparticulated precious and/or non-precious catalytic centers interacting with their support? Catalysts 6(9):145
120. Dresp S, Luo F, Schmack R, Kuhl S, Gliech M, Strasser P (2016) An efficient bifunctional two-component catalyst for oxygen reduction and oxygen evolution in reversible fuel cells, electrolyzers and rechargeable air electrodes. Energy Environ Sci 9(6):2020–2024. https://doi.org/10.1039/C6EE01046F

121. Banham D, Ye S, Pei K, J-i Ozaki, Kishimoto T, Imashiro Y (2015) A review of the stability and durability of non-precious metal catalysts for the oxygen reduction reaction in proton exchange membrane fuel cells. J Power Sources 285:334–348. https://doi.org/10.1016/j.jpowsour.2015.03.047
122. Trogadas P, Fuller TF, Strasser P (2014) Carbon as catalyst and support for electrochemical energy conversion. Carbon 75:5–42. https://doi.org/10.1016/j.carbon.2014.04.005
123. Domínguez C, Pérez-Alonso FJ, Gómez de la Fuente JL, Al-Thabaiti SA, Basahel SN, Alyoubi AO, Alshehri AA, Peña MA, Rojas S (2014) Influence of the electrolyte for the oxygen reduction reaction with Fe/N/C and Fe/N/CNT electrocatalysts. J Power Sources 271:87–96. https://doi.org/10.1016/j.jpowsour.2014.07.173
124. Zhao C, Li D, Feng Y (2013) Size-controlled hydrothermal synthesis and high electrocatalytic performance of CoS_2 nanocatalysts as non-precious metal cathode materials for fuel cells. J Mater Chem A 1(18):5741–5746. https://doi.org/10.1039/c3ta10296c
125. Othman R, Dicks AL, Zhu Z (2012) Non precious metal catalysts for the PEM fuel cell cathode. Int J Hydrogen Energy 37(1):357–372. https://doi.org/10.1016/j.ijhydene.2011.08.095
126. Lai L, Potts JR, Zhan D, Wang L, Poh CK, Tang C, Gong H, Shen Z, Lin J, Ruoff RS (2012) Exploration of the active center structure of nitrogen-doped graphene-based catalysts for oxygen reduction reaction. Energy Environ Sci 5(7):7936–7942. https://doi.org/10.1039/c2ee21802j
127. Feng Y, Alonso-Vante N (2012) Carbon-supported cubic $CoSe_2$ catalysts for oxygen reduction reaction in alkaline medium. Electrochim Acta 72:129–133. https://doi.org/10.1016/j.electacta.2012.04.003
128. Chen Z, Higgins D, Yu A, Zhang L, Zhang J (2011) A review on non-precious metal electrocatalysts for PEM fuel cells. Energy Environ Sci 4(9):3167–3192
129. Ishihara A, Ohgi Y, Matsuzawa K, Mitsushima S, K-i Ota (2010) Progress in non-precious metal oxide-based cathode for polymer electrolyte fuel cells. Electrochim Acta 55(27):8005–8012. https://doi.org/10.1016/j.electacta.2010.03.002
130. Feng Y, Alonso-Vante N (2008) Nonprecious metal catalysts for the molecular oxygen-reduction reaction. Phys Stat Sol (b) 245(9):1792–1806
131. Fu S, Zhu C, Song J, Du D, Lin Y Metal-organic framework-derived non-precious metal nanocatalysts for oxygen reduction reaction. Adv Energy Mater:1700363-n/a. https://doi.org/10.1002/aenm.201700363
132. Kong D, Cha JJ, Wang H, Lee HR, Cui Y (2013) First-row transition metal dichalcogenide catalysts for hydrogen evolution reaction. Energy Environ Sci 6(12):3553–3558. https://doi.org/10.1039/C3EE42413H
133. Wan S, Hu J, Li G-D, Yang L, Liu Y, Gao R, Li X, Zou X (2017) Nano-netlike carbon fibers decorated with highly dispersed $CoSe_2$ nanoparticles as efficient hydrogen evolution electrocatalysts. J Alloys Compd 702:611–618. https://doi.org/10.1016/j.jallcom.2017.01.207
134. Huang H, Huang W, Yang Z, Huang J, Lin J, Liu W, Liu Y (2017) Strongly coupled MoS_2 nanoflake-carbon nanotube nanocomposite as an excellent electrocatalyst for hydrogen evolution reaction. J Mater Chem A. https://doi.org/10.1039/C6TA09612C
135. Xu J, Cui J, Guo C, Zhao Z, Jiang R, Xu S, Zhuang Z, Huang Y, Wang L, Li Y (2016) Ultrasmall Cu_7S_4@MoS_2 Hetero-Nanoframes with Abundant Active Edge Sites for Ultrahigh-Performance Hydrogen Evolution. Angew Chem Int Edition:n/a-n/a. https://doi.org/10.1002/anie.201600686
136. Deng H, Zhang C, Xie Y, Tumlin T, Giri L, Karna SP, Lin J (2016) Laser induced MoS_2/carbon hybrids for hydrogen evolution reaction catalysts. J Mater Chem A 4(18):6824–6830. https://doi.org/10.1039/C5TA09322H
137. Zou X, Zhang Y (2015) Noble metal-free hydrogen evolution catalysts for water splitting. Chem Soc Rev 44(15):5148–5180. https://doi.org/10.1039/C4CS00448E
138. Zeng M, Li Y (2015) Recent advances in heterogeneous electrocatalysts for the hydrogen evolution reaction. J Mater Chem A 3(29):14942–14962. https://doi.org/10.1039/C5TA02974K

139. Lu Q, Hutchings GS, Yu W, Zhou Y, Forest RV, Tao R, Rosen J, Yonemoto BT, Cao Z, Zheng H, Xiao JQ, Jiao F, Chen JG (2015) Highly porous non-precious bimetallic electrocatalysts for efficient hydrogen evolution. Nat Commun 6. https://doi.org/10.1038/ncomms7567
140. Kibsgaard J, Jaramillo TF, Besenbacher F (2014) Building an appropriate active-site motif into a hydrogen-evolution catalyst with thiomolybdate $[Mo_3S_{13}]^{2-}$ clusters. Nat Chem 6 (3):248–253. https://doi.org/10.1038/nchem.1853 http://www.nature.com/nchem/journal/v6/n3/abs/nchem.1853.html#supplementary-information
141. Hou Y, Laursen AB, Zhang J, Zhang G, Zhu Y, Wang X, Dahl S, Chorkendorff I (2013) Layered nanojunctions for hydrogen-evolution catalysis. Angew Chem Int Ed 52(13):3621–3625. https://doi.org/10.1002/anie.201210294
142. Bonde J, Moses PG, Jaramillo TF, Nørskov JK, Chorkendorff I (2009) Hydrogen evolution on nano-particulate transition metal sulfides. Faraday Discuss 140:219–231
143. Jaramillo TF, Jorgensen KP, Bonde J, Nielsen JH, Horch S, Chorkendorff I (2007) Identification of active edge sites for electrochemical H_2 evolution from MoS_2 nanocatalysts. Science 317(5834):100–102. https://doi.org/10.1126/science.1141483
144. Hinnemann B, Moses PG, Bonde J, Jørgensen KP, Nielsen JH, Horch S, Chorkendorff I, Nørskov JK (2005) Biomimetic hydrogen evolution: MoS2 nanoparticles as catalyst for hydrogen evolution. J Am Chem Soc 127(15):5308–5309. https://doi.org/10.1021/ja0504690
145. Ganesan V, Ramasamy P, Kim J (2017) Hierarchical $Ni_{3.5}Co_{5.5}S_8$ nanosheet-assembled hollow nanocages: Superior electrocatalyst towards oxygen evolution reaction. Int J Hydrogen Energy 42(9):5985–5992. https://doi.org/10.1016/j.ijhydene.2016.12.060
146. Hong WT, Risch M, Stoerzinger KA, Grimaud A, Suntivich J, Shao-Horn Y (2015) Toward the rational design of non-precious transition metal oxides for oxygen electrocatalysis. Energy Environ Sci 8(5):1404–1427. https://doi.org/10.1039/C4EE03869J
147. Sheng W, Bivens AP, Myint M, Zhuang Z, Forest RV, Fang Q, Chen JG, Yan Y (2014) Non-precious metal electrocatalysts with high activity for hydrogen oxidation reaction in alkaline electrolytes. Energy Environ Sci 7:1719–1724. https://doi.org/10.1039/c3ee43899f
148. Raj IA, Vasu KI (1990) Transition metal-based hydrogen electrodes in alkaline solution—electrocatalysis on nickel based binary alloy coatings. J Appl Electrochem 20(1):32–38. https://doi.org/10.1007/bf01012468
149. Brown DE, Mahmood MN, Man MCM, Turner AK (1984) Preparation and characterization of low overvoltage transition metal alloy electrocatalysts for hydrogen evolution in alkaline solutions. Electrochim Acta 29(11):1551–1556. https://doi.org/10.1016/0013-4686(84)85008-2
150. Yin Y, Zhang Y, Gao T, Yao T, Zhang X, Han J, Wang X, Zhang Z, Xu P, Zhang P, Cao X, Song B, Jin S (2017) Synergistic phase and disorder engineering in 1T-$MoSe_2$ nanosheets for enhanced hydrogen-evolution reaction. Adv Mater 29 (28):1700311-n/a. https://doi.org/10.1002/adma.201700311
151. Eftekhari A (2017) Molybdenum diselenide ($MoSe_2$) for energy storage, catalysis, and optoelectronics. Appl Mater Today 8:1–17. https://doi.org/10.1016/j.apmt.2017.01.006
152. Guo W, Chen Y, Wang L, Xu J, Zeng D, Peng D-L (2017) Colloidal synthesis of $MoSe_2$ nanonetworks and nanoflowers with efficient electrocatalytic hydrogen-evolution activity. Electrochim Acta 231:69–76. https://doi.org/10.1016/j.electacta.2017.02.048
153. Feng Q, Duan K, Xie H, Xue M, Du Y, Wang C (2016) Electrocatalytic Hydrogen Evolution Reaction of 2H $MoSe_2$ Nanoflowers and 2H-$MoSe_2$/α-MoO_3 Heterostucture. Electrochim Acta 222:499–504. https://doi.org/10.1016/j.electacta.2016.11.002
154. Qu B, Li C, Zhu C, Wang S, Zhang X, Chen Y (2016) Growth of $MoSe_2$ nanosheets with small size and expanded spaces of (002) plane on the surfaces of porous N-doped carbon nanotubes for hydrogen production. Nanoscale 8(38):16886–16893. https://doi.org/10.1039/C6NR04619C

155. Jiang M, Zhang J, Wu M, Jian W, Xue H, Ng T-W, Lee C-S, Xu J (2016) Synthesis of 1T-$MoSe_2$ ultrathin nanosheets with an expanded interlayer spacing of 1.17 nm for efficient hydrogen evolution reaction. J Mater Chem A 4 (39):14949–14953. https://doi.org/10.1039/c6ta07020e
156. Tang H, Dou K, Kaun C-C, Kuang Q, Yang S (2014) $MoSe_2$ nanosheets and their graphene hybrids: synthesis, characterization and hydrogen evolution reaction studies. J Mater Chem A 2(2):360–364. https://doi.org/10.1039/C3TA13584E
157. Karfa P, Madhuri R, Sharma PK, Tiwari A (2017) Designing of transition metal dichalcogenides based different shaped trifunctional electrocatalyst through “adjourn-reaction” scheme. Nano Energy 33:98–109. https://doi.org/10.1016/j.nanoen.2017.01.012
158. Wang X, Chen Y, Zheng B, Qi F, He J, Li P, Zhang W (2016) Few-layered WSe_2 nanoflowers anchored on graphene nanosheets: a highly efficient and stable electrocatalyst for hydrogen evolution. Electrochim Acta 222:1293–1299. https://doi.org/10.1016/j.electacta.2016.11.104
159. Wang X, Chen Y, Qi F, Zheng B, He J, Li Q, Li P, Zhang W, Li Y (2016) Interwoven WSe_2/CNTs hybrid network: a highly efficient and stable electrocatalyst for hydrogen evolution. Electrochem Commun 72:74–78. https://doi.org/10.1016/j.elecom.2016.09.007
160. Wang X, Chen Y, Zheng B, Qi F, He J, Li Q, Li P, Zhang W (2017) Graphene-like WSe_2 nanosheets for efficient and stable hydrogen evolution. J Alloys Compd 691:698–704. https://doi.org/10.1016/j.jallcom.2016.08.305
161. McKone JR, Pieterick AP, Gray HB, Lewis NS (2012) Hydrogen evolution from Pt/Ru-Coated p-Type WSe_2 photocathodes. J Am Chem Soc 135(1):223–231. https://doi.org/10.1021/ja308581g
162. Xiao H, Wang S, Wang C, Li Y, Zhang H, Wang Z, Zhou Y, An C, Zhang J (2016) Lamellar structured $CoSe_2$ nanosheets directly arrayed on Ti plate as an efficient electrochemical catalyst for hydrogen evolution. Electrochim Acta 217:156–162. https://doi.org/10.1016/j.electacta.2016.09.043
163. Zhou Y, Xiao H, Zhang S, Li Y, Wang S, Wang Z, An C, Zhang J (2017) Interlayer expanded lamellar $CoSe_2$ on carbon paper as highly efficient and stable overall water splitting electrodes. Electrochim Acta 241:106–115. https://doi.org/10.1016/j.electacta.2017.04.084
164. Zhou W, Lu J, Zhou K, Yang L, Ke Y, Tang Z, Chen S (2016) $CoSe_2$ nanoparticles embedded defective carbon nanotubes derived from MOFs as efficient electrocatalyst for hydrogen evolution reaction. Nano Energy 28:143–150. https://doi.org/10.1016/j.nanoen.2016.08.040
165. Lin J, He J, Qi F, Zheng B, Wang X, Yu B, Zhou K, Zhang W, Li Y, Chen Y (2017) In-situ selenization of co-based metal-organic frameworks as a highly efficient electrocatalyst for hydrogen evolution reaction. Electrochim Acta 247:258–264. https://doi.org/10.1016/j.electacta.2017.06.179
166. Kim JK, Park GD, Kim JH, Park S-K, Kang YC (2017) Rational design and synthesis of extremely efficient macroporous $CoSe_2$–CNT composite microspheres for hydrogen evolution reaction. Small 13 (27):1700068-n/a. https://doi.org/10.1002/smll.201700068
167. Guo Y, Shang C, Wang E (2017) An efficient CoS_2/$CoSe_2$ hybrid catalyst for electrocatalytic hydrogen evolution. J Mater Chem A 5(6):2504–2507. https://doi.org/10.1039/C6TA09983A
168. Lee C-P, Chen W-F, Billo T, Lin Y-G, Fu F-Y, Samireddi S, Lee C-H, Hwang J-S, Chen K-H, Chen L-C (2016) Beaded stream-like $CoSe_2$ nanoneedle array for efficient hydrogen evolution electrocatalysis. J Mater Chem A 4(12):4553–4561. https://doi.org/10.1039/C6TA00464D
169. Zhou H, Wang Y, He R, Yu F, Sun J, Wang F, Lan Y, Ren Z, Chen S (2016) One-step synthesis of self-supported porous $NiSe_2$/Ni hybrid foam: An efficient 3D electrode for hydrogen evolution reaction. Nano Energy 20:29–36. https://doi.org/10.1016/j.nanoen.2015.12.008

170. Ge Y, Gao S-P, Dong P, Baines R, Ajayan PM, Ye M, Shen J (2017) Insight into the hydrogen evolution reaction of nickel dichalcogenide nanosheets: activities related to non-metal ligands. Nanoscale 9(17):5538–5544. https://doi.org/10.1039/C6NR09977G
171. Wang F, Li Y, Shifa TA, Liu K, Wang F, Wang Z, Xu P, Wang Q, He J (2016) Selenium-Enriched nickel selenide nanosheets as a robust electrocatalyst for hydrogen generation. Angew Chem Int Ed 55(24):6919–6924. https://doi.org/10.1002/anie.201602802
172. Liang J, Yang Y, Zhang J, Wu J, Dong P, Yuan J, Zhang G, Lou J (2015) Metal diselenide nanoparticles as highly active and stable electrocatalysts for the hydrogen evolution reaction. Nanoscale 7(36):14813–14816. https://doi.org/10.1039/C5NR03724G
173. Theerthagiri J, Sudha R, Premnath K, Arunachalam P, Madhavan J, Al-Mayouf AM (2017) Growth of iron diselenide nanorods on graphene oxide nanosheets as advanced electrocatalyst for hydrogen evolution reaction. Int J Hydrogen Energy 42(18):13020–13030. https://doi.org/10.1016/j.ijhydene.2017.04.042
174. Zhang Y, Liu K, Wang F, Shifa TA, Wen Y, Wang F, Xu K, Wang Z, Jiang C, He J (2017) Dendritic growth of monolayer ternary $WS_{2(1-x)}Se_{2x}$ flakes for enhanced hydrogen evolution reaction. Nanoscale 9(17):5641–5647. https://doi.org/10.1039/C7NR00895C
175. Lim WY, Hong M, Ho GW (2016) In situ photo-assisted deposition and photocatalysis of $ZnIn_2S_4$/transition metal chalcogenides for enhanced degradation and hydrogen evolution under visible light. Dalton Trans 45(2):552–560. https://doi.org/10.1039/C5DT03775A
176. Barber J, Tran PD (2013) From natural to artificial photosynthesis. J R Soc Interface 10 (81):1–16. https://doi.org/10.1098/rsif.2012.0984
177. Kanan MW, Surendranath Y, Nocera DG (2009) Cobalt-phosphate oxygen-evolving compound. Chem Soc Rev 38(1):109–114. https://doi.org/10.1039/B802885K
178. Tributsch H, Bennett JC (1977) Electrochemistry and photochemistry of MoS_2 layer crystals. I J Electroanal Chem 81(1):97–111. https://doi.org/10.1016/S0022-0728(77)80363-X
179. Lukowski MA, Daniel AS, Meng F, Forticaux A, Li L, Jin S (2013) Enhanced hydrogen evolution catalysis from chemically exfoliated metallic MoS_2 nanosheets. J Am Chem Soc 135(28):10274–10277. https://doi.org/10.1021/ja404523s
180. Lauritsen JV, Kibsgaard J, Helveg S, Topsoe H, Clausen BS, Laegsgaard E, Besenbacher F (2007) Size-dependent structure of MoS_2 nanocrystals. Nat Nano 2(1):53–58
181. Kibsgaard J, Chen Z, Reinecke BN, Jaramillo TF (2012) Engineering the surface structure of MoS_2 to preferentially expose active edge sites for electrocatalysis. Nat Mater 11 (11):963–969. doi:http://www.nature.com/nmat/journal/v11/n11/abs/nmat3439.html#supplementary-information
182. Xie J, Zhang J, Li S, Grote F, Zhang X, Zhang H, Wang R, Lei Y, Pan B, Xie Y (2013) Controllable disorder engineering in oxygen-incorporated MoS_2 ultrathin nanosheets for efficient hydrogen evolution. J Am Chem Soc 135(47):17881–17888. https://doi.org/10.1021/ja408329q
183. Merki D, Fierro S, Vrubel H, Hu X (2011) Amorphous molybdenum sulfide films as catalysts for electrochemical hydrogen production in water. Chem Sci 2(7):1262–1267. https://doi.org/10.1039/C1SC00117E
184. Li Y, Yu Y, Huang Y, Nielsen RA, Goddard WA, Li Y, Cao L (2015) Engineering the composition and crystallinity of molybdenum sulfide for high-performance electrocatalytic hydrogen evolution. ACS Catalysis 5(1):448–455. https://doi.org/10.1021/cs501635v
185. Lassalle-Kaiser B, Merki D, Vrubel H, Gul S, Yachandra VK, Hu X, Yano J (2015) Evidence from in situ X-ray absorption spectroscopy for the involvement of terminal disulfide in the reduction of protons by an amorphous molybdenum sulfide electrocatalyst. J Am Chem Soc 137(1):314–321. https://doi.org/10.1021/ja510328m
186. Jaegermann W, Schmeisser D (1986) Reactivity of layer type transition metal chalcogenides towards oxidation. Surf Sci 165(1):143–160. https://doi.org/10.1016/0039-6028(86)90666-7
187. Lukowski MA, Daniel AS, English CR, Meng F, Forticaux A, Hamers RJ, Jin S (2014) Highly active hydrogen evolution catalysis from metallic WS_2 nanosheets. Energy Environ Sci 7(8):2608–2613. https://doi.org/10.1039/C4EE01329H

188. Mayorga-Martinez CC, Ambrosi A, Eng AYS, Sofer Z, Pumera M (2015) Transition metal dichalcogenides (MoS_2, $MoSe_2$, WS_2 and WSe_2) exfoliation technique has strong influence upon their capacitance. Electrochem Commun 56:24–28. https://doi.org/10.1016/j.elecom.2015.03.017
189. Escalera-López D, Griffin R, Isaacs M, Wilson K, Palmer RE, Rees NV (2017) Electrochemical sulfidation of WS_2 nanoarrays: Strong dependence of hydrogen evolution activity on transition metal sulfide surface composition. Electrochem Commun 81:106–111. https://doi.org/10.1016/j.elecom.2017.06.016
190. Shang X, Yan K-L, Liu Z-Z, Lu S-S, Dong B, Chi J-Q, Li X, Liu Y-R, Chai Y-M, Liu C-G (2017) Oxidized carbon fiber supported vertical WS_2 nanosheets arrays as efficient 3 D nanostructure electrocatalyts for hydrogen evolution reaction. Appl Surf Sci 402:120–128. https://doi.org/10.1016/j.apsusc.2017.01.059
191. Vrubel H, Hu X (2013) Growth and activation of an amorphous molybdenum sulfide hydrogen evolving catalyst. ACS Catalysis 3(9):2002–2011. https://doi.org/10.1021/cs400441u
192. Chang Y-H, Lin C-T, Chen T-Y, Hsu C-L, Lee Y-H, Zhang W, Wei K-H, Li L-J (2013) Highly efficient electrocatalytic hydrogen production by MoS_x grown on graphene-protected 3D Ni foams. Adv Mater 25(5):756–760. https://doi.org/10.1002/adma.201202920
193. Li DJ, Maiti UN, Lim J, Choi DS, Lee WJ, Oh Y, Lee GY, Kim SO (2014) Molybdenum sulfide/n-doped cnt forest hybrid catalysts for high-performance hydrogen evolution reaction. Nano Lett 14(3):1228–1233. https://doi.org/10.1021/nl404108a
194. Li Y, Wang H, Xie L, Liang Y, Hong G, Dai H (2011) MoS_2 nanoparticles grown on graphene: an advanced catalyst for the hydrogen evolution reaction. J Am Chem Soc 133(19):7296–7299. https://doi.org/10.1021/ja201269b
195. He HY (2017) One-step assembly of 2H-1T MoS_2:Cu/reduced graphene oxide nanosheets for highly efficient hydrogen evolution. Scientific Reports 7:45608. https://doi.org/10.1038/srep45608 https://www.nature.com/articles/srep45608#supplementary-information
196. Voiry D, Salehi M, Silva R, Fujita T, Chen M, Asefa T, Shenoy VB, Eda G, Chhowalla M (2013) Conducting MoS_2 nanosheets as catalysts for hydrogen evolution reaction. Nano Lett 13(12):6222–6227. https://doi.org/10.1021/nl403661s
197. Benson J, Li M, Wang S, Wang P, Papakonstantinou P (2015) Electrocatalytic hydrogen evolution reaction on edges of a few layer molybdenum disulfide nanodots. ACS Appl Mater Interfaces 7(25):14113–14122. https://doi.org/10.1021/acsami.5b03399
198. Ye G, Gong Y, Lin J, Li B, He Y, Pantelides ST, Zhou W, Vajtai R, Ajayan PM (2016) Defects engineered monolayer mos_2 for improved hydrogen evolution reaction. Nano Lett 16(2):1097–1103. https://doi.org/10.1021/acs.nanolett.5b04331
199. Guo B, Yu K, Li H, Song H, Zhang Y, Lei X, Fu H, Tan Y, Zhu Z (2016) Hollow structured micro/nano mos_2 spheres for high electrocatalytic activity hydrogen evolution reaction. ACS Appl Mater Interfaces 8(8):5517–5525. https://doi.org/10.1021/acsami.5b10252
200. Deng J, Li H, Wang S, Ding D, Chen M, Liu C, Tian Z, Novoselov KS, Ma C, Deng D, Bao X (2017) Multiscale structural and electronic control of molybdenum disulfide foam for highly efficient hydrogen production. Nature Communications 8:14430. https://doi.org/10.1038/ncomms14430. https://www.nature.com/articles/ncomms14430#supplementary-information
201. Dai X, Du K, Li Z, Liu M, Ma Y, Sun H, Zhang X, Yang Y (2015) Co-Doped MoS_2 nanosheets with the dominant comos phase coated on carbon as an excellent electrocatalyst for hydrogen evolution. ACS Appl Mater Interfaces 7(49):27242–27253. https://doi.org/10.1021/acsami.5b08420
202. Escalera-López D, Niu Y, Yin J, Cooke K, Rees NV, Palmer RE (2016) Enhancement of the hydrogen evolution reaction from ni-mos_2 hybrid nanoclusters. ACS Catalysis 6(9):6008–6017. https://doi.org/10.1021/acscatal.6b01274
203. Ennaoui A, Fiechter S, Pettenkofer C, Alonso-Vante N, Buker K, Bronold M, Hopfner C, Tributsch H (1993) Iron disulfide for solar energy conversion. Sol Energy Mater Sol Cells 29(4):289–370

204. Brostigen G, Kjekshus A (1970) Bonding schemes for compounds with the pyrite, marcasite, and arsenopyrite type structures. Acta Chem Scand 24:2993–3012. https://doi.org/10.3891/acta.chem.scand.24-2993
205. Jaegermann W, Tributsch H (1988) Interfacial properties of semiconducting transition metal chalcogenides. Prog Surf Sci 29(1–2):1–167. https://doi.org/10.1016/0079-6816(88)90015-9
206. Faber MS, Dziedzic R, Lukowski MA, Kaiser NS, Ding Q, Jin S (2014) High-performance electrocatalysis using metallic cobalt pyrite (CoS_2) Micro- and Nanostructures. J Am Chem Soc 136(28):10053–10061. https://doi.org/10.1021/ja504099w
207. Leighton C, Manno M, Cady A, Freeland JW, Wang L, Umemoto K, Wentzcovitch RM, Chen TY, Chien CL, Kuhns PL, Hoch MJR, Reyes AP, Moulton WG, Dahlberg ED, Checkelsky J, Eckert J (2007) Composition controlled spin polarization in $Co_{1-x}Fe_xS_2$ alloys. J Phys: Condens Matter 31:315219
208. Samad L, Cabán-Acevedo M, Shearer MJ, Park K, Hamers RJ, Jin S (2015) Direct chemical vapor deposition synthesis of phase-pure iron pyrite (FeS_2) thin films. Chem Mater 27(8):3108–3114. https://doi.org/10.1021/acs.chemmater.5b00664
209. Caban-Acevedo M, Stone ML, Schmidt JR, Thomas JG, Ding Q, Chang H-C, Tsai M-L, He J-H, Jin S (2015) Efficient hydrogen evolution catalysis using ternary pyrite-type cobalt phosphosulphide. Nat Mater 14(12):1245–1251. https://doi.org/10.1038/nmat4410. http://www.nature.com/nmat/journal/v14/n12/abs/nmat4410.html#supplementary-information
210. Tongay S, Zhou J, Ataca C, Lo K, Matthews TS, Li J, Grossman JC, Wu J (2012) Thermally driven crossover from indirect toward direct bandgap in 2D semiconductors: $MoSe_2$ versus MoS_2. Nano Lett 12(11):5576–5580. https://doi.org/10.1021/nl302584w
211. Tsai C, Chan K, Abild-Pedersen F, Norskov JK (2014) Active edge sites in $MoSe_2$ and WSe_2 catalysts for the hydrogen evolution reaction: a density functional study. Phys Chem Chem Phys 16(26):13156–13164. https://doi.org/10.1039/C4CP01237B
212. Kong D, Wang H, Cha JJ, Pasta M, Koski KJ, Yao J, Cui Y (2013) Synthesis of MoS_2 and $MoSe_2$ films with vertically aligned layers. Nano Lett 13(3):1341–1347. https://doi.org/10.1021/nl400258t
213. Gholamvand Z, McAteer D, Backes C, McEvoy N, Harvey A, Berner NC, Hanlon D, Bradley C, Godwin I, Rovetta A, Lyons MEG, Duesberg GS, Coleman JN (2016) Comparison of liquid exfoliated transition metal dichalcogenides reveals $MoSe_2$ to be the most effective hydrogen evolution catalyst. Nanoscale 8(10):5737–5749. https://doi.org/10.1039/C5NR08553E
214. Bastide S, Lévy-Clément C, Albu-Yaron A, Boucher AC, Alonso-Vante N (2000) $MoSe_2$ nanocrystallites synthesized at low temperature via a chemical solution route. Electrochem Solid-State Lett 3(9):450–451
215. Kim Y, Tiwari AP, Prakash O, Lee H (2017) Activation of ternary transition metal chalcogenide basal planes through chemical strain for the hydrogen evolution reaction. ChemPlusChem 82(5):785–791. https://doi.org/10.1002/cplu.201700164
216. Park S-K, Park GD, Ko D, Kang YC, Piao Y (2017) Aerosol synthesis of molybdenum diselenide–reduced graphene oxide composite with empty nanovoids and enhanced hydrogen evolution reaction performances. Chem Eng J 315:355–363. https://doi.org/10.1016/j.cej.2017.01.032
217. Basu M, Zhang Z-W, Chen C-J, Chen P-T, Yang K-C, Ma C-G, Lin CC, Hu S-F, Liu R-S (2015) Heterostructure of Si and $CoSe_2$: a promising photocathode based on a non-noble metal catalyst for photoelectrochemical hydrogen evolution. Angew Chem Int Ed 54(21):6211–6216. https://doi.org/10.1002/anie.201502573
218. Basu M, Zhang Z-W, Chen C-J, Lu T-H, Hu S-F, Liu R-S (2016) $CoSe_2$ embedded in C_3N_4: an efficient photocathode for photoelectrochemical water splitting. ACS Appl Mater Interfaces 8(40):26690–26696. https://doi.org/10.1021/acsami.6b06520
219. Luo Y, Alonso-Vante N (2017) Application of Metal Organic Framework (MOF) in the electrocatalytic process. In: Electrochemistry: Volume 14, vol 14. The Royal Society of Chemistry, pp 194–256. https://doi.org/10.1039/9781782622727-00194

220. Lu S, Pan J, Huang A, Zhuang L, Lu J (2008) Alkaline polymer electrolyte fuel cells completely free from noble metal catalysts. Proc Natl Acad Sci 105(52):20611–20614. https://doi.org/10.1073/pnas.0810041106
221. Hu Q, Li G, Pan J, Tan L, Lu J, Zhuang L (2013) Alkaline polymer electrolyte fuel cell with Ni-based anode and Co-based cathode. Int J Hydrogen Energy 38(36):16264–16268. https://doi.org/10.1016/j.ijhydene.2013.09.125
222. Baresel D, Sarholz W, Scharner P, Schmitz J (1974) Transition metal chalcogenides as oxygen catalysts for fuel cells. Ber Bunsen-Ges 78(6):608–611. https://doi.org/10.1002/bbpc.19740780616
223. Behret H, Binder H, Sandstede G (1975) Electrocatalytic oxygen reduction with thiospinels and other sulphides of transition metals. Electrochim Acta 20(2):111–117
224. Kitayama H, Yoshio I, Ichino T, Osaka T (1983) Oxygen electroreduction on cobalt sulphides in alkaline solution. Waseda Daigaku Rikogaku Kenkyusho Hokoku/Bulletin of Science and Engineering Research Laboratory 104:9–16
225. Feng Y, He T, Alonso-Vante N (2008) In situ Free-Surfactant Synthesis and ORR-Electrochemistry of Carbon-Supported Co_3S_4 and $CoSe_2$ Nanoparticles. Chem Mater 20(1):26–28
226. Cherevko S, Geiger S, Kasian O, Kulyk N, Grote J-P, Savan A, Shrestha BR, Merzlikin S, Breitbach B, Ludwig A, Mayrhofer KJJ (2016) Oxygen and hydrogen evolution reactions on Ru, RuO_2, Ir, and IrO_2 thin film electrodes in acidic and alkaline electrolytes: a comparative study on activity and stability. Catal Today 262:170–180. https://doi.org/10.1016/j.cattod.2015.08.014
227. Trotochaud L, Boettcher SW (2014) Precise oxygen evolution catalysts: Status and opportunities. Scripta Mater 74:25–32. https://doi.org/10.1016/j.scriptamat.2013.07.019
228. Chen W, Wang H, Li Y, Liu Y, Sun J, Lee S, Lee J-S, Cui Y (2015) In situ electrochemical oxidation tuning of transition metal disulfides to oxides for enhanced water oxidation. ACS Central Sci 1(5):244–251. https://doi.org/10.1021/acscentsci.5b00227
229. Wang H, Li Z, Li G, Peng F, Yu H (2015) Co_3S_4/NCNTs: A catalyst for oxygen evolution reaction. Catal Today 245:74–78. https://doi.org/10.1016/j.cattod.2014.06.006
230. Liu T, Liang Y, Liu Q, Sun X, He Y, Asiri AM (2015) Electrodeposition of cobalt-sulfide nanosheets film as an efficient electrocatalyst for oxygen evolution reaction. Electrochem Commun 60:92–96. https://doi.org/10.1016/j.elecom.2015.08.011
231. Sidik RA, Anderson AB (2006) Co_9S_8 as a catalyst for electroreduction of O_2: quantum chemistry predictions. J Phys Chem B 110(2):936–941
232. Zhu L, Susac D, Teo M, Wong KC, Wong PC, Parsons RR, Bizzotto D, Mitchell KAR, Campbell SA (2008) Investigation of CoS_2-based thin films as model catalysts for the oxygen reduction reaction. J Catal 258(1):235–242. https://doi.org/10.1016/j.jcat.2008.06.016
233. Feng Y, He T, Alonso-Vante N (2009) Oxygen reduction reaction on carbon-supported $CoSe_2$ nanoparticles in an acidic medium. Electrochim Acta 54(22):5252–5256. https://doi.org/10.1016/j.electacta.2009.03.052
234. Jirkovský JS, Björling A, Ahlberg E (2012) Reduction of oxygen on dispersed nanocrystalline CoS_2. J Phys Chem C 116(46):24436–24444. https://doi.org/10.1021/jp307669k
235. Tiwari AP, Kim D, Kim Y, Lee H (2017) Bifunctional oxygen electrocatalysis through chemical bonding of transition metal chalcogenides on conductive carbons. Adv Energy Mater:1602217-n/a. https://doi.org/10.1002/aenm.201602217
236. Chua XJ, Luxa J, Eng AYS, Tan SM, Sofer Z, Pumera M (2016) Negative electrocatalytic effects of p-doping niobium and tantalum on MoS_2 and WS_2 for the hydrogen evolution reaction and oxygen reduction reaction. ACS Catalysis 6(9):5724–5734. https://doi.org/10.1021/acscatal.6b01593
237. Wang T, Gao D, Zhuo J, Zhu Z, Papakonstantinou P, Li Y, Li M (2013) Size-dependent enhancement of electrocatalytic oxygen-reduction and hydrogen-evolution performance of MoS_2 particles. Chem—A Eur J 19(36):11939–11948. https://doi.org/10.1002/chem.201301406

238. Huang H, Feng X, Du C, Song W (2015) High-quality phosphorus-doped MoS_2 ultrathin nanosheets with amenable ORR catalytic activity. Chem Commun 51(37):7903–7906. https://doi.org/10.1039/C5CC01841B
239. Huang H, Feng X, Du C, Wu S, Song W (2015) Incorporated oxygen in MoS_2 ultrathin nanosheets for efficient ORR catalysis. J Mater Chem A 3(31):16050–16056. https://doi.org/10.1039/C5TA01600B
240. Zhao K, Gu W, Zhao L, Zhang C, Peng W, Xian Y (2015) MoS_2/Nitrogen-doped graphene as efficient electrocatalyst for oxygen reduction reaction. Electrochim Acta 169:142–149. https://doi.org/10.1016/j.electacta.2015.04.044
241. Yuan K, Zhuang X, Fu H, Brunklaus G, Forster M, Chen Y, Feng X, Scherf U (2016) Two-dimensional core-shelled porous hybrids as highly efficient catalysts for oxygen reduction reaction. Angew Chem Int Ed 55(24):6858–6863. https://doi.org/10.1002/anie.201600850
242. Arunchander A, Peera SG, Sahu AK (2017) Synthesis of flower-like molybdenum sulfide/graphene hybrid as an efficient oxygen reduction electrocatalyst for anion exchange membrane fuel cells. J Power Sources 353:104–114. https://doi.org/10.1016/j.jpowsour.2017.03.149
243. Zhou J, Xiao H, Zhou B, Huang F, Zhou S, Xiao W, Wang D (2015) Hierarchical MoS_2–rGO nanosheets with high MoS_2 loading with enhanced electro-catalytic performance. Appl Surf Sci 358:152–158. https://doi.org/10.1016/j.apsusc.2015.07.187
244. Zuo L-X, Jiang L-P, Zhu J-J (2017) A facile sonochemical route for the synthesis of MoS_2/Pd composites for highly efficient oxygen reduction reaction. Ultrason Sonochem 35:681–688. https://doi.org/10.1016/j.ultsonch.2016.02.006
245. Zhang H, Tian Y, Zhao J, Cai Q, Chen Z (2017) Small dopants make big differences: enhanced electrocatalytic performance of MoS2 monolayer for oxygen reduction reaction (ORR) by N– and P-Doping. Electrochim Acta 225:543–550. https://doi.org/10.1016/j.electacta.2016.12.144
246. Susac D, Zhu L, Teo M, Sode A, Wong KC, Wong PC, Parsons RR, Bizzotto D, Mitchell KAR, Campbell SA (2007) Characterization of FeS_2-based thin films as model catalysts for the oxygen reduction reaction. J Phys Chem C 111(50):18715–18723
247. Hibble SJ, Rice DA, Almond MJ, Mohammad KAH, Pearse SP, Sagar JR (1992) Preparation of new selenium-rich selenides, $CrSe_3$, $MoSe_{\sim 5}$, $WSe_{\sim 6\text{-}7}$, and $ReSe_{\sim 6\text{-}7}$, and known selenides, by the reaction of metal carbonyls with selenium. J Mater Chem 2(12):1237–1240. https://doi.org/10.1039/JM9920201237
248. Sato H, Nagasaki F, Kani Y, Senba S, Ueda Y, Kimura A, Taniguchi M (2001) Electronic structure of $CoSe_2$ studied by photoemission spectroscopy using synchrotron radiation. Solid State Commun 118(11):563–567. https://doi.org/10.1016/S0038-1098(01)00172-7
249. Alonso-Vante N, Feng Y, He T (2010) Carbon-supported $CoSe_2$ nanoparticles for oxygen reduction and hydrogen evolution in acid environments. USA Patent 20100233070
250. Feng YJ, He T, Alonso-Vante N (2010) Carbon-supported $CoSe_2$ nanoparticles for oxygen reduction reaction in acid medium. Fuel Cells 10(1):77–83
251. Lee K, Zhang L, Zhang JJ (2007) Ternary non-noble metal chalcogenide (W-Co-Se) as electrocatalyst for oxygen reduction reaction. Electrochem Commun 9(7):1704–1708
252. Lefevre M, Proietti E, Jaouen F, Dodelet J-P (2009) Iron-based catalysts with improved oxygen reduction activity in polymer electrolyte fuel cells. Science 324(5923):71–74. https://doi.org/10.1126/science.1170051
253. Nekooi P, Akbari M, Amini MK (2010) CoSe nanoparticles prepared by the microwave-assisted polyol method as an alcohol and formic acid tolerant oxygen reduction catalyst. Int J Hydrogen Energy 35(12):6392–6398. https://doi.org/10.1016/j.ijhydene.2010.03.134
254. Li H, Gao D, Cheng X (2014) Simple microwave preparation of high activity Se-rich $CoSe_2$/C for oxygen reduction reaction. Electrochim Acta 138:232–239. https://doi.org/10.1016/j.electacta.2014.06.065

255. Zhu L, Teo M, Wong PC, Wong KC, Narita I, Ernst F, Mitchell KAR, Campbell SA (2010) Synthesis, characterization of a $CoSe_2$ catalyst for the oxygen reduction reaction. Appl Catal A-Gen 386(1–2):157–165. https://doi.org/10.1016/j.apcata.2010.07.048
256. Gao M-R, Gao Q, Jiang J, Cui C-H, Yao W-T, Yu S-H (2011) A methanol-tolerant Pt/$CoSe_2$ nanobelt cathode catalyst for direct methanol fuel cells. Angew Chem Int Ed 50(21):4905–4908. https://doi.org/10.1002/anie.201007036
257. Chao Y-S, Tsai D-S, Wu A-P, Tseng L-W, Huang Y-S (2013) Cobalt selenide electrocatalyst supported by nitrogen-doped carbon and its stable activity toward oxygen reduction reaction. Int J Hydrogen Energy 38(14):5655–5664. https://doi.org/10.1016/j.ijhydene.2013.03.006
258. Wu R, Xue Y, Liu B, Zhou K, Wei J, Chan SH (2016) Cobalt diselenide nanoparticles embedded within porous carbon polyhedra as advanced electrocatalyst for oxygen reduction reaction. J Power Sources 330:132–139. https://doi.org/10.1016/j.jpowsour.2016.09.001
259. Unni SM, Mora-Hernandez JM, Kurungot S, Alonso-Vante N (2015) $CoSe_2$ supported on nitrogen-doped carbon nanohorns as a methanol-tolerant cathode for air-breathing microlaminar flow fuel cells. ChemElectroChem 2(9):1339–1345. https://doi.org/10.1002/celc.201500154
260. García-Rosado IJ, Uribe-Calderón J, Alonso-Vante N (2017) Nitrogen-doped reduced graphite oxide as a support for cose electrocatalyst for oxygen reduction reaction in alkaline media. J Electrochem Soc 164(6):F658–F666. https://doi.org/10.1149/2.1531706jes
261. Eng AYS, Ambrosi A, Sofer Z, Šimek P, Pumera M (2014) Electrochemistry of transition metal dichalcogenides: strong dependence on the metal-to-chalcogen composition and exfoliation method. ACS Nano 8(12):12185–12198. https://doi.org/10.1021/nn503832j
262. Guo J, Shi Y, Bai X, Wang X, Ma T (2015) Atomically thin $MoSe_2$/graphene and WSe_2/graphene nanosheets for the highly efficient oxygen reduction reaction. J Mater Chem A 3(48):24397–24404. https://doi.org/10.1039/C5TA06909B
263. Zhao D, Zhang S, Yin G, Du C, Wang Z, Wei J (2013) Tungsten doped Co–Se nanocomposites as an efficient non precious metal catalyst for oxygen reduction. Electrochim Acta 91:179–184. https://doi.org/10.1016/j.electacta.2013.01.001
264. Pan S, Cai Z, Duan Y, Yang L, Tang B, Jing B, Dai Y, Xu X, Zou J (2017) Tungsten diselenide/porous carbon with sufficient active edge-sites as a co-catalyst/Pt-support favoring excellent tolerance to methanol-crossover for oxygen reduction reaction in acidic medium. Appl Catal B: Environ 219:18–29. https://doi.org/10.1016/j.apcatb.2017.07.011
265. Karfa P, Madhuri R, Sharma PK (2017) Multifunctional fluorescent chalcogenide hybrid nanodots ($MoSe_2$:CdS and WSe_2:CdS) as electro catalyst (for oxygen reduction/oxygen evolution reactions) and sensing probe for lead. J Mater Chem A. https://doi.org/10.1039/C6TA08172J
266. Iwakura C, Hirao K, Tamura H (1977) Anodic evolution of oxygen on ruthenium in acidic solutions. Electrochim Acta 22(4):329–334. https://doi.org/10.1016/0013-4686(77)85082-2
267. Hutchings R, Müller K, Kötz R, Stucki S (1984) A structural investigation of stabilized oxygen evolution catalysts. J Mater Sci 19(12):3987–3994. https://doi.org/10.1007/bf00980762
268. Matsumoto Y, Sato E (1986) Electrocatalytic properties of transition metal oxides for oxygen evolution reaction. Mater Chem Phys 14(5):397–426. https://doi.org/10.1016/0254-0584(86)90045-3
269. Alonso-Vante N, Colell H, Stimming U, Tributsch H (1993) Anomalous low-temperature kinetic effects for oxygen evolution on RuO_2 and Pt electrodes. J Phys Chem 97(29):7381–7384
270. Rossmeisl J, Qu ZW, Zhu H, Kroes GJ, Nørskov JK (2007) Electrolysis of water on oxide surfaces. J Electroanal Chem 607(1–2):83–89
271. McCrory CCL, Jung S, Peters JC, Jaramillo TF (2013) Benchmarking heterogeneous electrocatalysts for the oxygen evolution reaction. J Am Chem Soc 135(45):16977–16987. https://doi.org/10.1021/ja407115p

272. García-Mota M, Bajdich M, Viswanathan V, Vojvodic A, Bell AT, Nørskov JK (2012) Importance of correlation in determining electrocatalytic oxygen evolution activity on cobalt oxides. J Phys Chem C 116(39):21077–21082. https://doi.org/10.1021/jp306303y
273. He Y, Zhang J, He G, Han X, Zheng X, Zhong C, Hu W, Deng Y (2017) Ultrathin Co_3O_4 nanofilm as an efficient bifunctional catalyst for oxygen evolution and reduction reaction in rechargeable zinc-air batteries. Nanoscale 9(25):8623–8630. https://doi.org/10.1039/C7NR02385E
274. Zhang C, Berlinguette CP, Trudel S (2016) Water oxidation catalysis: an amorphous quaternary Ba-Sr-Co-Fe oxide as a promising electrocatalyst for the oxygen-evolution reaction. Chem Commun 52(7):1513–1516. https://doi.org/10.1039/C5CC09361A
275. Smith RDL, Prévot MS, Fagan RD, Trudel S, Berlinguette CP (2013) Water oxidation catalysis: electrocatalytic response to metal stoichiometry in amorphous metal oxide films containing iron, cobalt, and nickel. J Am Chem Soc. https://doi.org/10.1021/ja403102j
276. Subbaraman R, Tripkovic D, Chang K-C, Strmcnik D, Paulikas AP, Hirunsit P, Chan M, Greeley J, Stamenkovic V, Markovic NM (2012) Trends in activity for the water electrolyser reactions on 3d M(Ni,Co,Fe,Mn) hydr(oxy)oxide catalysts. Nat Mater 11 (6):550–557. doi: http://www.nature.com/nmat/journal/v11/n6/abs/nmat3313.html#supplementary-information
277. Hou C-C, Cao S, Fu W-F, Chen Y (2015) Ultrafine CoP nanoparticles supported on carbon nanotubes as highly active electrocatalyst for both oxygen and hydrogen evolution in basic media. ACS Appl Mater Interfaces 7(51):28412–28419. https://doi.org/10.1021/acsami.5b09207
278. Kanan MW, Nocera DG (2008) In situ formation of an oxygen-evolving catalyst in neutral water containing phosphate and Co2+. Science 321(5892):1072–1075. https://doi.org/10.1126/science.1162018
279. Bursell M, Pirjamali M, Kiros Y (2002) $La_{0.6}Ca_{0.4}CoO_3$, $La_{0.1}Ca_{0.9}MnO_3$ and $LaNiO_3$ as bifunctional oxygen electrodes. Electrochim Acta 47(10):1651–1660. https://doi.org/10.1016/s0013-4686(02)00002-6
280. Suntivich J, May KJ, Gasteiger HA, Goodenough JB, Shao-Horn Y (2011) A perovskite oxide optimized for oxygen evolution catalysis from molecular orbital principles. Science 334(6061):1383–1385. https://doi.org/10.1126/science.1212858
281. Han B, Stoerzinger KA, Tileli V, Gamalski AD, Stach EA, Shao-Horn Y (2016) Nanoscale structural oscillations in perovskite oxides induced by oxygen evolution. Nat Mater advance online publication. https://doi.org/10.1038/nmat4764. http://www.nature.com/nmat/journal/vaop/ncurrent/abs/nmat4764.html#supplementary-information
282. Lee JG, Hwang J, Hwang HJ, Jeon OS, Jang J, Kwon O, Lee Y, Han B, Shul Y-G (2016) A new family of perovskite catalysts for oxygen-evolution reaction in alkaline media: $BaNiO_3$ and $BaNi_{0.83}O_{2.5}$. J Am Chem Soc 138(10):3541–3547. https://doi.org/10.1021/jacs.6b00036
283. Gao M-R, Zheng Y-R, Jiang J, Yu S-H (2017) Pyrite-type nanomaterials for advanced electrocatalysis. Acc Chem Res. https://doi.org/10.1021/acs.accounts.7b00187
284. Gao M-R, Xu Y-F, Jiang J, Zheng Y-R, Yu S-H (2012) Water oxidation electrocatalyzed by an efficient $Mn_3O_4/CoSe_2$ nanocomposite. J Am Chem Soc 134(6):2930–2933. https://doi.org/10.1021/ja211526y
285. Gao M-R, Cao X, Gao Q, Xu Y-F, Zheng Y-R, Jiang J, Yu S-H (2014) Nitrogen-doped graphene supported $cose_2$ nanobelt composite catalyst for efficient water oxidation. ACS Nano 8(4):3970–3978. https://doi.org/10.1021/nn500880v
286. Gao Q, Huang C-Q, Ju Y-M, Gao M-R, Liu J-W, An D, Cui C-H, Zheng Y-R, Li W-X, Yu S-H (2017) Phase-selective syntheses of cobalt telluride nanofleeces for efficient oxygen evolution catalysts. Angew Chem Int Ed 56(27):7769–7773. https://doi.org/10.1002/anie.201701998
287. Zheng Y-R, Gao M-R, Gao Q, Li H-H, Xu J, Wu Z-Y, Yu S-H (2015) An efficient $CeO_2/CoSe_2$ nanobelt composite for electrochemical water oxidation. Small 11(2):182–188. https://doi.org/10.1002/smll.201401423

288. Xu X, Song F, Hu X (2016) A nickel iron diselenide-derived efficient oxygen-evolution catalyst. Nature Commun 7:12324. https://doi.org/10.1038/ncomms12324. https://www.nature.com/articles/ncomms12324#supplementary-information
289. Gerken JB, Shaner SE, Masse RC, Porubsky NJ, Stahl SS (2014) A survey of diverse earth abundant oxygen evolution electrocatalysts showing enhanced activity from Ni-Fe oxides containing a third metal. Energy Environ Sci 7(7):2376–2382. https://doi.org/10.1039/C4EE00436A
290. Liang Y, Liu Q, Luo Y, Sun X, He Y, Asiri AM (2016) $Zn_{0.76}Co_{0.24}S/CoS_2$ nanowires array for efficient electrochemical splitting of water. Electrochim Acta 190:360–364. https://doi.org/10.1016/j.electacta.2015.12.153
291. Zhang K, Zhang L, Chen X, He X, Wang X, Dong S, Han P, Zhang C, Wang S, Gu L, Cui G (2012) Mesoporous cobalt molybdenum nitride: a highly active bifunctional electrocatalyst and its application in lithium–O2 batteries. J Phys Chem C 117(2):858–865. https://doi.org/10.1021/jp310571y
292. Zhong H, Tian R, Gong X, Li D, Tang P, Alonso-Vante N, Feng Y (2017) Advanced bifunctional electrocatalyst generated through cobalt phthalocyanine tetrasulfonate intercalated Ni_2Fe-layered double hydroxides for a laminar flow unitized regenerative micro-cell. J Power Sources 361:21–30. https://doi.org/10.1016/j.jpowsour.2017.06.057
293. Liu Y, Jiang H, Zhu Y, Yang X, Li C (2016) Transition metals (Fe Co, and Ni) encapsulated in nitrogen-doped carbon nanotubes as bi-functional catalysts for oxygen electrode reactions. J Mater Chem A 4(5):1694–1701. https://doi.org/10.1039/C5TA10551J
294. Liu Q, Jin J, Zhang J (2013) $NiCo_2S_4$@graphene as a bifunctional electrocatalyst for oxygen reduction and evolution reactions. ACS Appl Mater Interfaces 5(11):5002–5008. https://doi.org/10.1021/am4007897
295. Shen M, Ruan C, Chen Y, Jiang C, Ai K, Lu L (2015) Covalent entrapment of cobalt-iron sulfides in N-doped mesoporous carbon: extraordinary bifunctional electrocatalysts for oxygen reduction and evolution reactions. ACS Appl Mater Interfaces 7(2):1207–1218. https://doi.org/10.1021/am507033x
296. Jin C, Lu F, Cao X, Yang Z, Yang R (2013) Facile synthesis and excellent electrochemical properties of $NiCo_2O_4$ spinel nanowire arrays as a bifunctional catalyst for the oxygen reduction and evolution reaction. J Mater Chem A 1(39):12170–12177. https://doi.org/10.1039/C3TA12118F
297. Prabu M, Ketpang K, Shanmugam S (2014) Hierarchical nanostructured $NiCo_2O_4$ as an efficient bifunctional non-precious metal catalyst for rechargeable zinc-air batteries. Nanoscale 6(6):3173–3181. https://doi.org/10.1039/C3NR05835B

Chapter 5
Effect of Supports on Catalytic Centers

5.1 Introduction

The electrocatalytic power of nanodivided materials can be well utilized if they are highly dispersed onto conducting supports. These latter are a key component in the energy converting devices, since they play a critical role in the catalytic center durability and activity providing pathways for electrons and mass transfer. However, to achieve a higher durability, the electrocatalyst nanoparticles should be strongly coupled with conducting support. Moreover, the strong interaction of the electrocatalyst nanoparticles with the conducting support prevents the nanoparticles agglomeration, aggregation, and leaching out during the catalytic reaction. Some factors, like geometric (metal–metal distance) and ligand (*d*-band vacancy) effects, have been considered to be responsible for the electrocatalytic enhancement of the ORR process on Platinum-based electrocatalysts [1, 2]. The phenomenon that favors similar process is the so-called strong metal support interaction, known as the "SMSI" effect. Indeed, this effect was put in evidence by Tauster et al. [3, 4] when detecting an altered chemisorption property of some noble metals reduced on oxide (TiO_2) surface. This effect was defined as the formation of an interfacial bond occurring between the metal and the metal of the oxide's surface. Generally speaking, the strong coupling, between the metal nanoparticles and a conducting support, encompasses the enhancement of the catalytic activity through faster electron transport during the reaction, influencing favorably the current-potential characteristics shown in Fig. 5.1. This figure summarizes various processes of technical relevance in energy converting and storage devices; namely: oxygen reduction reaction (ORR), oxygen evolution reaction (OER), Hydrogen evolution reaction (HER), small organics oxidation, noted as fuel oxidation reaction (FOR), and the carbon dioxide reduction (CO_2-R). In this context, since carbon black (a popular support) cannot meet the harsh corrosion requirements, usually, encountered in, e.g., fuel cells, or electrolyzers, great efforts have been done so far to develop alternative support materials: graphitized carbon, metal oxides, and carbon–oxide nanocomposites.

N. Alonso-Vante, *Chalcogenide Materials for Energy Conversion*, Nanostructure Science and Technology, https://doi.org/10.1007/978-3-319-89612-0_5

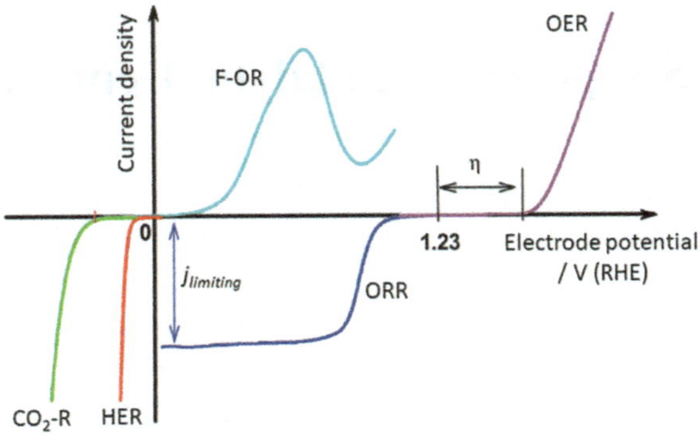

Fig. 5.1 Current-potential characteristics of multi-electron charge transfer electrocatalytic processes that take place in energy and storage devices: fuel cells and electrolyzers. FOR: fuel oxidation reaction; HER: hydrogen oxidation reaction; CO_2-R: carbon dioxide reduction reaction; ORR: oxygen reduction reaction; OER: oxygen evolution reaction

5.2 The Effects Encountered Between the Catalytic Center and the Oxide Support

The strong metal support interaction induces an electronic perturbation effect of the metal [5] which has been known in catalysis since the 1960s thanks to the work of Schwab [6] and Solymosi [7]. Interfacial contact between a metal and a support can result in charge redistribution at the interface. The electronic interactions are governed by fundamental principles; e.g., energy minimization and the continuity of the electric potential in a solid. A metal in contact with a support, depending on the degree of interaction, either weak (W) or strong (S), a metal support interaction (WMSI or SMSI) can occur. These phenomena are often invoked to explain catalytic effects resulting from electron transfer between metals and supports. Therefore, to further alter the catalytic activity by controlling the electron transfer, the metal-adsorbate interface should be modified. Consequently, the adsorption energy, W_{ad}, must be the contribution of the free surface energy of the support (if an oxide–) γ_{oxide}, the metal interacting with the support, γ_{metal}, and the interface (metal/oxide, Fig. 5.2a) as interface, $\gamma_{interface}$, written as:

$$W_{ad} = \gamma_{metal} + \gamma_{oxide} - \gamma_{interface} \tag{5.1}$$

The $\gamma_{interface}$ in Eq. (5.1) must strongly influence the growth of NPs or overlayers on the support. A heterojunction between the metal NPs and the support, e.g., n-type semiconductors, CeO_2, ZnO, SnO_2, TiO_2, can be formed, e.g., Ru NPs onto Titania [5], Fig. 5.2b. Its physics and chemistry define the contact nature, cf. Chap. 3, Figs. 3.3 and 3.5. It is important to recall that a reaction between the two contacting solids creates an interfacial layer which, at the end, will dominate the Schottky barrier height, e.g., a metal onto silicon produces a metal silicide, and

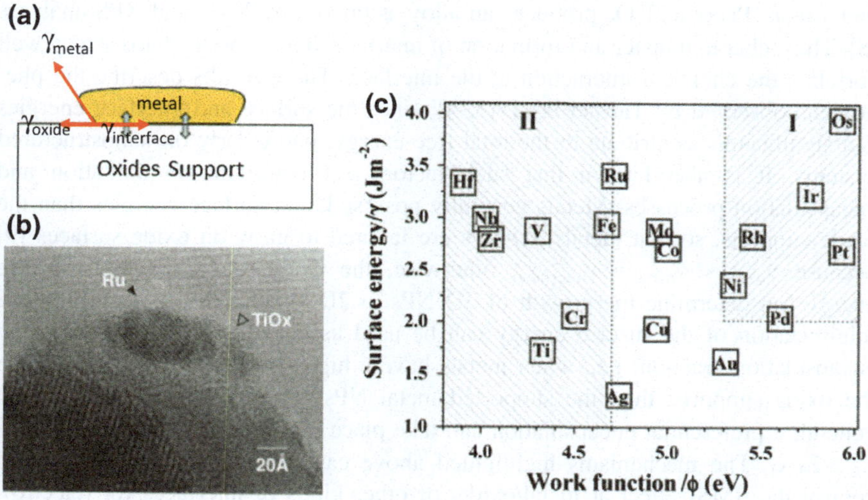

Fig. 5.2 **a** Schematics of an oxide-supported metal interface showing the energetic parameters, γ. **b** A transmission electron microscopy (TEM) picture showing the Ru NPs covered by fractions of amorphous Titania, showing additionally the metal/oxide junctions reported as active catalytic sites for the synthesis of hydrocarbon from hydrogen and carbon monoxide. Adapted from Ref. [5]. **c** The surface energy as a function of the work function, ϕ, of various transition metals. In Region I the encapsulation is expected ($\phi > 5.3$ eV and $\gamma > 2$ J/m^2). In region II ($\phi < 4.7$ eV), the oxidation of metals on Titania is possible. Reproduced from Ref. [10]. Copyright © 2005, American Chemical Society

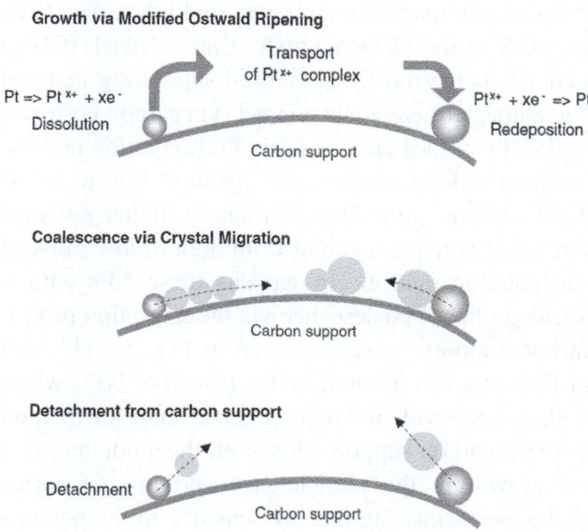

Fig. 5.3 **a** Scheme showing instability mechanisms for Pt NPs: Ostwald ripening; coalescence; and detachment due to corrosion, occurring in low temperature fuel cells. Adapted from Ref. [11]. Copyright © 2007, Springer Science-Business Media, LLC

metal (e.g., Pt) onto TiO_2 produces an alloy as proved by XRD and XPS analyses [8]. Thus, charge transfer and formation of interface dipole and the barrier may well correlate the chemical interaction at the interface. These results describe the phenomena observed by Tauster et al. [3, 4], since the surface and interface energies can significantly contribute to the total free energy, particularly on nanostructured systems. It is also known that such factors can contribute to oxidation and encapsulation processes. Metals generally possess larger surface energies than the oxide supports, so that metals 3D-NPs are favored to grow on oxide surfaces [9] because $\gamma_{oxide} < \gamma_{metal} + \gamma_{interface}$, otherwise, the decrease of the total surface energy can determine the growth of 3D NPs to 2D structures. In this regard, the minimization of the surface energy can be used as the driving force to favor an encapsulation reaction; i.e., when metals have a high surface energy compared to the oxide supports, then the supported metal NPs can be encapsulated [10]. In general, a preferential encapsulation can take place if the work function or the $E_{F,Ox} > E_{F,M}$. The mechanisms highlighted above can provide an idea the way to control the SMSI effect at metal/oxide, or other kinds of interfaces for (electro)-catalytic purposes to take place. As an example, Fig. 5.2c displays the surface energy of various well-oriented and polycrystalline metallic surfaces as a function of the metal work junctions. In Fig. 5.2c, region I (>5.3 eV) NPs' encapsulation is expected, whereas in region II (<4.7 eV) NPs' oxidation is possible on Titania.

5.3 Carbon-Supports

The state-of-the-art polymer electrolyte fuel cells (PEFCs) employ platinum (Pt) or Pt-based nanoparticles supported on high surface area carbon as electrocatalysts for the ORR at the PEFC cathode. Carbon black (CB), e.g., from Vulcan XC-72 and Ketjen®, is the most widely used supporting materials for catalysts' nanoparticles. The damage of electrical contact (via carbon corrosion or Pt particle detachment) as well as Pt surface area loss (via Pt dissolution into ionomer, particle agglomeration, surface blocking, or Ostwald ripening) lead to a loss in the ORR activity in aged PGM cathodes in acid media and at higher potentials. Indeed, it has been recognized that potential cycling from high to low moves the carbon- "attached" Pt NPs into solution, and electrodeposits these NPs with a new redistribution (Ostwald ripening effect); coalescence via the migration process, and/or detachment from the carbon support, as schematized in Fig. 5.3 [11, 12]. Experimentally, the loss of surface area is attributed to the growth of NPs, whereas the coarsening of particles leads to spherical, and non-spherical shapes which remain in contact, and or detach from the carbon support. However, the modeling of such mechanisms indicated that the growth in the particle size and loss in surface area are primarily due to coalescence/sintering that are sensitive to the potential limits and the initial particle size distribution [12]. One certainly has to recall that during the potential cycling process, the crossover effect of platinum ions causes NPs' precipitation in the ion-conductor of the membrane electrode assembly (MEA) by the influx of hydrogen ($H_2 + Pt^{x+} \rightarrow Pt + 2H^+$) from the anode side of the PEMFC. These facts put in clear that these phenomena contribute to the loss of the electrochemical activity of fuel cell electrodes and that the nature of the interaction of the support with the catalytic center is an important issue if voltages are driven higher than 1.1 V/RHE, since the carbon as support primarily corrodes. The electrochemical reaction of carbon corrosion is expressed as [13]:

$$C + 2H_2O \rightarrow CO_2 + 4H^+ + 4e^- \quad E = 0.207\,V \qquad (5.2)$$

Thermodynamically, carbon is not stable at the potential at which the cathode operates. However, the corrosion kinetics at the cathode electrode potential is very slow, and certainly due to carbon oxide and hydroxides intermediates that occur at higher potentials than 0.9 V/RHE, where C* is a carbon defect site [14].

$$C^* \text{-C-OH} + H_2O \rightarrow C^* + CO_2 + 3H^+ + 3e^- \quad E = 0.95\,V \qquad (5.3)$$

This scheme seems appropriate and explains experimental results obtained by the differential mass spectrometry [15]. The DEMS result is in agreement with recent data that considered a model based on the cathode catalyst layer represented as three interfaces: zone A (Nafion/carbon interface), zone B (platinum/Nafion interface), and zone C (platinum/carbon interface (see Fig. 5.4b). It turns out clearly that the presence of Pt NPs promotes the carbon corrosion from 0.8 V [14], a value slightly lower than that of carbon alone, as detected by DEMS [15]. Moreover,

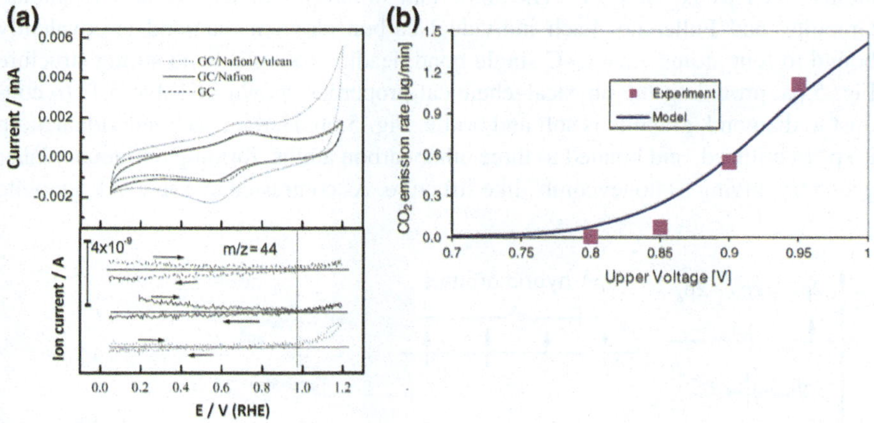

Fig. 5.4 **a** Cyclic voltammetry of carbon (glassy carbon-GC, Vulcan XC-72) electrodes in 0.5 M H_2SO_4, scan rate 5 mV/s. The Vulcan and Nafion® were deposited onto GC. The simultaneous detection of the mass signal for carbon dioxide (m/z = 44) is shown in the bottom panel. Reprinted from Ref. [15]. Copyright © 2002 Wiley-VCH Verlag GmbH & Co. KGaA, Weinheim. **b** Generated carbon dioxide emission, based on a three interfaces model: (Nafion/carbon interface); (platinum/Nafion interface); and (platinum/carbon interface) as a function the applied electrode voltage. Reprinted from Ref. [14]

corrosion experiments on Ketjen Black using DEMS revealed that the CO_2 detection limit was at 1.35 V/RHE [16]. Therefore, the enhancement of the ORR activity, and stability of carbon black supported Pt NPs must be based on the design of the catalytic center and on the presence of oxygen functional groups (e.g., >C=O, –OH; –COOH) at the carbon surfaces.

5.3.1 *Carbon, a Versatile Element*

Graphite, carbon fiber (CF), glassy carbon (GC), carbon blacks (CBs), diamond, and carbon nanotubes (CNTs) are related materials. Depending on the parameters used in the synthesis method (e.g., CVD), the amorphous form can be found, e.g., on CNTs. The electronic structure of carbon contains six-electrons distributed in 1 s^2, 2 s^2, $2p_x^1$, $2p_y^1$, $2p_z$, Fig. 5.5. In carbon, the preferred tetravalence is explained by the hybridization sp^3. Since the difference in energy is low for *p* orbitals, these orbitals mix to form four equivalent hybridized orbitals, i.e., the 2s-orbital mix with a lower number 2p-orbitals to form sp^2- or sp-hybrid orbitals, with the spatial arrangement depicted in Fig. 5.5. Three-dimensional (3-D) tetrahedral, trigonal planar and linear structures are obtained. Consequently, in sp- and sp^2-hybridized orbitals there are two- or one- p-orbital not participating in the hybridization process, enabling an additional π-bond in contrast to the σ-bonds. The physical properties of the various forms of carbon are summarized in Table 5.1.

The sp^3 hybridized atom allows the carbon atom to form bonds with other kind of atoms, such as N, S, O, P. The important allotropes of carbon are: Diamond, Graphite, and Fullerene. Each individual carbon atom in diamond is covalently bound to four atoms via a C–C single bond, leading to a giant and strong structure Fig. 5.6a, providing the physical–chemical properties shown in Table 5.1. In contrast to diamond, graphite is soft and brittle, Fig. 5.6b. Herein, each individual atom is sp^2 hybridized, and bonded to three other carbon atoms, forming a trigonal planar geometry, giving a "honeycomb" like structure. As contrasted in Table 5.1, graphite

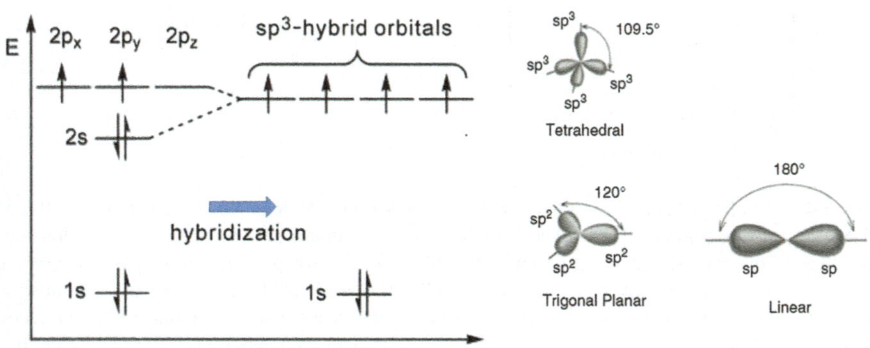

Fig. 5.5 Diagram showing the atomic orbitals and sp^3- sp^2, sp-hybridization of carbon

Table 5.1 Physical chemical properties of some carbons

Materials	Density/g cm^{-3}	Hardness/GPa	% sp^3	Gap energy/eV	Refs.
Diamond	3.515	100	100	5.5	[17, 18]
Graphite	2.267	N/A	0	0.04	
Glassy carbon	1.3–1.55	2–3	≈0	0.01	
CNT	0.8–1.2	N/A	N/A	0.3–2.0	[19]
[a]Vulcan XC-72	0.096	N/A	N/A	≈0	[20]
[b]Ketjenblack EC300J	0.125–0.145	N/A	N/A	≈0	[c]

[a]Surface area: 250 m^2/g
[b]Surface: 416 m^2/g
[c]AkzoNobel-datasheet

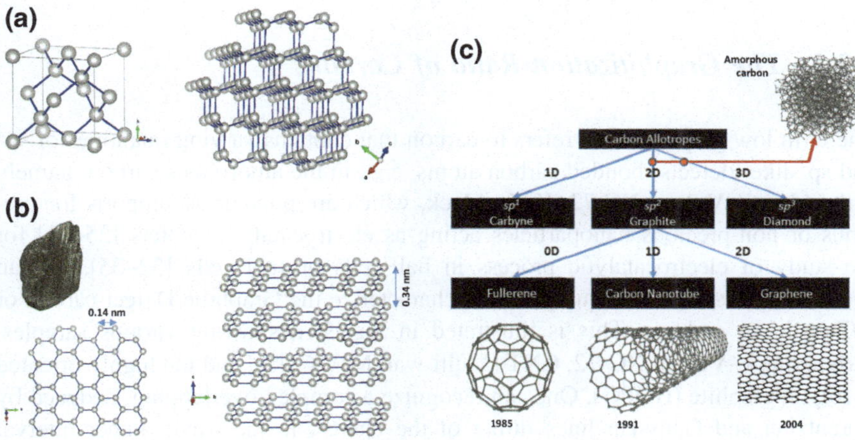

Fig. 5.6 Some carbon allotropes used as catalysts' supports in electrocatalysis. **a** and **b** show the two natural allotropes: diamond and graphite. These structures were generated via the VESTA program. **c** Besides the natural carbon allotropes other allotropes came into the stage from 1985 (Fullerenes); 1991 (Carbon Nanotubes, CNTs), and the rediscovered Graphene in 2004. Amorphous carbon (e.g., Vulcan XC-72) is constituted of sp^2 and sp^3-like defect carbons

is an electron-conducting material due to the fact that the remaining electron from each carbon atom is delocalized within the structure. The structure of graphite consists of sheets lying one on top of the other interacting with a Van der Waals force of attraction between each layer. Thermodynamically speaking, graphite is more stable than diamond, so that conversion of graphite to diamond was done under very harsh conditions of temperature (3000 K) and pressure (125 kbar) [21]. Fullerene, the third major allotrope of carbon, together with CNTs is a very interesting nanostructure material. The fullerene's general formula is (C_n, $n \geq 60$). This material, discovered in the early 1980s [22], is a large spherical molecule also known as "Bucky-balls." The strong electronic effect on such a molecule is based on the topology effect produced by the changes in their geometry. C_{60} is constituted by 12 pentagons in the graphene lattice plus 20 hexagons, Fig. 5.6c. One application of this molecule was described, herein, in Sect. 3.3.1.2.2, as a reservoir of electrons (c.f. Fig. 3.20a). The natural allotropes and other ones developed since 1985, i.e., Fullerenes [22], CNTs [23], and graphene sheets [24] are summarized in Fig. 5.6c. The amorphous carbon, a popular support used in electrocatalysis (e.g., carbon black–Vulcan XC-72 (280 m^2/g); Ketjenblack (400–1100 m^2/g) is also shown. This is a material that contains nanodomains of sp^2 and highly disordered sp^3-like carbons (C–C), cf. GC and Vulcan carbon, Table 5.1. It is important to recall that the dimensionality (0D–3D) of the nanoobjects determines the materials' properties. The isolation and the otherwise covalent functionalization of 2D graphene opened the potential interest in this material, as electrocatalytic support of chalcogenide materials for the HER, HOR, ORR, and OER, (cf. Chap. 4).

5.3.2 The Graphitization Ratio of Carbons

The term low graphitization refers to carbon that contains varying quantities of sp^2 and sp^3-like (defects) bonded carbon atoms, e.g., in the amorphous carbon, namely carbon black–Vulcan XC-72, Ketjenblack, which are popular as supports for precious or non-precious nanoparticles acting as electrocatalytic centers [25–31] for the study of electrocatalytic process in half-cells or fuel cells [32–35]. Raman spectroscopy is the technique per se to characterize the Graphitic/Defect-carbon or G/D ratio of carbons. This is illustrated in Fig. 5.7a showing various samples: carbon black–Vulcan XC-72, CNT (Multi-wall MW-CNT), and the highly oriented pyrolytic graphite (HOPG). One can recognize a complex overlapping, deduced by Lorentzian and Gaussian lines fitting of the curves, in the wavenumber interval between 1100 and 1800 cm^{-1}. Two important Raman lines D and G emerge showing the high degree of disorder in carbon black (Vulcan XC-72) and those of the MW-CNTs with interstitial disorder along the c-axis between the planes, notified by the band centered at 1580 cm^{-1}. In this example, the huge number of defects (C–C) in MW-CNT-1 with respect to MW-CNT-2 reflects the quality of the chemical vapor deposition (CVD) technique of synthesis used. The best fits finally provide the integrated intensity ratios of D versus G bands [36]. The G band in the Vulcan XC-72 shows a slight shift, to lower Raman wavenumbers, stressing the high degree of disorder in the sample. Therefore, the analysis of the Intensity I_D/I_G ratio allows the knowledge of the defects/sp^2 contribution, on the one hand, and on the other hand, to quantify the in-plane crystallite size, L_a [37] variation in the carbon material. A parameter that can be determined by:

$$L_a(nm) = 2.4 \times 10^{-10} \lambda_{laser}^4 \frac{I_G}{I_D} \tag{5.4}$$

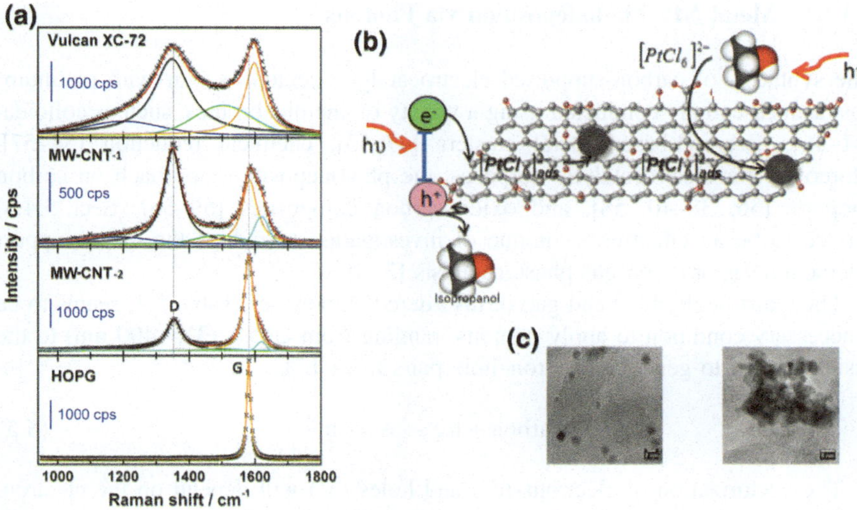

Fig. 5.7 **a** Raman spectra of Vulcan XC-72, MW-CNT-1, MW-CNT-2, and HOPG. Figure adapted from Ref. [36] with permission. Copyright © 2012, American Chemical Society. **b** Electron–hole pairs generation by UV-light to deposit Pt nanoparticles (NPs) onto graphitic surface or generation of Pt NPs via the photochemical pathway. **c** Selected TEM of a Pt NPS deposited via a chemical route (carbonyl)-left; and via the photodeposition-right. Figures (**b**) and (**c**) adapted from Ref. [39] with permission. Copyright © 2013 Elsevier B.V. All rights reserved

For the specific cases shown in Fig. 5.7, the evaluated I_D/I_G ratios were, respectively, 2.1; 1.7; 0.5 for Vulcan XC-72, MW-CNT-1, and MW-CNT-2, with the corresponding L_a, deduced from Eq. (5.4), of 8; 9.6; and 31.5 nm [36].

5.3.3 *The Support Interaction of Metal Nanoparticles onto Graphitic Domains*

The question is: how sp^3-like (defects) versus sp^2 domains, present on moderately disordered carbon, can affect the catalytic center? The requirement is to bring in close interaction the metal atoms with the sp^2-hybrid orbitals. This phenomenon can be attained, either chemically or photochemically, via UV–VIS photons, to selectively photodeposit the metallic NPs onto the sites where electron–hole pairs are generated and generate the *d*-π hybridization, a condition necessary for a strong metal support interaction [36, 38–40].

5.3.3.1 Metal NPs Photodeposition via Photons

The synthesis of carbon-supported electrocatalytic centers (so far with platinum) has been essentially conducted using a variety of chemical routes, such as colloidal [41–46], carbonyl-based metal clusters [47–53], chemical reduction [54–57]. Moreover, for electrocatalytic purposes, the photodeposition approach on carbon supports [36, 38–40, 58], and oxide–carbon composites [59–66] (Sect. 5.4.3) proved to be an effective technique of investigation to induce the metal support interaction, reported on gas phase catalysis [3, 4].

The quasi-negligible band gap of disordered carbons, cf. Table 5.1, seems to be a necessary condition to apply photons, ranging from UV-A (315–400 nm) to the visible region to generate electron–hole pairs ($e^- + h^+$).

$$\text{Carbon} + h\nu \rightarrow e^- + h^+ \tag{5.5}$$

The vectorization of electrons (e^-) and holes (h^+) will depend on the electron, and hole scavengers present in the reaction medium, in order to avoid ($e^- + h^+$) recombination. To photodeposit Pt NPs, the hole scavenger used was isopropanol (propan-2-ol), and the electron scavenger a platinum salt: $[PtCl_6]^{2-}$. One cannot neglect the fact that the UV-irradiation can be absorbed by the metal complex to produce an excited state of the metal complex: $[PtCl_6]^{2-*}$. This species can react with isopropanol to form a radical. Consequently, various reactions can be present during the photodeposition process, namely the (i) the radical one in the presence of isopropanol–water solution [67], Eq. (5.6):

$$[Pt^{IV}Cl]^{2-}(C_2H_6CHOH + H_2O) + h\nu \rightarrow [Pt^{II}Cl]^{2-} \rightarrow Pt^0 \tag{5.6}$$

And (ii) the electrochemical reaction via the generated conduction band electrons, Eq. (5.7):

$$\begin{aligned} &[Pt^{IV}Cl]^{2-}_{ad} + 2e^-_{cb} \rightarrow [Pt^{II}Cl]^{2-}_{ad} + 2Cl^- \\ &[Pt^{II}Cl]^{2-}_{ad} + 2e^-_{cb} \rightarrow Pt^0 + 4Cl^- \end{aligned} \tag{5.7}$$

The efficiency of reaction (5.7) is determined by the hole scavenging property of isopropanol, Eq. (5.8):

$$C_2H_6CHOH + 2h^+ \rightarrow C_2H_6CO + 2H^+ \tag{5.8}$$

It is to be noted that the photooxidation of isopropanol is complex, so that we assume that the initial product is acetone (C_2H_6CO), Eq. (5.8). Moreover, the subsequent reaction of acetone photooxidation can lead to the formation of formic acid, as reported on titanium dioxide surfaces [68]. Figure 5.7b summarizes both processes, (i) metal nanoparticles deposition via the conduction band electrons (left), and (ii) via the radical pathway (right). Figure 5.7c contrasts the transmission electron microscopy (TEM) pictures of Pt NPs generated onto MW-CNT-2, as an example. The nanoparticles' morphology generated by the chemical deposition (e.g., carbonyl route), and by the photodeposition method is different. In the latter method, the agglomeration effect is generally observed, indicating concentrated domains of sp^2 regions, where reactions (5.7) and (5.8) are favored. As it will be discussed below, such sites or domains are at the origin of the strong metal support interaction.

5.3.3.2 Electrochemical Probing the Electronic Interaction

The electrochemical CO stripping is used to probe the electronic surface modification of nanoparticulated electrocatalytic centers [38, 69, 70]. In acid electrolyte, the global electrochemical oxidation of carbon monoxide can be written as follows:

$$CO + H_2O \rightarrow CO_2 + 2H^+ + 2e^- \tag{5.9}$$

This reaction can be thought as a set of two steps, namely

$$CO_{ads} + OH_{ads} \rightarrow CO_2 + H^+ + e^- \tag{5.10}$$

where OH_{ads} indicates a partial oxidation of the metal:

$$M + H_2O \rightarrow M\text{-}OH_{ads} + H^+ + e^- \tag{5.11}$$

On platinum surface, the adsorption phenomenon of CO is accompanied by the back donation of d-orbital of platinum in the $2\pi^*$ antibonding orbital of CO, a phenomenon known as the Blyholder–Bagus back-donation mechanism [71, 72]. In Fig. 5.8a, the oxidation of adsorbed CO sets on at 0.69 V/RHE and peaks at 0.77 V/RHE on Pt/Vulcan XC-72. On Pt/HOPG this process is shifted to lower applied electrode potential by $\Delta E = -90$ mV (see arrow in the figure). Indeed, on a platinum surface with a higher electron density the process of oxidation sets on at 0.6 V/RHE and peaks at 0.68 V/RHE, with respect to Pt/ Vulcan XC-72. This shift is significant for a one-metal center catalyst; in the meantime, it puts in clear evidence the role of the support's nature: the graphitic domains. Independently of the I_D/I_G ratio, cf. Figure 5.7a, the use of photons can favor the sp^2 domains to nuclei metal nanoparticles, as demonstrated on Pt/MW-CNT-2. The CO stripping on photodeposited Pt NPs onto MW-CNT-2 shows the contribution of the two carbon domains: sp^2 and sp^3-like (defects), see Fig. 5.8a. Such changes in the electrochemical oxidation have to be associated with a different adsorption energy of the CO molecule induced by the effect of the support. The strength of such an interaction will depend on the domain extension of sp^2 in nanodivided carbons, such as MW-CNTs. It was demonstrated recently that the stability of the catalytic centers can be enhanced, Fig. 5.8b [73]. The understanding of this phenomenon can open ways to better chemically stabilize one of the best catalytic centers (and

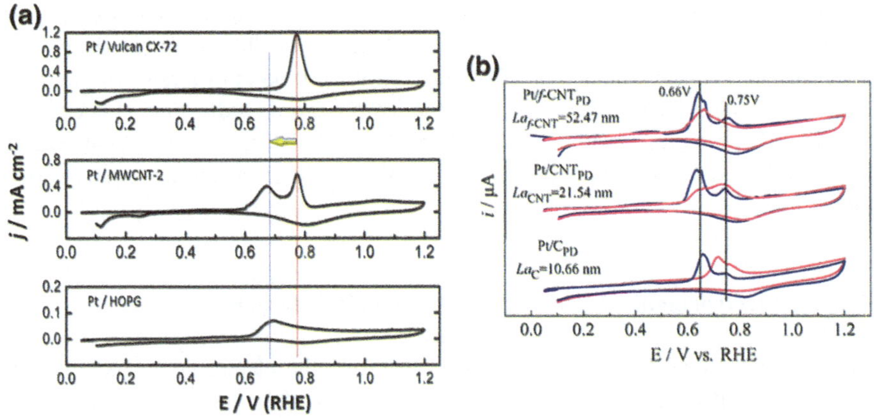

Fig. 5.8 **a** CO stripping, in 0.5 M H_2SO_4, on Pt (NPs) chemically deposited on amorphous carbon (upper panel); multi-wall carbon nanotubes, MW-CNT, (middle panel); and on highly oriented pyrolytic graphite, HOPG, (bottom panel): Figures modified from Ref. [36] with permission. Copyright © 2012, American Chemical Society. **b** The CO stripping, in $HClO_4$, on Pt (NPs) photodeposited on amorphous carbon, and on carbon nanotubes (CNT) having different L_a (crystallite domains). Figures adapted from Ref. [73] with permission. Copyright © 2016, American Chemical Society

non-precious ones), at the same time enhancing, e.g., the ORR kinetics, and hindering the agglomeration of nanoparticles.

The doping of graphene with non-metal atoms, e.g., N, B, P strongly alters its electronic surface charge, so that this material, so-called metal-free electrocatalyst has shown excellent electrocatalytic activity toward the ORR in alkaline medium [74–77]. Alike the photodeposited Pt on MW-CNT [73] the platinum deposited chemically onto N-doped graphene [40, 78] is also affected by the support interaction, as revealed by the chemical signature of Pt 4f XPS signals, see Fig. 5.9. On both examples, the Pt 4f binding energy shifts, testifying the change of the electronic properties of Pt NPs, via the hybridization, on one hand, with sp^2 domains of carbon, Fig. 5.9a, and, on the other hand, with pyridinic N-moieties leading to the formation of platinum-nitride species, Fig. 5.9b. The interaction is shown in Fig. 5.9, see insets. This phenomenon explains the lower oxidation potential of adsorbed CO during the CO stripping experiments, cf. Fig. 5.8. Therefore, these facts give credence to the electrochemical activation of various important processes, such as ORR, OER, HOR, and HER on precious, as well as on non-precious catalytic centers, such as the chalcogenides of Co, Mo, W, discussed in Chap. 4.

5.3.3.3 Effect of Carbon Supports' Nature on Chalcogenide Centers

The integration of chalcogenides with highly conducting supports with large surface area, e.g., CNTs, RGO, was used to develop efficient hybrid electrocatalysts.

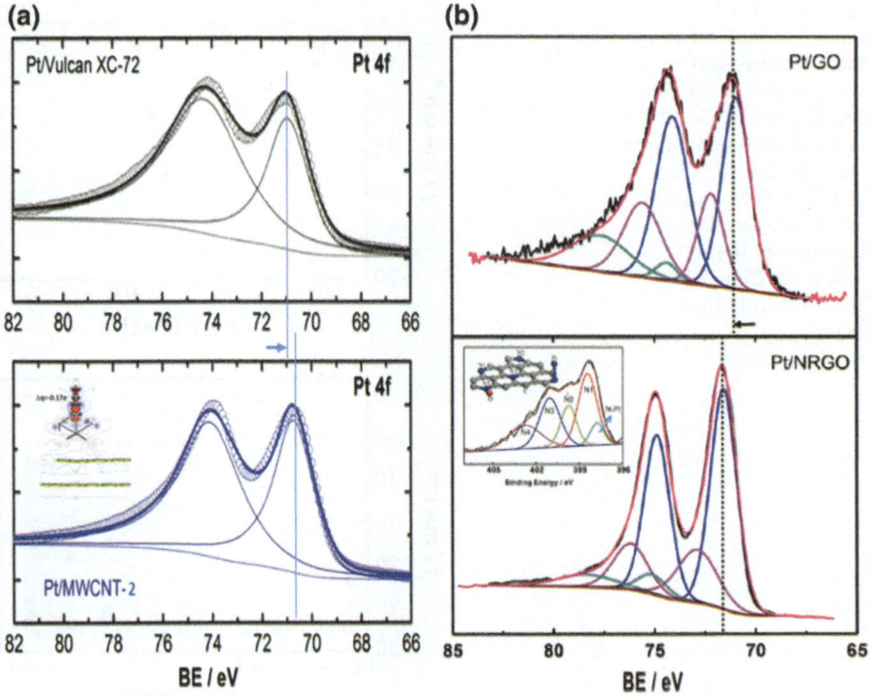

Fig. 5.9 **a** Pt 4f XPS spectra on Pt NPs supported onto Vulcan XC-72, and MW-CNT-2. The inset in the figure shows the calculated charge accumulation (blue) and depletion (red) is at the origin of the downshift of the binding energy (see arrow). Figures adapted from Ref. [38]. with permission. Copyright © 2013, American Chemical Society. **b** Pt 4f XPS spectra on Pt NPs supported onto NRGO, and GO. In the inset the XPS spectra of the N 1 s bands is shown. Herein, the interaction of coordinated Pt–N upshifts the binding energy (see arrow). Charge transfer is from the metal cluster to the nitrogen moieties. Figure adapted from Ref. [78] with permission. Copyright © 2012 Elsevier B.V. All rights reserved

Particularly for the HER, cf. Figs. 4.14, 4.15, 4.16 and 4.19; ORR, cf. Figs. 4.23 and 4.27, and the oxygen evolution reaction (OER) (cf. Fig. 4.29). As an example, the solvothermal synthesis of MoS_2 NPs on RGO (reduced graphene oxide) induced the chemical, and electrical coupling that turned out to be beneficial for the HER activity and stability even beyond 1000 cycles for the system: MoS_2/RGO [79]. As discussed in Chap. 4 (Sect. 4.3.1.1), the morphology of this hybrid system played a pivotal role, since the synthetic chemical route employed developed structures with an important amount of edges stacked onto the support (RGO), see inset image in Fig. 5.10 showing superior electrocatalytic activity in HER as compared to sole MoS_2. The support interaction, aside from the activity, confers to the hybrid system the enhanced chemical stability. Furthermore, exploring the coupling with carbon nanotubes as support, and changing the electronic property of pyrite (FeS_2) by doping with cobalt, the CNT supported $Fe_{0.9}Co_{0.1}S_2$/CNT outperformed the HER activity of FeS_2/CNT [80]. In addition to the coupling of pyrite

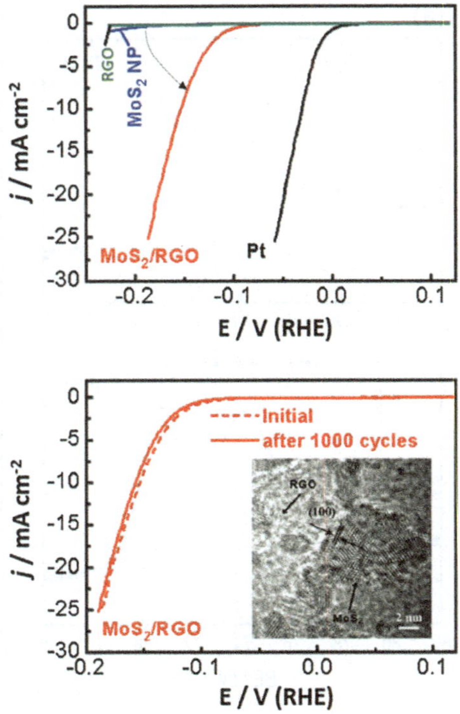

Fig. 5.10 HER current-potential curves on various electrocatalysts as shown in the caption. The durability test of the HER is shown on the hybrid material (MoS_2/RGO). The inset shows the TEM image of the hybrid material. Figures adapted from Ref. [79], with permission. Copyright © 2011, American Chemical Society

with CNT, the doping effect by cobalt further reduced the energy barrier on hydrogen adsorption. Further evidence of the effect of the carbon support onto cobalt selenide ($CoSe_2$) was verified in the electrocatalytic activity enhancement of this non-precious center toward the ORR in alkaline medium when supported onto functionalized CNT (f-CNT) as compared to carbon black [81]. Additionally, in acid medium, it was reported that this system, $CoSe_2$/f-CNT, was highly active, again as compared to carbon black, oxide–carbon composite, and nitrogen-doped carbon supports [81]. Carbon nanohorns (CNHs) are also a low graphitic material, showing horn-shaped carbon nanoparticles constituting a peculiar morphology: dahlia-like [82, 83] which can be further doped with nitrogen (NCNH). The ORR activity, in alkaline medium, showed that NCNH supported cobalt diselenide ($CoSe_2$/NCNH) was more efficient than $CoSe_2$/CNH, $CoSe_2$/C and NCNH [84]. XPS data on $CoSe_2$/NCNH further revealed that such an ORR enhancement can be the result of the catalyst support interaction since a charge transfer phenomenon from N atoms to $CoSe_2$ nanoparticles was put in evidence. The interaction of cobalt sulfide ($Co_{1-x}S$) on graphene (rGO) and/or MW-CNT was also reported to be beneficial for the ORR in alkaline medium. As a result, the electrocatalytic activity of $Co_{1-x}S$/MW-CNT outperformed that of $Co_{1-x}S$/rGO, owing, apparently, to the difference of the conductivity of the support [85]. Furthermore, chalcogenide compounds, the coupling of graphene with Fe–Ni-layered double hydroxide to form

FeNi–rGO–LDH also improved the activity and stability toward the oxygen evolution reaction (OER) in alkaline medium [86]. The strategy to modify carbon-based materials (e.g., rGO, MW-CNT, CNH) by doping with nitrogen has been stimulating to capitalize other catalytic centers, e.g., Co_3O_4 nanoparticles [87] for the ORR, and NiCo alloys [88] for the oxygen evolution reaction interacting with doped-N of the graphene (NGr) in alkaline medium.

5.4 Oxide and Oxide–Carbon Nanocomposites

Due to a severe corrosion of commonly used amorphous carbon supports, oxides, and oxide–carbon composites have been envisaged as electrocatalytic nanoparticle supports for the cornerstone energy conversion processes, such as the ORR and OER (cf. Fig. 5.1). The non-carbon materials are corrosion-resistant, potential strong metal support interaction (cf. Sect. 5.2), and stable porous structure. However, the low conductivity of most oxides can be overcome via doping of titanium oxides [89–96], tungsten oxide [97–99], tin oxide [100–103], and/or by chemically interacting with carbon to form a oxide–carbon composites [36, 64, 104–108].

5.4.1 Selected Metal Oxides of Ti, Sn, and W

Among the large band-gap oxide semiconductors, TiO_2 (Anatase phase 3.2 eV; Rutile phase 3.0 eV), SnO_2 (3.5 eV), and WO_3 (2.7 eV), (see red asterisks in Fig. 5.11) have been employed as heterogenous photocatalyst materials [109]. The figure at the bottom of Fig. 5.11a indicates the upper limit photon energy (3.09 eV) of the visible region. Therefore, for photoelectrochemical conversion, cf. Chap. 3, Sect. 3.2.3, the photon absorption of most oxide semiconductors is limited to 2–5% STH (solar to hydrogen) efficiency, assuming AM1.5 photons (cf. Fig. 3.9). Moreover, TiO_2, SnO_2, and WO_3 are stable in aqueous solutions, as compared to unstable ones, e.g., ZnO, Cu_2O [110], and have been employed, essentially, as electrocatalytic supports for Pt-based catalytic centers. The oxides as supports should be in the nanostructured form, otherwise, the requirement is to have a large area, which can be obtained using the soft chemistry route of synthesis or low temperature solution process, e.g., sol–gel. TEM of metal oxide mesoporous materials, Fig. 5.11b, can be used to observe for sol–gel synthesized Anatase (TiO_2) [111], hydrolysis of Na_2WO_4 with HCl to obtain the amorphous phase of $WO_3{\cdot}2H_2O$ [112], and microwave-assisted polyol method with $SnCl_4$. $5H_2O$ in tri-ethylene glycol as solvent to obtain SnO_2 [113].

The origin of the natural n-type conductivity of, e.g., Ti-, Sn oxides is, apparently, not determined by the oxygen vacancies, as it used to be thought [114], but caused by the presence of impurities. Such an impurity has been detected to be due to

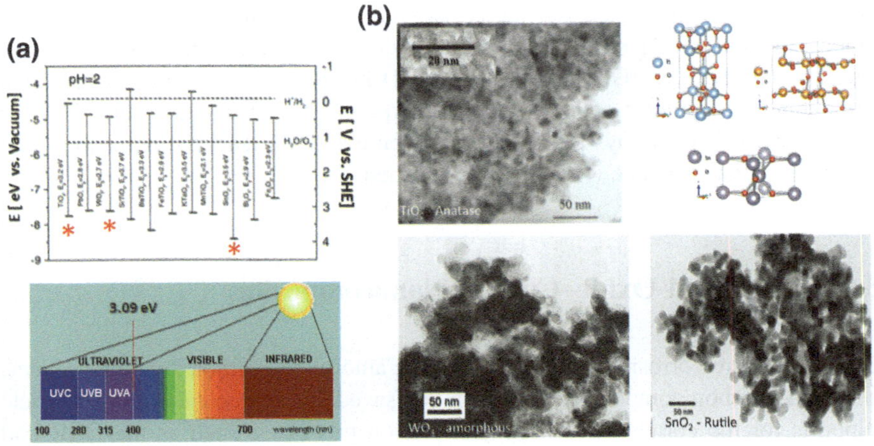

Fig. 5.11 **a** Energy bands positions of various oxide semiconductors with respect the SHE electrochemical scale, and the vacuum reference energy scale. The hydrogen and oxygen redox couples are positioned with respect to a pH of 2. The bottom panel shows the electromagnetic spectrum of the solar light. **b** TEM images of various oxide in nanoparticulated form: TiO_2 (anatase) [111], amorphous WO_3, and SnO_2 (rutile), with their corresponding unit cells generated by the program Vesta

hydrogen, acting electrically, in interstitial form or in substitutional sites for oxygen in the oxides. The presence of hydrogen has been predicted to act as shallow donor in oxide materials [115, 116], whereas the oxygen vacancies are deep donors in ZnO, and SnO_2, and certainly for TiO_2 too [114]. In the literature devoted to catalysis (based on TiO_2 [117–123], and electrocatalysis (based on TiO_2 [106, 107, 124–127], on WO_3 [104, 128, 129], and on SnO_2 [100–103]) some "pure" or not-intentionally doped nanostructured oxide materials have been tested as supports. To illustrate the effect of the oxide support interaction, we take as a model system nanostructured platinum deposited physically via the physical vapor deposition (PVD) onto Titania, reported by Hayden et al. [123]. Herein, different thicknesses of platinum were obtained onto Titania. The effect of the various "averaged equivalent" Pt thicknesses is reflected by the surface electrochemistry behavior during the carbon monoxide stripping, Fig. 5.12a. This result is contrasted with another example, where the method used was the electron–hole generation via UV photons to photodeposit Pt NPs onto nanostructured n-TiO_2, in the author's laboratory, Fig. 5.12b. In the former approach, lower Pt thickness (or coverage) performs the CO oxidation at higher electrode potential concomitant to the negative shift in the platinum reduction peak, discussed in Fig. 5 of the Ref. [123], and ascribed to the thermodynamic and kinetic effects as result of the interaction of oxygen with "small" platinum coverage. A phenomenon observed on carbon used as support [69, 130] and attributed to particle size effect. Moreover, this complex interplay, that contributes to the strength of adsorption energy of CO and oxygen, is the result of various factors, namely surface micro-strains and support interaction as reported by DFT calculation on Pd [131], and latter experimentally measured on this metal in the nanoparticulated form

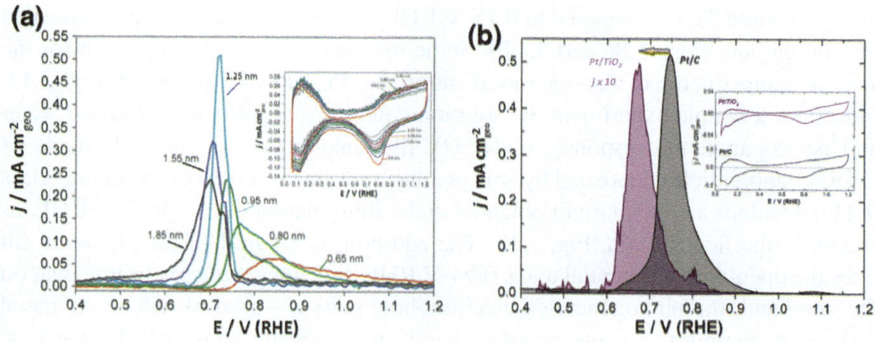

Fig. 5.12 CO stripping, in 0.5M $HClO_4$, on platinum thin films grown onto TiO_x supports with their corresponding cyclic voltammograms. Figure adapted from Ref. [123], with permission. Copyright © 2009, Royal Society of Chemistry. **b** The similar process recorded on photodeposited platinum onto nanoparticulated TiO_2 (Anatase). The inset shows the corresponding cyclic voltammograms in the same acid solution

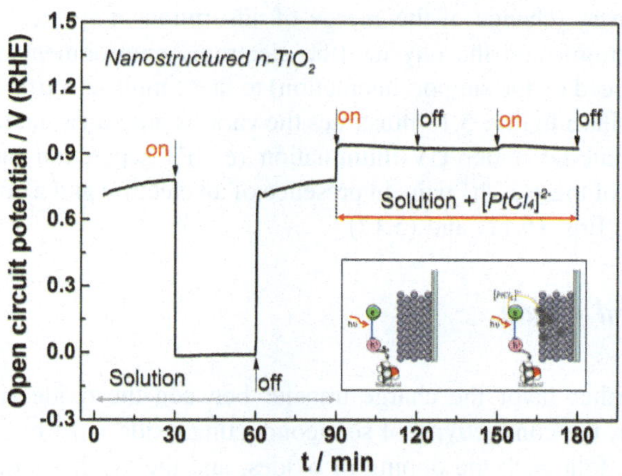

Fig. 5.13 Photodeposition process, in water containing 20 vol% isopropanol, of nanoparticulated platinum onto nanostructured TiO_2 spin-coated onto a conducting glass (FTO. During the firsts 90 min, the electron–hole separation takes place under UV-illumination. In presence of platinum salt (1.16×10^{-3} M), the open-circuit potential is barely modified by the on–off illumination, due to the charge transfer process to reduce the metal cations, and to oxidize the isopropanol molecule. The inset shows schematically the charge photogenerated separation and charge photogenerated transfer that occurs on the nanostructured titanium oxide. Courtesy of CA Campos-Roldan

onto various metal supports [132]. Unfortunately, the results depicted in Fig. 5.12a were not described or put in evidence by a surface spectroscopy tool. Nevertheless, one should keep in mind that two factors are at the source of the negative oxidation potential shift of carbon monoxide, used as a molecular probe, on carbon, cf. Fig. 5.8a, and on titanium dioxide Fig. 5.12, since the CO oxidation on PVD deposited Pt centers between 0.7 and 0.74 V/RHE (Fig. 5.12a, and 0.67 V/RHE on nanostructured TiO_2, compared to 0.75 V/RHE determined on chemically deposited Pt onto carbon, Figs. 5.8a and 5.12b. Some insights as to the driving force for the surface nanostructured Titania modification by Pt NPs are given in Fig. 5.13. Indeed, in a simple experiment the solution consisting of nitrogen outgassed water and isopropanol, the response, under UV-illumination, of an electrode made of n-TiO_2 nanoparticles deposited by spin coating, and sintered onto a conducting glass (FTO) develops an open-circuit potential under illumination of ca. −0.77 V/RHE, as shown in the first 90 min, Fig. 5.13. The addition of 2.4 mM of the platinum salt fixes the open-circuit potential to ca. 0.94 V/RHE. It is interesting to observe that, on the one hand, the photogenerated electron–hole pairs are used to reduce the metal cations of platinum, and on the other hand, the holes are consumed by the hole scavenger. The resulting semiconducting oxide surface modification by platinum is depicted in the cyclic voltammetry at 50 mV/s in the inset of Fig. 5.12b and compared to a commercial Pt/C material under the same conditions in 0.1 M $HClO_4$. However, in spite of the similarity observed on the cyclic voltammetry features, the electronic property of platinum is modified as testified by the CO stripping experiment. Therefore, the strong interaction of a catalytic center with the support put into play the chemistry (change of the energy of adsorption of species induced by the support interaction), and the physics (the electronic environment of the catalytic center also induced by the support interaction) to favor multi-electron charge transfer processes. The inset in Fig. 5.13 illustrates the various processes, namely the driving force of the material under UV-illumination (e^--h^+) separation and the transfer (vectorization) of these e^--h^+ pairs in presence of an electron and a hole acceptor, cf. Fig. 3.10d, and Eqs. (5.11) and (5.12).

5.4.2 *Doped Oxide*

In order to further favor the charge transfer between the oxide support and the catalytic center, the conductivity of semiconducting oxide must be enhanced. Two strategies were followed: the doping of oxides, and the synthesis of oxide–carbon composites (see Sect. 5.4.3). The majority of studies reported so far are devoted to enhance the activity and durability of platinum supported onto TiO_2 (Anatase phase) for the ORR. Among the Titania dopant elements we can find: Cr-doped [126]; Nb-doped [89, 133]; W-doped; [96]; Mo-doped [134]; Ru-doped [135]; Ta-doped [94, 136]; Sn-doped [93]; Y-, Ce-doped [105]; N-,C-doped [125]; Ag-doped [137]. This latter to improve the photocatalytic reduction of nitrite to nitrogen selectivity.

For the same process and catalytic center, Tin Oxide (SnO_2) was also doped: Nb-doped [138]; Indium-doped (known as ITO) [139]. Sb-doped [140, 141] serving as a support for a chalcogenide (RuSe) [142]. Unlike photodeposition, shown in Figs. 5.12 and 5.13, so far all the above data show that the platinum catalytic center was deposited chemically, and the general trend observed on all these systems is the change of the surface chemistry induced by the electronic modification of the catalytic center, leading to ORR catalytic activity and stability enhancement in acid

Fig. 5.14 Graph summarizes the electrochemical active surface area (ECSA) degradation of Pt chemically deposited onto undoped oxides and doped oxides. Comparison is made with carbon black (CB) supports. Figure from Ref. [25], with permission. Copyright © 2015 Elsevier Ltd. All rights reserved

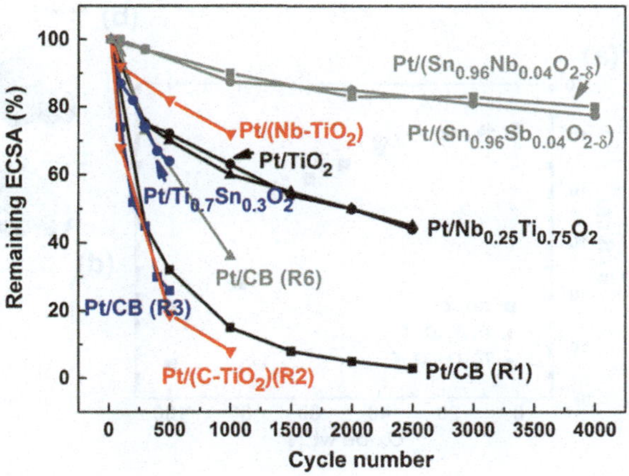

medium, Fig. 5.14, as a result of the electron charge transfer from support to platinum, inducing *d*-band vacancy of the catalytic center (Pt), as revealed by X-ray absorption near-edge spectroscopy (XANES) [134, 135]. From some literature data, the remaining electrochemical surface activity (ECSA) as a function of the number of cycles, summarized in Fig. 5.14 leads to a greater stability of the oxide supported metal catalytic center than that seen on carbon or carbon composite-supported electrocatalyst. Some recent data devoted to the SMSI phenomenon have been recently published [25, 31, 143, 144].

5.4.3 Oxide–Carbon Nanocomposites

The conductivity of semiconducting oxides can be also improved by the in situ synthesis of the oxide in presence of carbon [e.g., Carbon Vulcan XC-72; CNTs Reduced graphene oxide (RGO)] to form the oxide–carbon composites [8, 32–34, 61, 63, 64, 104, 145]. The relative powder conductivity of nanodivided oxide–carbon composites, obtained by soft chemistry (e.g., Sol–Gel), was measured by impedance spectroscopy [146] using a four-point probe [147], Fig. 5.15a. In spite of the intimate contact between the oxide nanoparticles and carbon, produced during the synthesis, the highest relative conductivity compared to pure carbon, is indeed maintained up to an oxide mass loading of ca. 40 wt%. For the measurements, the powder was submitted to a pressure of 25 lb/square-in. in the four-point cell. The addition of Y-, Ce-dopants modified the opto-electronic properties making of Y-doped TiO_2 less photoactive [64]. Yttrium increases the powder conductivity of ($Ti_{0.7}Y_{0.3}O_x$) by ca. two orders of magnitude compared to non-doped TiO_2. Therefore, the electronic conductivity of pure and/or doped oxide can be guaranteed by the presence of carbon. Indeed, the in situ synthesis of the oxide (Anatase) in presence of carbon, as depicted in Fig. 5.15b was performed using the hydrolysis of TIP (Titanium isopropoxide) chemical precursor with water at ca. 4 °C. Since the

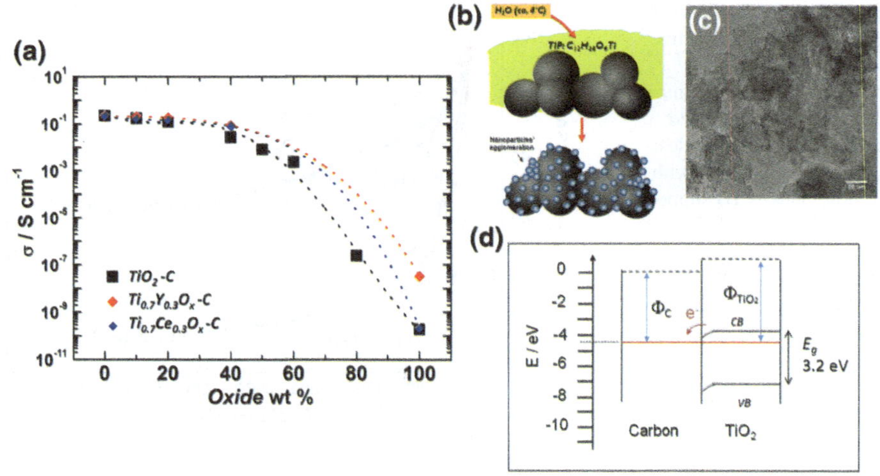

Fig. 5.15 **a** Relative electrical conductivity σ/S cm^{-1} for undoped-, and doped oxide–carbon nanocomposites in powder form. The powders were submitted to a torque of 25 lb/in.$^{-2}$ in a four-point cell. Courtesy of LA Estudillo-Wong. **b** The schematic synthesis of the oxide–carbon nanocomposite. **c** TEM of a 10 wt% TiO_2-C composite. **d** The energy scheme of the oxide–carbon heterojunction, showing the formation of an ohmic contact

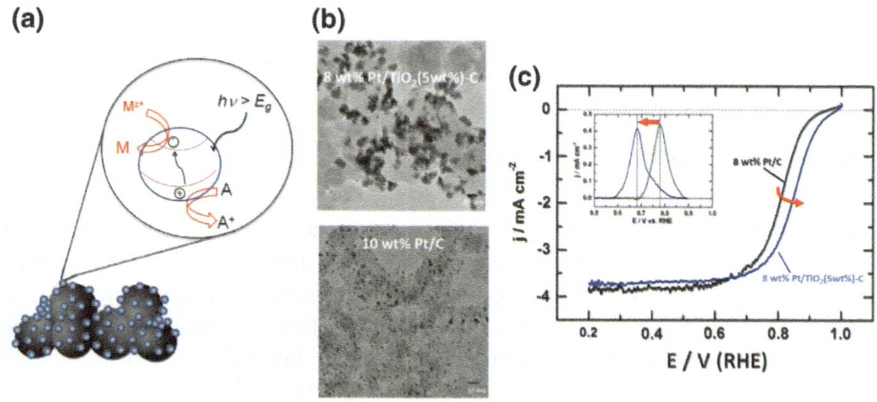

Fig. 5.16 **a** Scheme showing the selective photodeposition of nanoparticulated metals (e.g., Pt, Ru) under UV-irradiation on the oxide sites of the oxide–carbon nanocomposite. **b** TEM image showing 8 wt% Pt/5wt% TiO_2-C (upper); TEM image showing 10 wt% Pt/C, chemically deposited via carbonyl (bottom). **c** Current-potential curves for ORR in oxygen saturated electrolyte H_2SO_4 0.5M at 900 rpm on 8 wt% Pt/C, and 8 wt% Pt/5 wt% TiO_2/C. Figures (**b**) and (**c**) adapted from Ref. [8], with permission. Copyright © 2010 Elsevier B.V. All rights reserved

work function of carbon (4.3–4.6 eV) depends on the surface functionality [148], and for a non-stochiometric TiO_2, the work function is ca. 4.6 eV [149], it is reasonable to schematize the heterojunction carbon/oxide, as shown in Fig. 5.15d indicating an ohmic contact quality, at least responsible for the relative conductivity response up to an oxide mass loading of 40 wt%, Fig. 5.15a. The opportunity to selectively photodeposit metal nanoparticles onto the oxide sites of the nanocomposites can be realized via a photoreaction process [8], as indicated in Fig. 5.16; cf. inset in Fig. 5.13, where M^{z+} is a metal cation and A the hole scavenger, e.g., isopropanol. For similar electrocatalyst's mass loading, we note in the TEM images some differences regarding the metal distribution on the support. This is a clear indication of the photodeposition process, since the catalytic NPs are selectively deposited onto the oxide sites of the nanocomposite, as compared to the chemical deposition (Fig. 5.16b bottom) of the same catalytic center onto carbon. Moreover, differences are remarkable when comparing the interaction (so-called SMSI effect) generated by the photodeposition process between Pt NPs and the oxide sites, as compared to the chemical deposition route. Again, this SMSI effect can be easily visualized in the CO stripping process, see inset in Fig. 5.16c obtained on samples shown in Fig. 5.16b. The complex interplay at the interface between the catalytic center, and the support (oxide or carbon) will be discussed in Sect. 5.5. Meanwhile, the SMSI effect that produced the oxidation potential shift of adsorbed carbon monoxide monolayer is beneficial to enhance the kinetics of the ORR, since the sigmoidal curve recorded at 900 rpm in sulfuric acid saturated oxygen is shifted toward positive potential on the oxide supported electrocatalyst. The consequence of this phenomenon has already been described in Fig. 5.14.

5.5 The Rational Theoretical Approach

The examples provided above in Sects. 5.3 and 5.4 underline the importance of the support interaction, in heterogeneous (electro)catalysis, to stabilize and sustain the activation process. It was also discussed that the deposition of catalytic nanoparticles onto supports can be done using common chemical routes or, as developed in our group, using the photons to selectively deposit metal clusters onto the desired sites. In this section, we will focus on the last approach and discuss the consequence of the selective photodeposition of metal nanoparticles onto sp^2-carbon domains and oxide moieties that determine the change of the electronic structure toward the surface reactivity [150–152]. We take as the specific example platinum clusters. Other metal centers (precious or non-precious) seem to follow a similar trend.

5.5.1 Support Interaction on Carbon Materials

Figures 5.8a, b show that the carbon support, specifically the sp^2 domains, besides acting as an electronic conductor, it is positively affecting the electronic properties of the metal NPs, since it enhances their activity and stability toward, e.g., HER process. The density functional theory (DFT) calculations of the interaction between the metal clusters and the graphitic domain (sp^2 domains) revealed that the

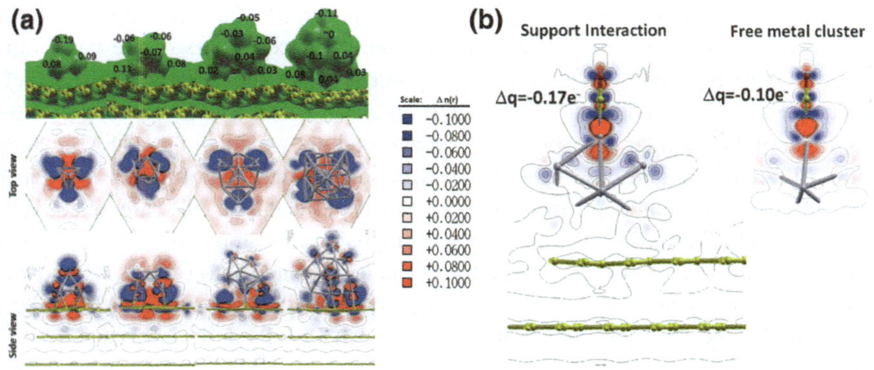

Fig. 5.17 **a** Charge iso-surface and Bader analysis of Pt_4, Pt_6, Pt_{13}, and Pt_{19} over graphite (upper panel). The charge accumulation (blue) and depletion (red)—Top view (middle panel), Side view (bottom panel). Figure (**a**) from Ref. [153] with permission. Copyright © 2013, Royal Society of Chemistry. **b** Free metal cluster and the cluster interacting with graphite showing the difference of electronic density between the graphite−Pt−CO system, and the electronic density of isolated graphite−Pt and CO. The scale for charge accumulation for (**a**) and (**b**) is shown at the center. Figure (**b**) adapted from Ref. [38], with permission. Copyright © 2013, American Chemical Society

presence of charge donation to the metal clusters depends on the adsorption strength of the metal clusters, so that this latter become polarized [153] (see Fig. 5.17). This figure shows, after metal cluster-graphite Bader analysis [154], a charge depletion at the metal cluster/graphite interface extended to the graphite second layer (see side view), as a result of a charge transfer to graphite. Negative charges accumulate predominantly at the top of all metal clusters: Pt_4, Pt_6, Pt_{13}, and Pt_{19}. This interaction leads to new bonding and antibonding states formation, revealing the hybridization of Pt–Graphite. This phenomenon is not only limited to the interaction area but to the surrounding graphite area is also affected (Fig. 5.17a—top view).

To further rationalize the support effect interaction toward the CO stripping process shown in Figs. 5.8 and 5.12, again the DFT simulation [155] brought light into this electrochemical process. The DFT calculations revealed that the extra charge on the Pt clusters on the top, cf. Fig. 5.17a, modifies the charge donated to the carbon monoxide molecule, as assessed by the XPS data (cf. Fig. 5.9 [38]). This charge modulation on the metal clusters monitored by the sp^2 carbon domain interaction is at the origin of the catalytic center stability, through surface-bound Pt NPs, shown during the process of oxidation/reaction in Fig. 5.8b, leading to an increase in back donation of Pt d-orbitals to $2\pi^*$ antibonding orbital of carbon monoxide, further producing a depletion of charge in the metal atom that interacts directly with CO, Fig. 5.17b. Otherwise, the charge transferred is inversely proportional to the adsorption energy of the carbon monoxide.

5.5.2 Support Interaction on Oxide Materials

The strong metal support interaction (SMSI) observed on oxide sites, undergoes also a complex interplay between the catalytic metal atoms and the metal of the oxide, since it results in the modification of the electronic properties of the catalytic center, as in the example shown in Fig. 5.12b for Pt NPs photodeposited onto Titania to form: Pt/TiO_2.

The deposition of size-selected metal nanoclusters on a well-defined oxide support was examined by DFT calculations, specifically with platinum metal clusters on rutile TiO_2 (110) surface [156]. This study showed that the structure at the metal/oxide interface is made of various contributions, namely, Pt–Pt, Pt–O, and Pt–Ti interactions, Fig. 5.18a. Moreover, the metal–metal bond between Pt and Ti can favor the change of the electronic structure of the catalytic center and can be

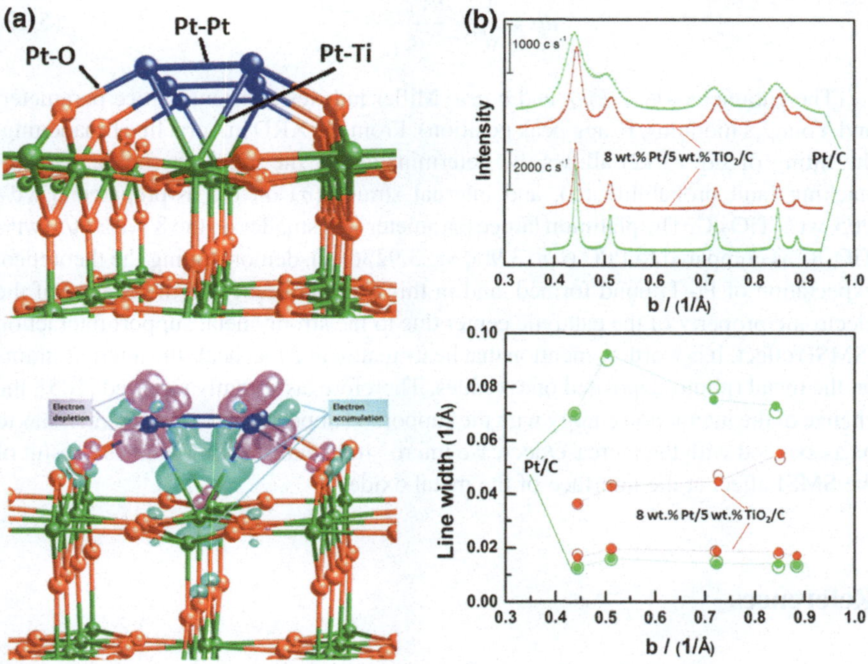

Fig. 5.18 Stable structure of Pt_6 on TiO_2 (110), with an edge-sharing bi-square on the oxide surface (upper panel), and the electron density difference between Pt_4/TiO_2 (110). The color code for the charge iso-surface is: electron accumulation (cyan iso-surface); electron depletion (purple). Figures adapted from Ref. [156], with permission. Copyright © 2012, American Chemical Society. **b** XRD patterns of as-prepared samples: 8 wt% Pt/C, and 8 wt% Pt/5 wt% TiO_2/C (XRD-upper curves); and after heat-treatment in air at 400 °C, 2 h (XRD-bottom curves); The corresponding Williamson–Hall plot is shown at the bottom panel. The open circles show the fitted experimental data using three free parameters: mean particle size <d>, stacking fault probability, α, and internal strains, σ. Figures (**b**) from Ref. [145], with permission. Copyright © 2010 Elsevier B.V. All rights reserved

thus associated with the charge transfer at the interfacial Pt–Ti nanoalloy formed. Indeed, the calculations showed that local density of states produces charge separation (electron density difference) as a result of a significant hybridization between Pt 5d with Ti 3p at the interface Pt clusters/oxide along the Pt–Ti bonds, as shown in Fig. 5.18a. Moreover, the action of building a strong interaction at the metal oxide interface can include phenomena related to strain and ligand arising from the impact of the support on intermetallic distances (nanoalloy formation) and electronic structure. The strain effect was found to influence the adsorption chemical properties of the surface and this concept can be used to tailor the catalytic activity of metals, which is significantly different from unstrained ones [157]. The experimental evidence was brought by the structure analysis of the photodeposited Pt NPs onto the anatase sites of the oxide–carbon composite [145], as shown in Fig. 5.18b. The XRD analyses on as-prepared and heat-treated (400 °C) samples are compared. The corresponding Williamson-Hall analysis was done, using the equation:

$$db = \frac{1}{d} + \frac{\alpha V_{hkl}}{a} + \frac{2\sigma b}{E_{hkl}} \tag{5.12}$$

(The parameters V_{hkl}, E_{hkl}, and *b are*: Miller indexes constant, lattice parameter, and Young's modulus, Bragg peak position). From the XRD integral line broadening, the fitting of Eq. (5.12) allows the determination of the mean crystallite size (*d*), stacking fault probability (α), and internal strains (σ) on the as-prepared 8 wt% Pt/5 wt% TiO_2/C. The platinum lattice parameter was smaller on the 8 wt% Pt/5 wt% TiO_2/C as compared to Pt/C (i.e., 3.900 vs. 3.9236 Å), demonstrating the theoretical expectation of Pt–Ti bond formed, and in this example testifying the change of the electronic property of the catalytic center due to the strong metal support interaction (SMSI) effect. It is worth to mention that heat-treatment effect heals the internal strains on the metal (photo)deposited onto oxides. Therefore, as recently reported [105], the change of the interaction energy with the support on doped or non-doped oxides has to be associated with the increase/decrease micro-strain and stacking fault as a result of the SMSI effect at the interface of the metal oxide.

References

1. Su L, Jia W, Li C-M, Lei Y (2014) Mechanisms for enhanced performance of platinum-based electrocatalysts in proton exchange membrane fuel cells. Chemsuschem 7 (2):361–378. https://doi.org/10.1002/cssc.201300823
2. Antolini E, Salgado JRC, Giz MJ, Gonzalez ER (2005) Effects of geometric and electronic factors on ORR activity of carbon supported Pt–Co electrocatalysts in PEM fuel cells. Int J Hydrogen Energy 30(11):1213–1220. https://doi.org/10.1016/j.ijhydene.2005.05.001
3. Tauster SJ, Fung SC, Baker RTK, Horsley JA (1981) Strong interactions in supported-metal catalysts. Science 211 (4487):1121–1125. doi:cuspal/science.211.4487.1121
4. Tauster SJ, Fung SC, Garten RL (1978) Strong metal-support interactions. Group 8 noble metals supported on TiO_2. J Am Chem Soc 100(1):170–175

5. Bell AT (2003) The impact of nanoscience on heterogeneous catalysis. Science 299 (5613):1688
6. Schwab G-M, Derleth H (1967) Inverse Mischkatalysatoren. Z Phys Chem 53(1–6):1. https://doi.org/10.1524/zpch.1967.53.1-6.001
7. Solymosi F (1968) Importance of the electric properties of supports in the carrier effect. Catal Rev 1(1):233–255. https://doi.org/10.1080/01614946808064705
8. Timperman L, Lewera A, Vogel W, Alonso-Vante N (2010) Nanostructured platinum becomes alloyed at oxide-composite substrate. Electrochem Commun 12(12):1772–1775
9. Bäumer M, Biener J, Madix RJ (1999) Growth, electronic properties and reactivity of vanadium deposited onto a thin alumina film. Surf Sci 432(3):189–198. https://doi.org/10.1016/S0039-6028(99)00400-8
10. Fu Q, Wagner T, Olliges S, Carstanjen H-D (2005) Metal−oxide interfacial reactions: encapsulation of Pd on TiO_2 (110). J Phys Chem B 109(2):944–951. https://doi.org/10.1021/jp046091u
11. Shao-Horn Y, Sheng WC, Chen S, Ferreira PJ, Holby EF, Morgan D (2007) Instability of supported platinum nanoparticles in low-temperature fuel cells. Top Catal 46(3):285–305. https://doi.org/10.1007/s11244-007-9000-0
12. Weber AZ, Borup RL, Darling RM, Das PK, Dursch TJ, Gu W, Harvey D, Kusoglu A, Litster S, Mench MM, Mukundan R, Owejan JP, Pharoah JG, Secanell M, Zenyuk IV (2014) A critical review of modeling transport phenomena in polymer-electrolyte fuel cells. J Electrochem Soc 161(12):F1254–F1299. https://doi.org/10.1149/2.0751412jes
13. Kinoshita K (1988) Carbon: electrochemical and physicochemical properties, 1st edn. Wiley-interscience, New York
14. Pandy A, Yang Z, Gummalla M, Atrazhev VV, Kuzminyh NY, Sultanov VI, Burlatsky S (2013) A carbon corrosion model to evaluate the effect of steady state and transient operation of a polymer electrolyte membrane fuel cell. J Electrochem Soc 160(9):F972–F979. https://doi.org/10.1149/2.036309jes
15. Manzo-Robledo A, Boucher AC, Pastor E, Alonso-Vante N (2002) Electro-oxidation of carbon monoxide and methanol on carbon-supported Pt–Sn nanoparticles: a DEMS study. Fuel Cells 2(2):109–116
16. Ashton SJ, Arenz M (2012) Comparative DEMS study on the electrochemical oxidation of carbon blacks. J Power Sour 217:392–399. https://doi.org/10.1016/j.jpowsour.2012.06.015
17. Aziz AA, Bakar SA, Rusop M (2010) Carbon nanostructured materials. In: Yahya N (ed) Carbon and oxide nanostructures, vol 5, Advanced structured materials. Springer, Berlin, Heidelberg, pp 165–193. https://doi.org/10.1007/8611_2010_14
18. Krueger A (2010) Carbon materials and nanotechnology. WILEY-VCH Verlag GmbH & Co. KGaA, Weinheim
19. Khatri I, Tetsuo S (2010) Carbon nanotubes towards polymer solar cell. In: Yahya N (ed) Carbon and oxide nanostructures, vol 5, Advanced structured materials. Springer, Berlin, Heidelberg, pp 101–123. https://doi.org/10.1007/8611_2010_16
20. McBreen J, Olender H, Srinivasan S, Kordesch KV (1981) Carbon supports for phosphoric acid fuel cell electrocatalysts: alternative materials and methods of evaluation. J Appl Electrochem 11(6):787–796. https://doi.org/10.1007/bf00615184
21. Bundy FP (1962) Direct conversion of graphite to diamond in static pressure apparatus. Science 137(3535):1057
22. Kroto HW, Heath JR, O'Brien SC, Curl RF, Smalley RE (1985) C60: Buckminsterfullerene. Nature 318:162. https://doi.org/10.1038/318162a0
23. Iijima S (1991) Helical microtubules of graphitic carbon. Nature 354:56. https://doi.org/10.1038/354056a0
24. Novoselov KS, Geim AK, Morozov SV, Jiang D, Zhang Y, Dubonos SV, Grigorieva IV, Firsov AA (2004) Electric field effect in atomically thin carbon films. Science 306 (5696):666–669. https://doi.org/10.1126/science.1102896

25. Luo Y, Alonso-Vante N (2015) The effect of support on advanced Pt-based cathodes towards the oxygen reduction reaction. State of the art. Electrochim Acta 179:108–118. https://doi.org/10.1016/j.electacta.2015.04.098
26. Lee J-S, Park GS, Lee HI, Kim ST, Cao R, Liu M, Cho J (2011) Ketjenblack carbon supported amorphous manganese oxides nanowires as highly efficient electrocatalyst for oxygen reduction reaction in alkaline solutions. Nano Lett 11(12):5362–5366. https://doi.org/10.1021/nl2029078
27. Jaouen F, Dodelet J-P (2007) Non-noble electrocatalysts for O_2 reduction: how does heat treatment affect their activity and structure? Part I. Model for carbon black gasification by NH_3: parametric calibration and electrochemical validation. J Phys Chem C 111 (16):5963–5970
28. Charreteur F, Jaouen F, Ruggeri S, Dodelet J-P (2008) Fe/N/C non-precious catalysts for PEM fuel cells: influence of the structural parameters of pristine commercial carbon blacks on their activity for oxygen reduction. Electrochim Acta 53(6):2925–2938
29. Lefèvre M, Dodelet J-P (2008) Fe-based electrocatalysts made with microporous pristine carbon black supports for the reduction of oxygen in PEM fuel cells. Electrochim Acta 53 (28):8269–8276
30. Antolini E (2009) Carbon supports for low-temperature fuel cell catalysts. Appl Catal B: Environ 88(1–2):1–24. https://doi.org/10.1016/j.apcatb.2008.09.030
31. Du L, Shao Y, Sun J, Yin G, Liu J, Wang Y (2016) Advanced catalyst supports for PEM fuel cell cathodes. Nano Energy 29:314–322. https://doi.org/10.1016/j.nanoen.2016.03.016
32. Luo Y, Calvillo L, Daiguebonne C, Daletou MK, Granozzi G, Alonso-Vante N (2016) A highly efficient and stable oxygen reduction reaction on Pt/CeOx/C electrocatalyst obtained via a sacrificial precursor based on a metal-organic framework. Appl Catal B: Environ 189:39–50. https://doi.org/10.1016/j.apcatb.2016.02.028
33. Luo Y, Habrioux A, Calvillo L, Granozzi G, Alonso-Vante N (2014) Yttrium Oxide/Gadolinium Oxide-modified platinum nanoparticles as cathodes for the oxygen reduction reaction. ChemPhysChem 15(10):2136–2144. https://doi.org/10.1002/cphc.201400042
34. Luo Y, Habrioux A, Calvillo L, Granozzi G, Alonso-Vante N (2015) Thermally induced strains on the catalytic activity and stability of Pt–M_2O_3/C (M=Y or Gd) catalysts towards oxygen reduction reaction. ChemCatChem 7(10):1573–1582. https://doi.org/10.1002/cctc.201500130
35. Susac D, Sode A, Zhu L, Wong PC, Teo M, Bizzotto D, Mitchell KAR, Parsons RR, Campbell SA (2006) A methodology for investigating new nonprecious metal catalysts for PEM fuel cells. J Phys Chem B 110(22):10762–10770
36. Ma J, Habrioux A, Guignard N, Alonso-Vante N (2012) The functionalizing effect of increasingly graphitic carbon supports on carbon-supported and TiO_2–carbon composite-supported Pt nanoparticles. J Phys Chem C 116(41):21788–21794. https://doi.org/10.1021/jp304947y
37. Cancado LG, Takai K, Enoki T, Endo M, Kim YA, Mizusaki H, Jorio A, Coelho LN, Magalhaes-Paniago R, Pimenta MA (2006) General equation for the determination of the crystallite size L_a of nanographite by Raman spectroscopy. Appl Phys Lett 88(16):163106
38. Ma J, Habrioux A, Morais C, Lewera A, Vogel W, Verde-Gómez Y, Ramos-Sanchez G, Balbuena PB, Alonso-Vante N (2013) Spectroelectrochemical probing of the strong interaction between platinum nanoparticles and graphitic domains of carbon. ACS Catal 3 (9):1940–1950. https://doi.org/10.1021/cs4003222
39. Ma J, Habrioux A, Pisarek M, Lewera A, Alonso-Vante N (2013) Induced electronic modification of Pt nanoparticles deposited onto graphitic domains of carbon materials by UV irradiation. Electrochem Commun 29:12–16. https://doi.org/10.1016/j.elecom.2012.12.028
40. Ma J, Habrioux A, Luo Y, Ramos-Sanchez G, Calvillo L, Granozzi G, Balbuena PB, Alonso-Vante N (2015) Electronic interaction between platinum nanoparticles and nitrogen-doped reduced graphene oxide: effect on the oxygen reduction reaction. J Mater Chem A 3(22):11891–11904. https://doi.org/10.1039/C5TA01285F

41. Bönnemann H, Braun G, Brijoux W, Brinkmann R, Tilling AS, Seevogel K, Siepen K (1996) Nanoscale colloidal metals and alloys stabilized by solvents and surfactants: preparation and use as catalyst precursors. J Organomet Chem 520(1–2):143–162
42. Koffi RC, Coutanceau C, Garnier E, Léger JM, Lamy C (2005) Synthesis, characterization and electrocatalytic behaviour of non-alloyed PtCr methanol tolerant nanoelectrocatalysts for the oxygen reduction reaction (ORR). Electrochim Acta 50(20):4117–4127
43. Veisz B, Tóth L, Teschner D, Paál Z, Gyorffy N, Wild U, Schlögl R (2005) Palladium-platinum powder catalysts manufactured by colloid synthesis: I. Preparation and characterization. J Mol Catal A: Chem 238 (1–2):56–62
44. Luo J, Njoki PN, Lin Y, Mott D, Wang LY, Zhong CJ (2006) Characterization of carbon-supported AuPt nanoparticles for electrocatalytic methanol oxidation reaction. Langmuir 22(6):2892–2898
45. Hernandez-Fernandez P, Rojas S, Ocon P, GomezdelaFuente JL, SanFabian J, Sanza J, Pena MA, Garcia-Garcia FJ, Terreros P, Fierro JLG (2007) Influence of the preparation route of bimetallic Pt-Au nanoparticle electrocatalysts for the oxygen reduction reaction. J Phys Chem C 111(7):2913–2923
46. Speder J, Zana A, Spanos I, Kirkensgaard JJK, Mortensen K, Arenz M (2013) On the influence of the Pt to carbon ratio on the degradation of high surface area carbon supported PEM fuel cell electrocatalysts. Electrochem Commun 34:153–156. https://doi.org/10.1016/j.elecom.2013.06.001
47. Alonso-Vante N (2006) Carbonyl tailored electrocatalysts. Fuel Cells 6(3–4):182–189
48. Favry E, Wang D, Fantauzzi D, Anton J, Su DS, Jacob T, Alonso-Vante N (2011) Synthesis, electrochemical characterization and molecular dynamics studies of surface segregation of platinum nano-alloy electrocatalysts. Phys Chem Chem Phys 13(20):9201–9208
49. Gago Aldo S, Habrioux A, Alonso-Vante N (2012) Tailoring nanostructured catalysts for electrochemical energy conversion systems. ntrev 1(5):427. https://doi.org/10.1515/ntrev-2012-0013
50. Vogel W, Kaghazchi P, Jacob T, Alonso-Vante N (2007) Genesis of Ru_xSe_y nanoparticles by pyrolysis of $Ru_4Se_2(CO)_{11}$: a combined X-ray in situ and DFT study. J Phys Chem C 111:3908–3913
51. Yang H, Vogel W, Lamy C, Alonso-Vante N (2004) Structure and electrocatalytic activity of carbon-supported Pt-Ni alloy nanoparticles toward the oxygen reduction reaction. J Phys Chem B 108(30):11024–11034
52. Boucher AC, Alonso-Vante N, Dassenoy F, Vogel W (2003) Structural and electrochemical studies of Pt-Sn nanoparticulate catalysts. Langmuir 19(26):10885–10891
53. Alonso-Vante N (2008) Tailoring of metal cluster-like materials for the molecular oxygen reduction reaction. Pure Appl Chem 80(10):2103–2114
54. Deivaraj TC, Lee JY (2005) Preparation of carbon-supported PtRu nanoparticles for direct methanol fuel cell applications—a comparative study. J Power Sourc 142(1–2):43–49
55. Swider KE, Merzbacher CI, Hagans PL, Rolison DR (1997) Synthesis of ruthenium dioxide-titanium dioxide aerogels: redistribution of electrical properties on the nanoscale. Chem Mater 9(5):1248–1255
56. Elezovic NR, Babic BM, Radmilovic VR, Vracar LM, Krstajic NV (2009) Synthesis and characterization of MoO_x-Pt/C and TiO_x-Pt/C nano-catalysts for oxygen reduction. Electrochim Acta 54(9):2404–2409
57. Peng Z, Yang H (2009) Synthesis and oxygen reduction electrocatalytic property of Pt-on-Pd bimetallic heteronanostructures. J Am Chem Soc 131(22):7542–7543. https://doi.org/10.1021/ja902256a
58. Ma J, Habrioux A, Gago AS, Alonso-Vante N (2013) Towards understanding the essential role played by the platinum-support interaction on electrocatalytic activity. ECS Trans 45 (21):25–33. https://doi.org/10.1149/04521.0025ecst
59. Lin C-T, Huang HJ, Yang J-J, Shiao M-H A simple fabrication process of Pt-TiO_2 hybrid electrode for photo-assisted methanol fuel cells. Microelectron Eng In Press, Corrected Proof

60. Shironita S, Mori K, Shimizu T, Ohmichi T, Mimura N, Yamashita H (2008) Preparation of nano-sized platinum metal catalyst using photo-assisted deposition method on mesoporous silica including single-site photocatalyst. Appl Surf Sci 254(23):7604–7607
61. Timperman L, Feng YJ, Vogel W, Alonso-Vante N (2010) Substrate effect on oxygen reduction electrocatalysis. Electrochim Acta 55(26):7558–7563
62. Timperman L, Gago AS, Alonso-Vante N (2011) Oxygen reduction reaction increased tolerance and fuel cell performance of Pt and Ru_xSe_y onto oxide-carbon composites. J Power Sour 196(9):4290–4297
63. Ruiz Camacho B, Morais C, Valenzuela MA, Alonso-Vante N (2013) Enhancing oxygen reduction reaction activity and stability of platinum via oxide-carbon composites. Catal Today 202:36–43. https://doi.org/10.1016/j.cattod.2012.03.033
64. Estudillo-Wong LA, Luo Y, Díaz-Real JA, Alonso-Vante N (2016) Enhanced oxygen reduction reaction stability on platinum nanoparticles photo-deposited onto oxide-carbon composites. Appl Catal B: Environ 187:291–300. https://doi.org/10.1016/j.apcatb.2016.01.030
65. Ma J, Habrioux A, Alonso-Vante N (2013) Enhanced HER and ORR behavior on photodeposited Pt nanoparticles onto oxide–carbon composite. J Solid State Electrochem 17 (7):1913–1921. https://doi.org/10.1007/s10008-013-2046-y
66. Ma J, Valenzuela E, Gago AS, Rousseau J, Habrioux A, Alonso-Vante N (2014) Photohole trapping induced platinum cluster nucleation on the surface of TiO_2 nanoparticles. J Phys Chem C 118(2):1111–1117. https://doi.org/10.1021/jp410846k
67. Harada M, Einaga H (2006) Formation mechanism of Pt particles by photoreduction of Pt ions in polymer solutions. Langmuir 22(5):2371–2377
68. Bickley RI, Jayanty RKM (1974) Photo-adsorption and photo-catalysis on titanium dioxide surfaces. Photo-adsorption of oxygen and the photocatalyzed oxidation of isopropanol. Faraday Discuss Chem Soc 58:194–204. https://doi.org/10.1039/DC9745800194
69. Maillard F, Savinova ER, Simonov PA, Zaikovskii VI, Stimming U (2004) Infrared spectroscopic study of CO adsorption and electro-oxidation on carbon-supported Pt nanoparticles: Interparticle versus intraparticle heterogeneity. J Phys Chem B 108 (46):17893–17904
70. Herrero E, Chen Q-S, Hernandez J, Sun S-G, Feliu JM (2011) Effects of the surface mobility on the oxidation of adsorbed CO on platinum electrodes in alkaline media. The role of the adlayer and surface defects. Phys Chem Chem Phys 13(37):16762–16771. https://doi.org/10.1039/C1CP21909J
71. Blyholder G (1964) Molecular orbital view of chemisorbed carbon monoxide. J Phys Chem 68(10):2772–2778
72. Bagus PS, Pacchioni G (1992) The contribution of metal sp electrons to the chemisorption of CO: theoretical studies of CO on Li, Na, and Cu. Surf Sci 278(3):427–436. https://doi.org/10.1016/0039-6028(92)90678-Y
73. Campos-Roldán CA, Ramos-Sánchez G, Gonzalez-Huerta RG, Vargas García JR, Balbuena PB, Alonso-Vante N (2016) Influence of sp^3–sp^2 carbon nanodomains on metal/support interaction, catalyst durability, and catalytic activity for the oxygen reduction reaction. ACS Appl Mater Interfaces 8:23260–23269. https://doi.org/10.1021/acsami.6b06886
74. Qu L, Liu Y, Baek J-B, Dai L (2010) Nitrogen-doped graphene as efficient metal-free electrocatalyst for oxygen reduction in fuel cells. ACS Nano 4(3):1321–1326. https://doi.org/10.1021/nn901850u
75. Pan F, Jin J, Fu X, Liu Q, Zhang J (2013) Advanced oxygen reduction electrocatalyst based on nitrogen-doped graphene derived from edible sugar and urea. ACS Appl Mater Interfaces 5(21):11108–11114. https://doi.org/10.1021/am403340f
76. Fei H, Ye R, Ye G, Gong Y, Peng Z, Fan X, Samuel ELG, Ajayan PM, Tour JM (2014) Boron- and nitrogen-doped graphene quantum dots/graphene hybrid nanoplatelets as efficient electrocatalysts for oxygen reduction. ACS Nano 8(10):10837–10843. https://doi.org/10.1021/nn504637y

77. Wang Z, Cao X, Ping J, Wang Y, Lin T, Huang X, Ma Q, Wang F, He C, Zhang H (2015) Electrochemical doping of three-dimensional graphene networks used as efficient electrocatalysts for oxygen reduction reaction. Nanoscale 7(21):9394–9398. https://doi.org/10.1039/C4NR06631F
78. He D, Jiang Y, Lv H, Pan M, Mu S (2013) Nitrogen-doped reduced graphene oxide supports for noble metal catalysts with greatly enhanced activity and stability. Appl Catal B: Environ 132–133:379–388. https://doi.org/10.1016/j.apcatb.2012.12.005
79. Li Y, Wang H, Xie L, Liang Y, Hong G, Dai H (2011) MoS_2 nanoparticles grown on graphene: an advanced catalyst for the hydrogen evolution reaction. J Am Chem Soc 133 (19):7296–7299. https://doi.org/10.1021/ja201269b
80. Wang D-Y, Gong M, Chou H-L, Pan C-J, Chen H-A, Wu Y, Lin M-C, Guan M, Yang J, Chen C-W, Wang Y-L, Hwang B-J, Chen C-C, Dai H (2015) Highly active and stable hybrid catalyst of cobalt-doped FeS_2 nanosheets-carbon nanotubes for hydrogen evolution reaction. J Am Chem Soc 137(4):1587–1592. https://doi.org/10.1021/ja511572q
81. Sreekuttan MU, Campos-Roldan CA, Mora-Hernandez JM, Luo Y, Estudillo-Wong LA, Alonso-Vante N (2015) The effect of carbon-based substrates onto non-precious and precious electrocatalytic centers. ECS Trans 69(17):35–42. https://doi.org/10.1149/06917.0035ecst
82. Kasuya D, Yudasaka M, Takahashi K, Kokai F, Iijima S (2002) Selective production of single-wall carbon nanohorn aggregates and their formation mechanism. J Phys Chem B 106(19):4947–4951. https://doi.org/10.1021/jp020387n
83. Zhu S, Xu G (2010) Single-walled carbon nanohorns and their applications. Nanoscale 2 (12):2538–2549. https://doi.org/10.1039/C0NR00387E
84. Unni SM, Mora-Hernandez JM, Kurungot S, Alonso-Vante N (2015) $CoSe_2$ supported on nitrogen-doped carbon nanohorns as a methanol-tolerant cathode for air-breathing microlaminar flow fuel cells. ChemElectroChem 2(9):1339–1345. https://doi.org/10.1002/celc.201500154
85. Liang Y, Li Y, Wang H, Dai H (2013) Strongly coupled inorganic/nanocarbon hybrid materials for advanced electrocatalysis. J Am Chem Soc 135(6):2013–2036. https://doi.org/10.1021/ja3089923
86. Long X, Li J, Xiao S, Yan K, Wang Z, Chen H, Yang S (2014) A strongly coupled graphene and FeNi double hydroxide hybrid as an excellent electrocatalyst for the oxygen evolution reaction. Angew Chem Int Ed 53(29):7584–7588. https://doi.org/10.1002/anie.201402822
87. Singh SK, Dhavale VM, Kurungot S (2015) Surface-Tuned Co_3O_4 nanoparticles dispersed on nitrogen-doped graphene as an efficient cathode electrocatalyst for mechanical rechargeable zinc-air battery application. ACS Appl Mater Interfaces 7(38):21138–21149. https://doi.org/10.1021/acsami.5b04865
88. Singh SK, Kumar D, Dhavale VM, Pal S, Kurungot S (2016) Strategic preparation of efficient and durable NiCo alloy supported N-doped porous graphene as an oxygen evolution electrocatalyst: a theoretical and experimental investigation. Adv Mater Interfaces 3(20):n/a–n/a. https://doi.org/10.1002/admi.201600532
89. Gao Y, Liang Y, Chambers SA (1996) Thermal stability and the role of oxygen vacancy defects in strong metal support interaction—Pt on Nb-doped TiO_2(100). Surf Sci 365 (3):638–648. https://doi.org/10.1016/0039-6028(96)00763-7
90. Ioroi T, Siroma Z, Fujiwara N, S-i Yamazaki, Yasuda K (2005) Sub-stoichiometric titanium oxide-supported platinum electrocatalyst for polymer electrolyte fuel cells. Electrochem Commun 7(2):183–188. https://doi.org/10.1016/j.elecom.2004.12.007
91. Ioroi T, Senoh H, S-i Yamazaki, Siroma Z, Fujiwara N, Yasuda K (2008) Stability of corrosion-resistant magnéli-phase Ti_4O_7-supported PEMFC catalysts at high potentials. J Electrochem Soc 155(4):B321–B326. https://doi.org/10.1149/1.2833310
92. Shi F, Baker LR, Hervier A, Somorjai GA, Komvopoulos K (2013) Tuning the electronic structure of titanium oxide support to enhance the electrochemical activity of platinum nanoparticles. Nano Lett 13(9):4469–4474. https://doi.org/10.1021/nl402392u

93. Gao Y, Hou M, Shao Z, Zhang C, Qin X, Yi B (2014) Preparation and characterization of $Ti_{0.7}Sn_{0.3}O_2$ as catalyst support for oxygen reduction reaction. J Energy Chem 23(3):331–337. https://doi.org/10.1016/S2095-4956(14)60155-8
94. Kumar A, Ramani V (2014) Strong metal-support interactions enhance the activity and durability of platinum supported on tantalum-modified titanium dioxide electrocatalysts. ACS Catal 4(5):1516–1525. https://doi.org/10.1021/cs500116h
95. Bendova M, Gispert-Guirado F, Hassel AW, Llobet E, Mozalev A (2017) Solar water splitting on porous-alumina-assisted TiO_2-doped WO_x nanorod photoanodes: paradoxes and challenges. Nano Energy 33:72–87. https://doi.org/10.1016/j.nanoen.2017.01.029
96. Abadias G, Gago AS, Alonso-Vante N (2011) Structural and photoelectrochemical properties of $Ti_{1-x}W_xO_2$ thin films deposited by magnetron sputtering. Surf Coat Technol 205(2):S265–S270
97. Patil PS, Mujawar SH, Inamdar AI, Shinde PS, Deshmukh HP, Sadale SB (2005) Structural, electrical and optical properties of TiO_2 doped WO_3 thin films. Appl Surf Sci 252(5):1643–1650
98. Wang F, Di Valentin C, Pacchioni G (2012) Doping of WO_3 for photocatalytic water splitting: hints from density functional theory. J Phys Chem C 116(16):8901–8909. https://doi.org/10.1021/jp300867j
99. Yang Y, Zhan F, Li H, Liu W, Yu S (2017) In situ Sn-doped WO_3 films with enhanced photoelectrochemical performance for reducing CO_2 into formic acid. J Solid State Electrochem 21(8):2231–2240. https://doi.org/10.1007/s10008-017-3569-4
100. Zhang X, Zhu H, Guo Z, Wei Y, Wang F (2011) Sulfated SnO_2 modified multi-walled carbon nanotubes—a mixed proton–electron conducting support for Pt catalysts in direct ethanol fuel cells. J Power Sour 196(6):3048–3053. https://doi.org/10.1016/j.jpowsour.2010.11.129
101. Elezović NR, Babić BM, Radmilović VR, Krstajić NV (2013) Synthesis and characterization of Pt catalysts on SnO_2 based supports for oxygen reduction reaction. J Electrochem Soc 160(10):F1151–F1158. https://doi.org/10.1149/2.095310jes
102. Rabis A, Kramer D, Fabbri E, Worsdale M, Kötz R, Schmidt TJ (2014) Catalyzed SnO_2 thin films: theoretical and experimental insights into fabrication and electrocatalytic properties. J Phys Chem C 118(21):11292–11302. https://doi.org/10.1021/jp4120139
103. Monyoncho EA, Ntais S, Brazeau N, Wu J-J, Sun C-L, Baranova EA (2016) Role of the metal-oxide support in the catalytic activity of Pd nanoparticles for ethanol electrooxidation in alkaline media. ChemElectroChem 3(2):218–227. https://doi.org/10.1002/celc.201500432
104. Lewera A, Timperman L, Roguska A, Alonso-Vante N (2011) Metal-support interactions between nanosized Pt and metal oxides (WO_3 and TiO_2) studied using X-ray photoelectron spectroscopy. J Phys Chem C 115(41):20153–20159. https://doi.org/10.1021/jp2068446
105. Estudillo-Wong LA, Ramos-Sanchez G, Calvillo L, Granozzi G, Alonso-Vante N (2017) Support interaction effect of platinum nanoparticles on non-, Y-, Ce-doped anatase and its implication on the ORR in acid and alkaline media. ChemElectroChem 4:3264–3275. https://doi.org/10.1002/celc.201700715
106. Armstrong KJ, Elbaz L, Bauer E, Burrell AK, McCleskey TM, Brosha EL (2012) Nanoscale titania ceramic composite supports for PEM fuel cells. J Mater Res 27(15):2046–2054. https://doi.org/10.1557/jmr.2012.169
107. Rigdon WA, Huang X (2014) Carbon monoxide tolerant platinum electrocatalysts on niobium doped titania and carbon nanotube composite supports. J Power Sour 272:845–859. https://doi.org/10.1016/j.jpowsour.2014.09.054
108. Luo Y, Alonso-Vante N (2017) Application of metal organic framework (MOF) in the electrocatalytic process. In: Electrochemistry: volume 14. R Soc Chem, pp 194–256. https://doi.org/10.1039/9781782622727-00194
109. Kudo A, Miseki Y (2009) Heterogeneous photocatalyst materials for water splitting. Chem Soc Rev 38(1):253–278. https://doi.org/10.1039/B800489G
110. Hassel A, Schultze JW (2003) Passivity of metals, alloys, and semiconductors. In: Bard AJ, Frankel GS, Stratmann M (eds) Encyclopedia of electrochemistry. Corrosion and oxide films, vol 4. Wiley-VCH, pp 216–270

111. Lana Villarreal T (2004) Depuración de aguas por métodos fotoelectroquímicos. Ph.D. thesis, The University of Navarra and The University of Poitiers
112. Choi Y-G, Sakai G, Shimanoe K, Miura N, Yamazoe N (2002) Preparation of aqueous sols of tungsten oxide dihydrate from sodium tungstate by an ion-exchange method. Sens Actuators B: Chem 87(1):63–72. https://doi.org/10.1016/S0925-4005(02)00218-6
113. Bargougui R, Pichavant A, Hochepied JF, Berger MH, Gadri A, Ammar S (2016) Synthesis and characterization of SnO_2, TiO_2 and $Ti_{0.5}Sn_{0.5}O_2$ nanoparticles as efficient materials for photocatalytic activity. Opt Mater 58 (Supplement C):253–259. doi:https://doi.org/10.1016/j.optmat.2016.05.026
114. Janotti A, Varley JB, Lyons JL, Van de Walle CG (2012) Controlling the conductivity in oxide semiconductors. In: Wu J, Cao J, Han W-Q, Janotti A, Kim H-C (eds) Functional metal oxide nanostructures, vol 149 Springer series in materials science. Springer Berlin Heidelberg, Springer New York Dordrecht Heidelberg London, pp 23–35
115. Varley JB, Peelaers H, Janotti A, Walle CGVd (2011) Hydrogenated cation vacancies in semiconducting oxides. J Phys: Condens Matter 23(33):334212
116. Van de Walle CG (2003) Hydrogen as a shallow center in semiconductors and oxides. physica status solidi (b) 235(1):89–95. https://doi.org/10.1002/pssb.200301539
117. Łaniecki M, Małecka-Grycz M, Domka F (2000) Water–gas shift reaction over sulfided molybdenum catalysts: I. Alumina, titania and zirconia-supported catalysts. Appl Catal A-Gen 196(2):293–303. doi:https://doi.org/10.1016/S0926-860X(99)00480-9
118. Idakiev V, Yuan ZY, Tabakova T, Su BL (2005) Titanium oxide nanotubes as supports of nano-sized gold catalysts for low temperature water-gas shift reaction. Appl Catal A-Gen 281(1–2):149–155. https://doi.org/10.1016/j.apcata.2004.11.021
119. Ketchie WC, Maris EP, Davis RJ (2007) In-situ X-ray absorption spectroscopy of supported Ru catalysts in the aqueous phase. Chem Mater 19(14):3406–3411
120. Zhang R, Elzatahry AA, Al-Deyab SS, Zhao D (2012) Mesoporous titania: from synthesis to application. Nano Today 7(4):344–366. https://doi.org/10.1016/j.nantod.2012.06.012
121. Reddy BM, Khan A (2005) Recent advances on TiO_2-ZrO_2 mixed oxides as catalysts and catalyst supports. Catal Rev 47(2):257–296. https://doi.org/10.1081/CR-200057488
122. Huang S-Y, Ganesan P, Park S, Popov BN (2009) Development of a titanium dioxide-supported platinum catalyst with ultrahigh stability for polymer electrolyte membrane fuel cell applications. J Am Chem Soc. https://doi.org/10.1021/ja904810h
123. Hayden BE, Pletcher D, Suchsland J-P, Williams LJ (2009) The influence of Pt particle size on the surface oxidation of titania supported platinum. Phys Chem Chem Phys 11(10):1564–1570
124. Sharma S, Pollet BG (2012) Support materials for PEMFC and DMFC electrocatalysts—a review. J Power Sour 208:96–119. https://doi.org/10.1016/j.jpowsour.2012.02.011
125. Gebauer C, Jusys Z, Wassner M, Hüsing N, Behm RJ (2014) Membrane fuel cell cathode catalysts based on titanium oxide supported platinum nanoparticles. ChemPhysChem 15(10):2094–2107. https://doi.org/10.1002/cphc.201402019
126. Kim J-H, Chang S, Kim Y-T (2014) Compressive strain as the main origin of enhanced oxygen reduction reaction activity for Pt electrocatalysts on chromium-doped titania support. Appl Catal B: Environ 158–159:112–118. https://doi.org/10.1016/j.apcatb.2014.04.003
127. Savych I, Bernard d'Arbigny J, Subianto S, Cavaliere S, Jones DJ, Rozière J (2014) On the effect of non-carbon nanostructured supports on the stability of Pt nanoparticles during voltage cycling: a study of TiO_2 nanofibres. J Power Sour 257:147–155. https://doi.org/10.1016/j.jpowsour.2014.01.112
128. Dou M, Hou M, Zhang H, Li G, Lu W, Wei Z, Shao Z, Yi B (2012) A highly stable anode, carbon-free, catalyst support based on tungsten trioxide nanoclusters for proton-exchange membrane fuel cells. Chemsuschem 5(5):945–951. https://doi.org/10.1002/cssc.201100706
129. Liu Y, Shrestha S, Mustain WE (2012) Synthesis of nanosize tungsten oxide and its evaluation as an electrocatalyst support for oxygen reduction in acid media. ACS Catal 2(3):456–463. https://doi.org/10.1021/cs200657w

130. Maillard F, Schreier S, Hanzlik M, Savinova ER, Weinkauf S, Stimming U (2005) Influence of particle agglomeration on the catalytic activity of carbon-supported Pt nanoparticles in CO monolayer oxidation. Phys Chem Chem Phys 7(2):385–393. https://doi.org/10.1039/B411377B
131. Roudgar A, Groß A (2003) Local reactivity of thin Pd overlayers on Au single crystals. J Electroanal Chem 548:121–130. https://doi.org/10.1016/S0022-0728(03)00230-4
132. Hu S, Scudiero L, Ha S (2014) Electronic effect of Pd-transition metal bimetallic surfaces toward formic acid electrochemical oxidation. Electrochem Commun 38:107–109. https://doi.org/10.1016/j.elecom.2013.11.010
133. Chen G, Bare SR, Mallouk TE (2002) Development of supported bifunctional electrocatalysts for unitized regenerative fuel cells. J Electrochem Soc 149(8):A1092–A1099
134. Ho VTT, Pan C-J, Rick J, Su W-N, Hwang B-J (2011) Nanostructured $Ti_{0.7}Mo_{0.3}O_2$ support enhances electron transfer to Pt: high-performance catalyst for oxygen reduction reaction. J Am Chem Soc 133(30):11716–11724. https://doi.org/10.1021/ja2039562
135. Ho VTT, Pillai KC, Chou H-L, Pan C-J, Rick J, Su W-N, Hwang B-J, Lee J-F, Sheu H-S, Chuang W-T (2011) Robust non-carbon $Ti_{0.7}Ru_{0.3}O_2$ support with co-catalytic functionality for Pt: enhances catalytic activity and durability for fuel cells. Energy Environ Sci 4(10):4194–4200
136. Stassi A, Gatto I, Baglio V, Passalacqua E, Aricò AS (2013) Oxide-supported PtCo alloy catalyst for intermediate temperature polymer electrolyte fuel cells. Appl Catal B: Environ 142–143:15–24. https://doi.org/10.1016/j.apcatb.2013.05.008
137. Zhang F, Pi Y, Cui J, Yang Y, Zhang X, Guan N (2007) Unexpected selective photocatalytic reduction of nitrite to nitrogen on silver-doped titanium dioxide. J Phys Chem C 111 (9):3756–3761
138. Lee K-S, Park I-S, Cho Y-H, Jung D-S, Jung N, Park H-Y, Sung Y-E (2008) Electrocatalytic activity and stability of Pt supported on Sb-doped SnO_2 nanoparticles for direct alcohol fuel cells. J Catal 258(1):143–152
139. Zhao S, Wangstrom AE, Liu Y, Rigdon WA, Mustain WE (2015) Stability and activity of Pt/ITO electrocatalyst for oxygen reduction reaction in alkaline media. Electrochim Acta 157:175–182. https://doi.org/10.1016/j.electacta.2015.01.030
140. Fabbri E, Patru A, Rabis A, Kötz R, Schmidt TJ (2014) Advanced cathode materials for polymer electrolyte fuel cells based on Pt/metal oxides: from model electrodes to catalyst systems. CHIMIA Int J Chem 68(4):217–220. https://doi.org/10.2533/chimia.2014.217
141. Fabbri E, Rabis A, Kotz R, Schmidt TJ (2014) Pt nanoparticles supported on Sb-doped SnO_2 porous structures: developments and issues. Phys Chem Chem Phys 16(27):13672–13681. https://doi.org/10.1039/C4CP00238E
142. Malkhandi S, Yangang Y, Rao V, Bund A, Stimming U (2011) Synthesis and electrochemical study of antimony-doped tin oxide supported RuSe catalysts for oxygen reduction reaction. Electrocatalysis 2(1):20–23. https://doi.org/10.1007/s12678-010-0033-y
143. Hsieh B-J, Tsai M-C, Pan C-J, Su W-N, Rick J, Chou H-L, Lee J-F, Hwang B-J (2017) Tuning metal support interactions enhances the activity and durability of TiO_2-supported Pt nanocatalysts. Electrochim Acta 224 (Supplement C):452–459. doi:https://doi.org/10.1016/j.electacta.2016.12.020
144. Pan C-J, Tsai M-C, Su W-N, Rick J, Akalework NG, Agegnehu AK, Cheng S-Y, Hwang B-J (2017) Tuning/exploiting strong metal-support interaction (SMSI) in heterogeneous catalysis. J Taiwan Inst Chem Eng 74:154–186. https://doi.org/10.1016/j.jtice.2017.02.012
145. Vogel W, Timperman L, Alonso-Vante N (2010) Probing metal substrate interaction of Pt nanoparticles: structural XRD analysis and oxygen reduction reaction. Appl Catal A-Gen 377:167–173
146. Korshunov KV, Tsarev MV, Mokrushin VV, Shapovalov AM, Zabavin EV (2015) Application of impedance spectroscopy to study oxidized powders of titanium hydride. J Alloys Compd 645 Supplement 1:S140–S143. https://doi.org/10.1016/j.jallcom.2015.01.131

147. Espinola A, Miguel PM, Salles MR, Pinto AR (1986) Electrical properties of carbons—resistance of powder materials. Carbon 24(3):337–341. https://doi.org/10.1016/0008-6223(86)90235-6
148. Fabish TJ, Hair ML (1977) The dependence of the work function of carbon black on surface acidity. J Colloid Interface Sci 62(1):16–23. https://doi.org/10.1016/0021-9797(77)90060-1
149. Ioannides T, Verykios XE (1996) Charge transfer in metal catalysts supported on doped TiO_2: a theoretical approach based on metal-semiconductor contact theory. J Catal 161(2):560–569
150. Greeley J, Nørskov JK, Mavrikakis M (2002) Electronic structure and catalysis on metal surfaces. Annu Rev Phys Chem 53:319–348
151. Greeley J (2012) Computational studies of trends in electrocatalysis. ECS Trans 45(2):85–95. https://doi.org/10.1149/1.3701970
152. Greeley J, Markovic NM (2012) The road from animal electricity to green energy: combining experiment and theory in electrocatalysis. Energy Environ Sci 5(11):9246–9256. https://doi.org/10.1039/C2EE21754F
153. Ramos-Sanchez G, Balbuena PB (2013) Interactions of platinum clusters with a graphite substrate. Phys Chem Chem Phys 15:11950–11959. https://doi.org/10.1039/c3cp51791h
154. Bader RFW (1998) A bond path: a universal indicator of bonded interactions. J Phys Chem A 102(37):7314–7323. https://doi.org/10.1021/jp981794v
155. Ramos-Sánchez G, Balbuena PB (2014) CO adsorption on Pt clusters supported on graphite. J Electroanal Chem 716:23–30. https://doi.org/10.1016/j.jelechem.2013.09.025
156. D-e Jiang, Overbury SH, Dai S (2012) Structures and energetics of Pt clusters on TiO_2: interplay between metal-metal bonds and metal-oxygen bonds. J Phys Chem C 116(41):21880–21885. https://doi.org/10.1021/jp3072102
157. Mavrikakis M, Hammer B, Nørskov JK (1998) Effect of strain on the reactivity of metal surfaces. Phys Rev Lett 81(13):2819–2822

Chapter 6
Micro-fuel Cells

6.1 Introduction

The kinetic and thermodynamic principles of a fuel cell system were discussed in Chap. 2 (cf., e.g., Fig. 2.13). The micro-fuel cell (μFC) device obeys the same kinetic and thermodynamic concepts. In order to be integrated in micro-systems and nanosystems in the emerging technologies, led by the proliferation of MEMS/NEMS (micro- and nanoelectromechanical systems), μFCs must satisfy the requirement of energy, where the operating temperature between RT and 90 °C permits the use of either passive [1, 2] or active delivery of reactants [3]. These two-principal operating modes of μFCs can framework the elimination or the use of ancillary devices that provide the feeding of fuel in the μFC. The technique allows these systems to be operated in air-breathing conditions at the cathode providing a net power output in the energy balance of plant. However, the study of novel devices leads to the use of ancillary equipment to understand the engineering process in the micro-devices. Under this circumstance, micro-fluidic fuel cells (μFCs) appear to be a promising power source for portable electronic devices.

6.2 Micro-fluidic Fuel Cells

The micro-fluidic fuel cell is an alternative cell type that does not need a membrane, so-called self-styled membrane-less device. This latter indeed represents, besides the high cost, a bottleneck for processes such as fuel-crossover, and water management in conventional fuel cell systems (cf. Fig. 2.2). The way of functioning of the two systems is schematized in Fig. 6.1. The heart of the fuel cell is the membrane electrode assembly (MEA). The role of the membrane in this fuel cell is to separate fuel and oxidant, while maintaining a good ionic conductivity, Fig. 6.1a. In the micro-fuel cell (μFC), the laminar nature of the flow should be maintained, and

N. Alonso-Vante, *Chalcogenide Materials for Energy Conversion*, Nanostructure Science and Technology, https://doi.org/10.1007/978-3-319-89612-0_6

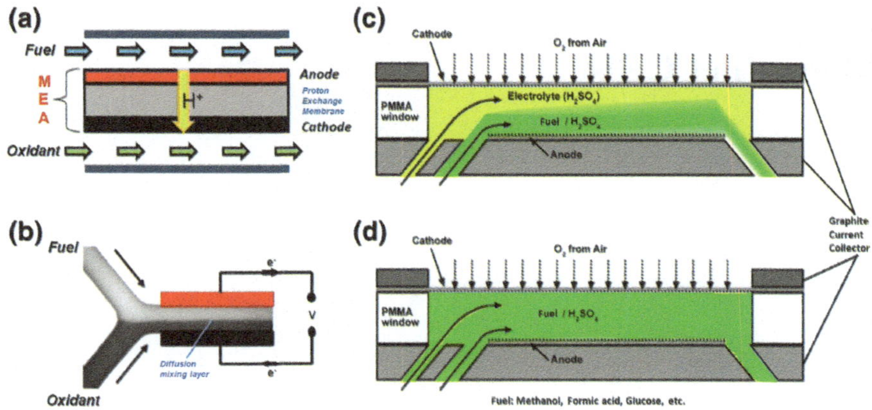

Fig. 6.1 A bridged working principles of fuel cells. **a** A membrane polymer acid fuel cell (PEM-FC); **b** Y-shaped micro-laminar flow fuel cell (μLFFC); **c** an air-breathing micro-laminar flow fuel cell (μLFFC); **d** an air-breathing mixed-reactant micro-laminar flow fuel cell (MR-μLFFC)

the system is named micro-laminar flow fuel cell (μLFFC). The flow regime (Reynolds conditions) in this system naturally separates the reactant streams, while also keeping a good ionic conductivity, Fig. 6.1b. Under these conditions, the μLFFC can work (i) either as a laminar flow fuel cell [4–13] (LFFC, Fig. 6.1c), or as (ii) a mixed-reactant flow fuel cell [9, 11, 12, 14], MRFFC, Fig. 6.1d. In the LFFC mode, the flow regime essentially avoids the crossover effect of fuel and oxidant (in the streams) that can affect the reaction selectivity at the anode or at the cathode interface because of a lack of tolerance at the anode or at the cathode materials (cf. Figs. 4.6 and 4.7). Besides these complexities regarding the nature of the catalytic centers of the electrode materials, engineering analyses have been put forward to optimize the design, and operational control using mathematical modeling. For example, Xuan et al. [15], using the semi-empirical Graetz–Damköhler analysis, arrived at the conclusion that the fuel utilization and the current density is strongly related to the electrode kinetics. Additionally, a co-laminar flow in the μFC limits the fuel utilization, and therefore, the μFC performance. Naturally, these calculations open ways as to the optimization of future μFC systems, from an engineering technical point of view concerning the materials science development.

6.2.1 Working Principle of LFFC

The fluid at the microscale level is submitted to laminar conditions, where the stream with the components moves in parallel without being mixed, Fig. 6.1b. The number of Reynolds (*Re*) characterizes the inertial forces with respect to the viscosity, Eq. (6.1):

$$Re = \frac{\rho U D_h}{\mu} \tag{6.1}$$

where ρ is the volumetric mass or density of the fluid, kg m^{-3}; U the average flow speed, m s^{-1}; D_h, the hydraulic diameter, m; and μ the kinematic viscosity, m^2 s^{-1}. Additionally, the parameter that represents the ratio between the transfer by convection and the transfer by diffusion is the Péclet number (*Pe*), Eq. (6.2)

$$Pe = \frac{U D_h}{D} \tag{6.2}$$

where D is the diffusion coefficient, m^2 s^{-1}. A well-defined parallel flow lines flow, with a laminar regime, is characterized by $Re < 2000$. Moreover, a high-effective *Pe* will characterize a decrease in diffusive mixing with the flow rate. Furthermore, the velocity of a particle of the fluid can be determined by the Navier–Stokes relationship, Eq. (6.3):

$$\rho\left(\frac{\partial \vec{u}}{\partial t} + \vec{u}.\nabla\vec{u}\right) = -\nabla p + \mu\nabla^2\vec{u} + \vec{f} \tag{6.3}$$

where p is the pressure, u is the velocity of a particle of the fluid, t is the temporal parameter, and f is the body force per unit of volume. Equation (6.3) allows to describe and predict the motion of the fluids at the microscale, i.e., micro-fluidics. In general, the instabilities in a fluid are directly associated to the non-linear term $(\vec{u} \cdot \nabla\vec{u})$ that represents the inertial mass transport by convection. For an incompressible fluid (ρ = constant), the conservation of mass follows Eq. (6.4) of continuity:

$$\left(\frac{\partial \rho}{\partial t} + \nabla \cdot (\rho\vec{u})\right) = 0 \tag{6.4}$$

The solution of Eq. (6.3) with (6.1) and (6.4) lead to the well-known parabolic, flat, and uniform profile of flow speed, without turbulence. The streams that flow through a channel of micro-metric dimensions ($D_h < 1000$ μm) behave differently from those that flow in large diameter tubes and can be used to replace FCs: H_2/O_2; CH_3OH/O_2.

The oxidant and fuel are conveyed along the micro-channel in the μLFFC without mixing, Fig. 6.1b, c. The streams containing the fuel (e.g., CH_3OH, CH_3CH_2OH, HCOOH, glucose) and the oxidant (O_2) are, respectively, named anolyte and catholyte. The protons diffuse through the liquid interface (diffusion mixing layer) between the two laminar flows [16–18]. Moreover, the stream that uses the O_2 from air corresponds to an air-breathing LFFC system [19], as schematized in Fig. 6.1c, d, in these membrane-less fuel cells.

If the μLFFC device uses methanol as fuel, this device is termed a micro-direct methanol fuel cell (μDMFC) and the crossover phenomenon is present at high alcohol concentration. The fuel can gradually cross from the catholyte to the anolyte through the diffusion mixing layer (cf. Fig. 6.1c). The width of this diffusion mixing layer can be calculated with Eq. (6.5) [16]:

$$\Delta x \approx \left(\frac{Dhy}{U}\right)^{1/3} \tag{6.5}$$

where h is the channel height, y is the distance of the fluid that flows downstream, and U is the average flow speed.

The simultaneous oxidation of the fuel and the reduction of molecular oxygen at the cathode generate a mixed-potential. This parameter lowers the overall performance of the FCs. A phenomenon which can be reduced by increasing the flow speed of the catholyte [19], or using a nanoporous separator placed between the streams [20–22]. The role of this latter is to reduce the fuel diffusion gradient, but increase the device's internal resistance.

6.2.2 *Micro-laminar Flow Fuel Cells—μLFFC*

Taking advantage of the hydrodynamics in the microscale dimensions, micro-laminar flow fuel cells (μLFFC) have been designed to operate without membranes. Table 6.1 summarizes various designs and their corresponding performances. An overview of data summarized in Table 6.1, without taking into consideration the vanadium-redox system [23], reveals that most of the μLFFC developed so far use oxygen, as a dissolved oxidant in the catholyte stream; and the fuels, such as, hydrogen, methanol, formic acid, dissolved in the anolyte stream (see Fig. 6.1b). The anode and cathode materials are essentially Pt-based materials. In acid medium, the use of fuels like methanol, and formic acid, can affect, by the crossover effect phenomenon, the micro-fuel cell performance, as indicated by the cell voltages data (electromotive force). The performance of low-temperature fuel cells is essentially focused on the cathode performance. Thus, we can read from Table 6.1, for platinum-based cathodes that the cell voltage spans between 0.6 and 0.62 V (methanol); 0.64 and 0.78 V (formic acid). For chalcogenide cathodes, the cell voltages are, namely on Ru_xSe_y 0.55 V (methanol), and 0.84 V (formic acid); on $CoSe_2$ 0.27 V (methanol); 0.49 V (formic acid); on Pt_xS_y 0.48 V (methanol), and on Pt_xSe_y 0.37 V (methanol). All chalcogenides framed in the Table are methanol tolerant cathode materials, in acid as well as in alkaline medium.

Aiming at optimizing the micro-laminar fuel cell performance, various designs (less- or more-sophisticated) have been considered, as shown in the first column of Table 6.1. The most common is the Y-shaped configuration, Fig. 6.1a. Figure 6.1c can be considered as Y-shaped with a self-breathing window for the cathode side. One notices that the diffusion mixing layer shortens at the end of the cell path, so that the crossover effect can take place. This cell can be easily converted to a

Table 6.1 Membrane-less micro-fuel cells systems

Micro-fuel cell design	Anode–cathode	Fuel/oxidant	[(a)]Cell voltage/V	[(a)]Current density @ P_{max} mA/cm^2	[(a)]P_{max} mW/cm^2	References
Membrane-less vanadium laminar flow cell	Au/graphite–Au/graphite	V(III)/V(II)/ V(V)/V(IV)	–	[(b)]35 (@ MCV of 1.1 V)	38	[23]
Y-shaped formic acid μFC	Pt/C–Pt/C	Formic Acid/O_2 + H_2SO_4	–	[(b)]0.4 (@ MCV of 0.4 V)	0.17	[4]
Y-shaped formic acid μFC–$KMnO_4$	Pt/C–Pt/C	Formic Acid/ H_2SO_4 + $KMnO_4$	–	[(b)]4.0 (@ MCV of 0.55 V)	2.4	[4]
Acidic Y-shaped membrane-less μFC	Unsupported Pt/Ru–Pt black	MeOH/O_2 + H_2SO_4	–	[(b)]6.5 (@ MCV of 0.41 V)	2.7	[24]
Alkaline Y-shaped membrane-less μFC	Unsupported Pt/Ru–Pt black	MeOH/O_2 + KOH	–	[(b)]3.8 (@ MCV of 0.56 V)	1.9	[24]
Mixed shaped membrane-less μFC	Unsupported Pt/Ru–Pt black	MeOH + KOH/O_2 + H_2SO_4	–	[(b)]43.8 (@ MCV of 0.26 V)	11.8	[24]
Planar membrane-less μFC	Pt/C–Pt/C	Formic Acid + H_2SO_4/ O_2 + H_2SO_4	0.64	–	0.18	[25]
Planar membrane-less dual electrolyte μFC	Pt/Ta–Pt/Ta	H_2 + KOH/O_2 + H_2SO_4	1.4	–	0.6	[26]
H_2O_2 dual electrolyte μFC	No data	H_2O_2 + NaOH/ H_2O_2 + H_2SO_4	0.7	–	24	[27]
Non-noble "Mixed-reactant" μFC	Ni hydroxide–Ag oxide	MeOH + H_2O_2 + KOH	–	[(b)]0.68 (@ MCV of 0.036 V)	[(*)]0.025	[28]
Orthogonal flow membrane-less μFC (1)	Pt black–Pt black	H_2 + H_2SO_4/O_2 + H_2SO_4	0.87	100	45	[29]
Orthogonal flow membrane-less μFC (2)	Nanoporous Pt–Nanoporous Au	MeOH + KOH/O_2 gas	0.6	152	46	[29]

(continued)

Table 6.1 (continued)

Micro-fuel cell design	Anode–cathode	Fuel/oxidant	[a]Cell voltage/V	[a]Current density @ P_{max} mA/cm^2	[a]P_{max} mW/cm^2	References
Methanol tolerant air-breathing μFC	PtRu black–Ru_xSe_y/carbon paper	5 M MeOH + H_2SO_4/ O_2 + H_2SO_4	0.55	–	3.86	[8]
Optimized planar μFC	Pd black–Pt black	Formic Acid + H_2SO_4/forced O_2 and H_2SO_4	0.85	170	55	[30]
Counter-flow μFC	Porous Carbon (E-tek)–Porous Carbon (E-tek)	V^{3+}/V^{2+}	1.27	9	5	[31]
Y-shaped μFC	Pd/MWCNT–Pt/C	0.1 M Formic Acid/ O_2 + H_2SO_4	0.9	8	3.3	[32]
Y-shaped metal chalcogenide μFC	Pd/C–Ru_xSe_y/C	0.5 M Formic Acid + H_2SO_4/ O_2 + H_2SO_4	0.84	11.4	1.9	[33]
Bridge-shaped μFC	Pd layer–No catalyst	Formic Acid + H_2SO_4/ $KMnO_4$ + H_2SO_4	1.35	29	25.5	[34]
Vanadium μFC stack	Toray Carbon Paper–Toray Carbon Paper	V^{2+} + H_2SO_4/VO_2^+ + H_2SO_4	1.1	17.5	7.5	[35]
Tolerant MeOH chalcogenides cathodes μFC (1)	PtRu/C–Pt_xS_y/C	MeOH + H_2SO_4/O_2 + H_2SO_4	0.48	26	4.0	[11]
Tolerant MeOH chalcogenides cathodes μFC (2)	PtRu/C–$CoSe_2$/C	MeOH + H_2SO_4/O_2 + H_2SO_4	0.27	2	0.23	[11]
Tolerant HCOOH chalcogenides cathodes μFC	Pd/C–$CoSe_2$/C	Formic Acid + H_2SO_4/ O_2 + H_2SO_4	0.49	6.2	1.04	[11]
μFC to evaluate bubbles formation	Pd/C–Pt/MWCNT's	Formic Acid + H_2SO_4/ O_2 + H_2SO_4	0.78	62	22.2	[36]
Ultra-thin polymer μFC	PtRu/C–Pt/C	MeOH + H_2SO_4/O_2 + H_2SO_4	0.62	63	10.9	[37]
Mixed-reactants μFC	PtRu/C–Pt_xSe_y/C	MeOH + H_2SO_4	0.37	19	3.7	[38]

(continued)

Table 6.1 (continued)

Micro-fuel cell design	Anode–cathode	Fuel/oxidant	[(a)]Cell voltage/V	[(a)]Current density @ P_{max} mA/cm^2	[(a)]P_{max} mW/cm^2	References
Two-phase flow μFC	Pt–Pt	H_2 bubbles and H_2SO_4/O_2 bubbles and H_2SO_4	0.88	16	3.8	[39]
Mixing promoters μFC	Cr–Pt	$AQ_2SH + H_2SO_4$/Iron sulfate + H_2SO_4	0.54	10.7	2.82	[40]
Tolerant MeOH chalcogenides cathodes μFC	PtRu/C–$CoSe_2$/NCNH	MeOH + KOH/O_2 + KOH	0.48	60	10.5	[14]
Non-precious μFC	Pt/C–CoSe/N-RGO	H_2/O_2 + KOH	0.9	150	49.7	[41]

Hatched zones refer to devices using chalcogenide cathode materials
[(a)]The cell voltage, current density, and cell power values were taken from the text or directly read from the graphics
[(b)]The current density values were taken at the maximum cell voltage (MCV)
(*) data obtained from the 10 μm design at a flow rate of 100 μL min^{-1}

mixed-reactant cell on condition that the cathode is tolerant to the fuel used, without affecting the oxygen reduction reaction (absence of mixed-potential).

6.2.3 Cathode Electrocatalysts and Fuel Tolerance

In Chap. 4, Sect. 4.2.2.2., the methanol crossover problem was addressed for the specific case of platinum electrodes. Sulfide [11, 42] and selenide [12, 43] coordinated surface atoms onto nanoparticulated platinum enhance the methanol tolerance during the oxygen reduction reaction (cf. Figs. 4.6b and 4.7). It turns out that a similar phenomenon was evaluated on Chevrel phase ($Mo_{4.2}Ru_{1.8}Se_8$) [44, 45], ruthenium selenide (Ru_xSe_y) [10, 46–50], and cobalt diselenide $CoSe_2$ materials [11, 51–54]. The formic acid fuel was also tested for tolerance, specifically on the Ru_xSe_y [10, 33, 50, 55] and CoSe [56] cathodes. Figure 6.2 summarizes the progress on cathode materials reported so far, as fuel tolerant chalcogenide materials for the micro-laminar flow fuel cell aiming at working in laminar or mixed-reactant mode. Some examples of the micro-direct formic acid flow fuel cell (μ-DFA-LF-FC), as well as the micro-direct methanol flow fuel cell (μ-DM-LF-FC) are discussed below.

6.2.3.1 Case Study of the Y-Shape Geometry

The performance comparison of μLFFCs, Fig. 6.1b, using as oxygen cathodes 20 wt%Ru_xSe_y/C (1.1 $mg_{Ru_xSe_y}/cm^2$), and 30 wt% Pt/C (1.1 mg_{Pt}/cm^2), is shown in Fig. 6.3. The formic acid oxidation at the anode for both systems was 30 wt% Pd/C (1.7 mg_{Pd}/cm^2). The anolyte and catholyte streams flowed at a rate of 0.1 and

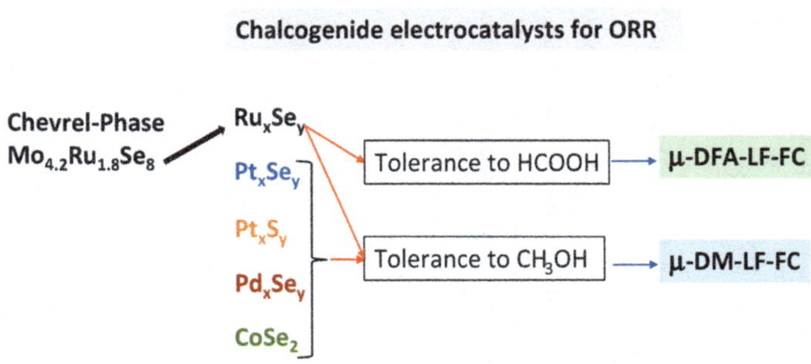

Fig. 6.2 A scheme showing at the origin a Chevrel phase compound (cf. Chap. 4, Sect. 4.2.2.1). The chalcogenide based on Pt, Pd, and Co metal centers investigated so far for the micro-fluidic fuel cell in laminar and mixed-reactant conditions. The earlier ORR investigation on the Chevrel phase led to the synthesis of Ru_xSe_y, with an extension to platinum, palladium, cobalt, rhodium, etc.

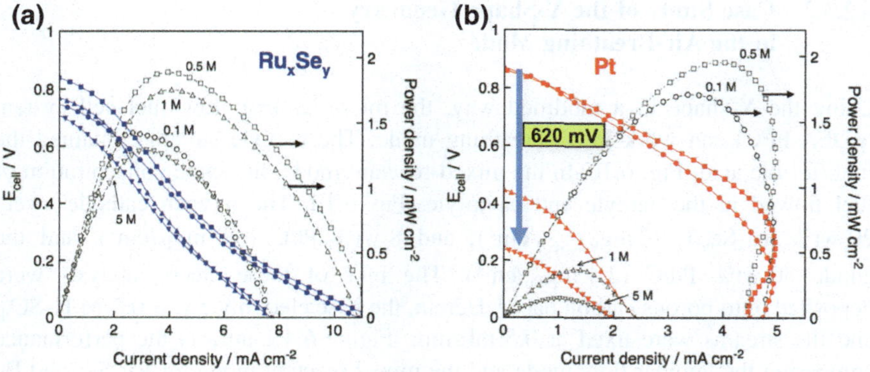

Fig. 6.3 Current-potential-power density curves in a Y-shaped LFFC using as cathodes **a** Ru_xSe_y/C (1.1 mg.cm^{-2}) and **b** Pt/C E-tek (1.1 mg.cm^{-2}). The anode for both systems was Pd/C E-tek (1.7 mg.cm^{-2}). At the anolyte, the flow rate was fixed at 1.2 mL min^{-1}, and at the catholyte 0.1 mL min^{-1}. The fuel at various concentrations was HCOOH (●) 0.1, (■) 0.5, (▲) 1, (▼) 5 M, at 25 °C. Figures (a) and (b) adapted from reference [33], with permission.

Table 6.2 Performance parameters on carbon supported Ru_xSe_y and Pt cathodes measured in a Y-shaped micro-laminar fuel cell

Ru_xSe_y				Pt		
HCOOH/M	Cell voltage/V	j_{max}/mA cm^{-2}	P_{max}/mW cm^{-2}	Cell voltage/V	j_{max}/mA cm^{-2}	P_{max}/mW cm^{-2}
0.1	0.77	7.73	1.43	0.87	4.73	1.71
0.5	0.84	11.44	1.89	0.86	4.44	1.97
1	0.75	10.89	1.76	0.46	2.40	0.31
5	0.71	7.73	1.30	0.24	1.95	0.14

Length of the channel of 45 mm, and a distance between electrodes of 2 mm. Base electrolyte 0.5 M H_2SO_4. Fuel stream: HCOOH; Oxidant stream: saturated Oxygen, at RT

1.2 mL/min, respectively. With the increase of the formic acid concentration from 0.1 to 0.5 M, there is an increase of the performance for both systems, where a maximum power density of 1.89 and 1.97 mW/cm^2 for Ru_xSe_y and Pt, respectively, is attained. Under this circumstance, the crossover negative effect is absent. Moreover, the micro-fuel cells' performance drops dramatically with the formic acid concentration of 1 to 5 M for Pt due to the diffusion of the anolyte in the diffusion mixing layer interface—crossover effect. As a result of this phenomenon, the cell voltage decreases by 0.62 V. This negative effect encountered at Pt is relatively negligible at the chalcogenide Ru_xSe_y cathode material, so that at the highest formic acid concentration the performance of the chalcogenide is 9.3 times that of Pt. Thus, this fact testifies the high tolerance to formic acid, as summarized in Table 6.2.

6.2.3.2 Case Study of the Y-Shape Geometry in the Air-Breathing Mode

Using the Y-shape in a modified way, the micro-laminar flow fuel cell system (μDFA-FFC) can work in air-breathing mode. The anolyte stream contained the fuel: formic acid, Fig. 6.1c. In the mixed-reactant mode, the same concentration of fuel flowed in the anolyte and catholyte, Fig. 6.1d. The oxygen cathodes were 20 wt% Ru_xSe_y/C ($2\,mg_{Ru_xSe_y}/cm^2$), and 8 wt% Pt/C ($0.8\,mg_{Pt}/cm^2$), and the anode 30 wt% Pd/C ($2.4\,mg_{Pd}/cm^2$). The inks of these electrocatalysts were deposited onto porous carbon paper. Herein, the base electrolyte was 0.5 M H_2SO_4, and the streams were fixed at 0.5 mL/min. Figure 6.4 contrasts the performance comparing the laminar flow mode and the mixed-reactant mode for Ru_xSe_y and Pt, respectively, using a fuel concentration of 5 M, Table 6.3. As discussed above in Sect. 6.2.3.1, the tolerance to formic acid of Ru_xSe_y is remarkable, since the maximum power density (P_{max}) drops 5%, whereas with Pt the P_{max} drops 35%. The MRFFC approach suggests a stream oversimplification in the system.

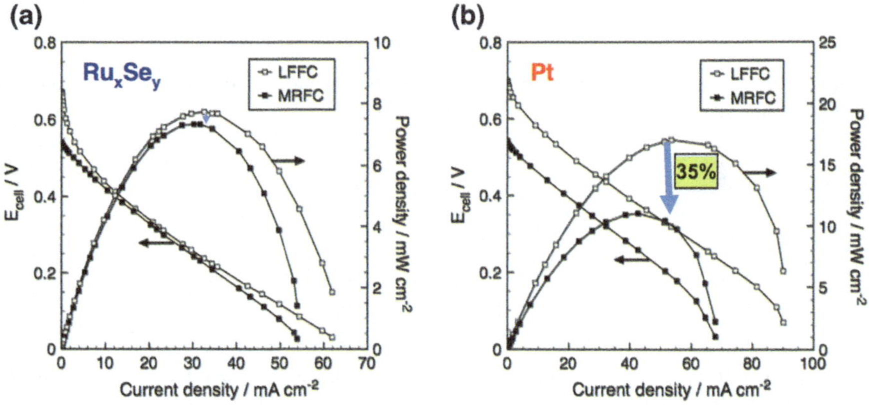

Fig. 6.4 Current-potential-power density curves in a self-breathing Y-shaped MRFC using as cathodes **a** Ru_xSe_y/C (2 mg cm^{-2}) and **b** Pt/C E-tek (0.8 mg cm^{-2}) using as fuel 5 M HCOOH, at 25 °C. The flow rate was fixed at 0.5 mL min^{-1}. Pd/C E-tek (2.4 mg.cm^{-2}) served as anode. Figures adapted from references [57], courtesy of Aldo S. Gago

Table 6.3 Performance parameters on carbon-supported Ru_xSe_y and Pt cathodes measured in a Y-shaped cell in the LFFC and MRFFC mode

Ru_xSe_y				Pt		
FC-Mode	Cell voltage/V	j_{max}/ mA cm^{-2}	P_{max}/mW cm^{-2}	Cell voltage/V	j_{max}/ mA cm^{-2}	P_{max}/mW cm^{-2}
LFFC	0.67	61.9	7.7	0.7	90.2	17.1
MRFFC	0.54	53.9	7.3	0.54	67.9	11.1

Length of the channel of 28 mm, and a distance between electrodes of 1.2 mm. Base electrolyte 0.5 M H_2SO_4. Fuel stream: 5 M HCOOH; Oxidant stream: Pure oxygen (1Bar), at RT

The data in Tables 6.2 and 6.3 clearly identify the performance improvement of the electrode materials. The use of (i) a porous separator between the electrodes (Millipore GSWP 0.22 μm), (ii) a porous carbon support for the electrocatalyst, and (iii) pure oxygen at the air-breathing side of cathode are positive factors for LFFC and MRFFC devices, Fig. 6.1c, d [57]. The ruthenium chalcogenide, Ru_xSe_y, is additionally tolerant to methanol [33, 46, 58–64]. The selectivity of ruthenium-based chalcogenides compared to platinum-based materials is due to the higher density of states near the Fermi level favoring the formation of Ru–OH species at more negative potential than platinum [65]. It is important to recall that on ruthenium oxide-like surface, the interaction of carbon monoxide is less compared to a metallic surface [66]. On the other hand, as reported by the density functional theory (DFT) calculations [67], the tolerance to methanol is attributed to a high dehydrogenation energy barrier on metal chalcogenides.

6.2.3.3 Multi-channel Geometry in the μFFFC

The study of cathode materials selectivity properties was further pursued via a multi-channel micro-fuel cell working as a micro-laminar fuel cell (μLFC), and as micro-mixed-reactant laminar fuel cell (μMR-LFC).

Figure 6.5 shows the fabrication, and assembly process of this kind of air-breathing fuel cell to measure either in the laminar flow fuel cell, provided by a nanoporous separator (3)–(5)–(7), or in the mixed-reactant flow fuel cell (4)–(6)–(7) mode. The cell in (7) shows the window for the air self-breathing mode at the cathode. The nanoporous separator in (3) placed between the SU-8 structures served

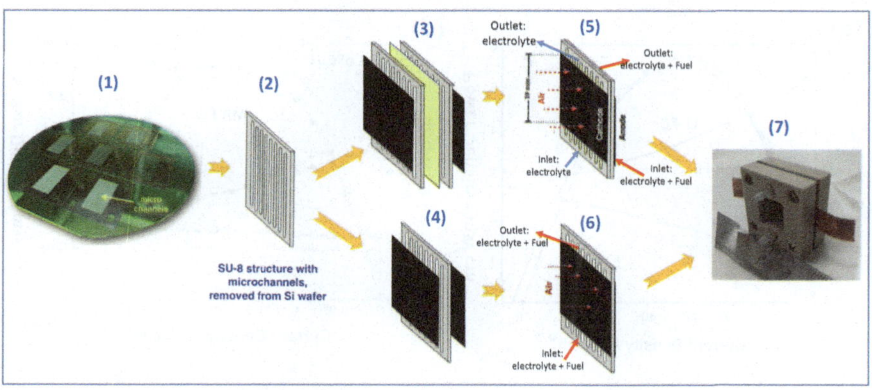

Fig. 6.5 Process of fabrication of the micro-fluidic fuel cell with micro-channels. (1): UV-print on silicon wafer; (2) the micro-channels (each channel was 250 mm high, 750 mm wide, and 20 mm long) with varying channel dimensions; (3) (5) the electrodes laminar flow fuel cell assembly with a nanoporous separator (polycarbonate filter Cyclopore, 0.1 mm pore size); (4) (6) the electrodes mixed-reactant assembly; (7) the micro-fuel cell assembly provided with two sheets of copper, in contact with the electrodes, as current collectors

to prevent the mixing of the streams in the micro-channels [11, 36]. With this arrangement, various transition metal chalcogenides were investigated, namely Pt_xS_y [11, 42], Pt_xSe_y [12, 43], Pd_xSe_y [68], $CoSe_2$ [11, 69], toward the methanol tolerance. The micro-fuel cell performance in the laminar flow with the separator and under the mixed-reactant conditions is summarized in Table 6.4.

The coordinated chalcogen (S, Se) on Pt, Pd enhances the methanol tolerance with respect to the only metal center in the mixed-reactant working conditions. Judging by the values of P_{max}, this parameter drops by 80% on the metal centers, whereas the methanol poisoning effect is significantly minimized by the Se, and S coordination onto the surface of Pt, and palladium, where the P_{max} drop attains at the most 30%. Figure 6.6 contrasts exemplarily the impact of the crossover effect of carbon supported Pt_xS_y, Pt_xSe_y, Pd_xSe_y, and $CoSe_2$, respectively. On this latter, it is worth to note the valuable effect of the support. Indeed, the nitrogen-doping and morphology of carbon (e.g., carbon nanohorns (CNH)) seem to be a primordial ingredient to favor the metal support interaction. It is also to be noted that the

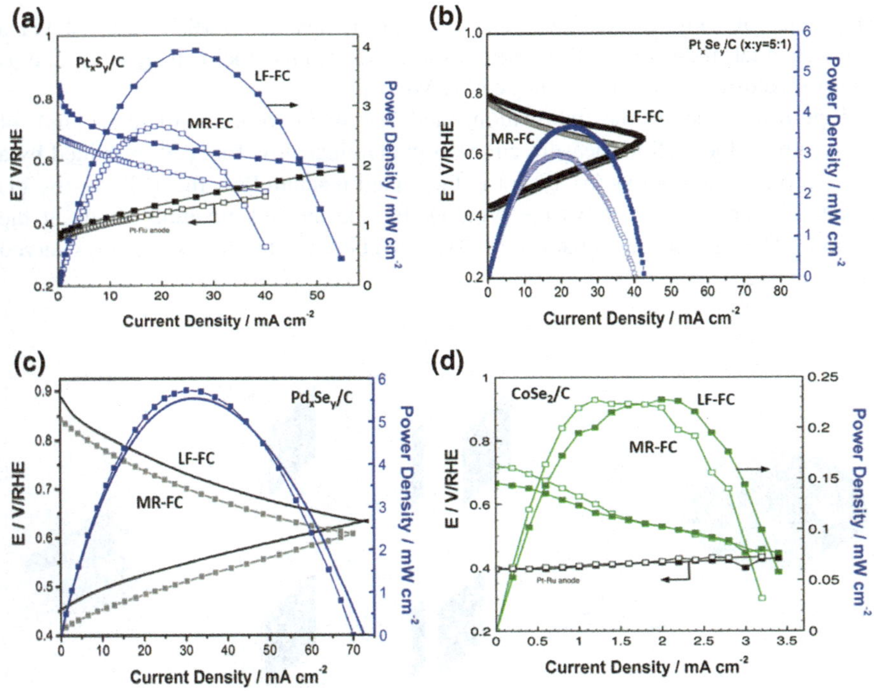

Fig. 6.6 Current-potential-power density curves in a self-breathing LFFC and MRFC using as cathodes: **a** Pt_xS_y/C, **b** Pt_xSe_y/C, **c** Pd_xSe_y/C, and **d** $CoSe_2$/C. Figures (a), (d) adapted from reference [11], with permission. Copyright © 2012 WILEY-VCH Verlag GmbH & Co. KGaA, Weinheim. Figure (b) adapted from reference [12], with permission. Copyright © 2014, Royal Society of Chemistry. Figure (c) adapted from reference [68]. Courtesy of JM Mora-Hernández. The anode on these systems was PtRu/C using as fuel 5 M CH_3OH at 25 °C. The flow rate was ca. 3.4 mL min^{-1}

micro-fuel cell device Fig. 6.5 (in position 7) allows to record the fuel (5 M methanol) oxidation at the anode (PtRu/C), as well as the impact (not shown here) of the fuel concentration.

6.2.4 The H_2/O_2 Micro-fuel Cell Platform

The micro-fuel cell system can serve as a H_2/O_2 platform tool to test the electrocatalytic properties on novel materials in real conditions using acid or alkaline streams, without the need of a membrane, required in a classical fuel cell system. Platinum-based electrodes in alkaline [70, 71] and acid [72–74] media have been investigated to assess the proof of concept. It was determined that in alkaline medium the performance of hydrogen-fueled room-temperature alkaline fuel cell is limited by the carbonate species present in the electrolyte [71] with less impact at the cathode side. It is known that the carbonate formation (CO_3^{2-}/HCO_3^-) reduces the OH^- concentration, with a concomitant reduction of the electrolyte conductivity, thus affecting the reaction kinetics, a problem in the traditional alkaline fuel cells (AFCs), that are limited to the use of high purity H_2 and O_2 gas streams with ultra-low CO_2 concentration. In the micro-fuel cell, however, the flowing electrolyte stream can be used as an effective means to prevent carbonate species accumulation.

6.2.4.1 H_2/O_2 μFFFC

Besides the device presented in Fig. 6.5, a H_2/O_2 functional system to work in acid or alkaline media is shown in Fig. 6.7a to evaluate the performance of the anode and the cathode with a reference electrode connected to the outlet of the system (not shown). A case study performed with a cobalt chalcogen (hexagonal phase structure) cathode in alkaline medium is presented, and, again, compared to platinum electrode. The study highlights the following aspects: The versatility of using the membrane-less device as a platform and to evaluate the effect of the catalytic center support (cf. Chap. 5). The hexagonal phase of cobalt selenide (CoSe) is the example shown in Fig. 6.7b, c. This electrocatalytic material was synthesized, in presence of the carbonaceous support, a solution containing $CoSO_4 \cdot 7H_2O$ and SeO_2. The reduction was completed by adding $NaBH_4$, powder filtered, dried, and heat treated at 600 °C. It was assessed by XRD measurements that the crystal phase structure was the hexagonal (Freboldite) material [41]. The carbonaceous materials used as supports were amorphous carbon (C), N–C and doped with 25% Polypyrrole, heat treated at 1000 °C (NC25(1000)); N-RGO and doped with 25% Polypyrrole, heat treated at 1000 °C (NG25(1000)). The measurements in alkaline medium using the H_2/O_2 micro-fuel cell indicate that the support effect is beneficial to enhance the ORR electrocatalytic activity of CoSe, Fig. 6.7b. The highest performance is obtained on CoSe/NG25(1000). The cathode and anode characteristics

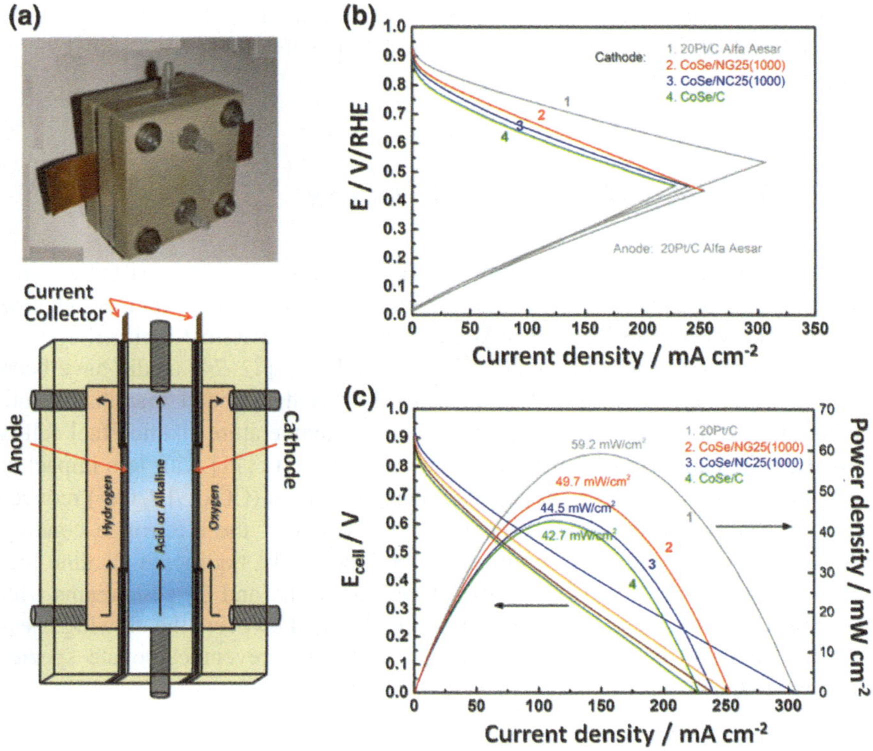

Fig. 6.7 **a** The H_2/O_2 micro-laminar fuel cell for acid or alkaline electrolytes. **b** The cathode and anode performance in H_2/O_2 micro-laminar fuel cell in 3 M KOH. (**c**) Cell voltage and power density performance deduced from (**b**). Figures (b) and (c) adapted and reprinted from reference [41]

translated into the micro-fuel cell performance, clearly deliver an enhancement of the power density of ca. 11% of the catalyst supported on NG25(1000) with respect to the NC25(1000) support. The results depicted in Fig. 6.7b, c clearly indicate that the specific morphology of the carbon supports favors the catalytic center coordination favorable to the catalytic center-support interaction discussed in Chap. 5.

6.3 Micro-fluidic Fuel Cells—What Else?

The sections above, in this chapter, were essentially devoted to the progress done in the use of chalcogenide materials in micro-fuel cell systems. The advantage, of this kind of materials (precious or non-precious centers), is their enhanced activity to perform the ORR reactions in presence of a fuel (formic acid, methanol). It is

Table 6.4 Micro-channel laminar micro-fuel cell systems

Cathode	Cell voltage/V	j_{max}/mA cm^{-2}	P_{max}/mW/cm^2	References
Pt/C	0.50	66	5.7	[12]
	0.19	27.7	1.2	
Pt_xS_y/C	0.49	54.2	3.9	[11]
	0.32	40	2.7	
Pt_xSe_y/C	0.39	44	3.7	[12]
	0.37	43.3	3	
Pd/C	0.49	138	13.2	[68]
	0.30	84.6	5	
Pd_xSe_y/C	0.63	72	5.7	
	0.43	69.8	5.6	
$CoSe_2$/C	0.32	3.2	0.29	[11]
	0.27	3.4	0.29	
$CoSe_2$/CNH	0.42	79.28	4.53	[14]
$CoSe_2$/NCNH	0.48	124.2	10.04	

Hatched zones refer to the mixed-reactant working mode using a 5 M CH_3OH

certainly important to point out that progress in the development of micro-fuel cells is being addressed to develop further anode–cathode micro-fuel cells based on non-precious metal center electrocatalysts. Additionally, the search of bifunctional electrocatalyst will permit to manufacture unitized micro-systems by means of a similar device depicted in Fig. 6.7a [13].

To the best knowledge, since the first report on chemical energy converting devices, in the microscale using the fluidic dynamics [23], research and development in this field of micro-fluidic fuel cell and battery technology for electrochemical energy conversion and storage has progressed. In this sense, various books [75, 76] and reviews have been devoted; e.g., to the development of micro-machined membrane-based fuel cells [77], micro-fluidic microbial fuel cells [78], and chalcogenides and portable devices [50, 79, 80].

References

1. Sabaté N, Esquivel JP, Santander J, Hauer JG, Verjulio RW, Gràcia I, Salleras M, Calaza C, Figueras E, Cané C, Fonseca L (2014) New approach for batch microfabrication of silicon-based micro fuel cells. Microsyst Technol 20(2):341–348. https://doi.org/10.1007/s00542-013-1781-4
2. Esquivel JP, Sabaté N, Santander J, Torres-Herrero N, Gràcia I, Ivanov P, Fonseca L, Cané C (2009) Influence of current collectors design on the performance of a silicon-based passive micro direct methanol fuel cell. J Power Sour 194(1):391–396. https://doi.org/10.1016/j.jpowsour.2009.04.065

3. Wang Y, Pham L, Vasconcellos GPSd, Madou M (2010) Fabrication and characterization of micro PEM fuel cells using pyrolyzed carbon current collector plates. J Power Sour 195 (15):4796–4803. https://doi.org/10.1016/j.jpowsour.2010.02.050
4. Choban ER, Markoski LJ, Wieckowski A, Kenis PJA (2004) Microfluidic fuel cell based on laminar flow. J Power Sour 128(1):54–60. https://doi.org/10.1016/j.jpowsour.2003.11.052
5. Choban ER, Waszczuk P, Kenis PJA (2005) Characterization of limiting factors in laminar flow-based membraneless microfuel cells. Electrochem Solid-State Lett 8(7):A348–A352
6. Jayashree RS, Gancs L, Choban ER, Primak A, Natarajan D, Markoski LJ, Kenis PJA (2005) Air-breathing laminar flow-based microfluidic fuel cell. J Am Chem Soc 127(48):16758–16759
7. Jayashree RS, Egas D, Spendelow JS, Natarajan D, Markoski LJ, Kenis PJA (2006) Air-breathing laminar flow-based direct methanol fuel cell with alkaline electrolyte. Electrochem Solid-State Lett 9(5):A252–A256. https://doi.org/10.1149/1.2185836
8. Whipple DT, Jayashree RS, Egas D, Alonso-Vante N, Kenis PJA (2009) Ruthenium cluster-like chalcogenide as a methanol tolerant cathode catalyst in air-breathing laminar flow fuel cells. Electrochim Acta 54(18):4384–4388
9. Gago AS, Timperman L, Alonso-Vante N (2011) Ru_xSe_y cluster-like as cathode catalyst in a formic acid laminar flow fuel cell. Paper presented at the European Fuel Cell Forum 2011, Kultur- und Kongresszentrum Luzern Lucerne/Switzerland, July 12
10. Timperman L, Gago AS, Alonso-Vante N (2011) Oxygen reduction reaction increased tolerance and fuel cell performance of Pt and Ru_xSe_y onto oxide-carbon composites. J Power Sour 196(9):4290–4297
11. Gago AS, Gochi-Ponce Y, Feng Y-J, Esquivel JP, Sabaté N, Santander J, Alonso-Vante N (2012) Tolerant chalcogenide cathodes of membraneless micro fuel cells. Chemsuschem 5(8):1488–1494. https://doi.org/10.1002/cssc.201200009
12. Ma J, Habrioux A, Morais C, Alonso-Vante N (2014) Electronic modification of Pt via Ti and Se as tolerant cathodes in air-breathing methanol microfluidic fuel cells. Phys Chem Chem Phys 16:13820–13826. https://doi.org/10.1039/c3cp54564d
13. Zhong H, Tian R, Gong X, Li D, Tang P, Alonso-Vante N, Feng Y (2017) Advanced bifunctional electrocatalyst generated through cobalt phthalocyanine tetrasulfonate intercalated Ni_2Fe-layered double hydroxides for a laminar flow unitized regenerative micro-cell. J Power Sour 361:21–30. https://doi.org/10.1016/j.jpowsour.2017.06.057
14. Unni SM, Mora-Hernandez JM, Kurungot S, Alonso-Vante N (2015) $CoSe_2$ supported on nitrogen-doped carbon nanohorns as a methanol-tolerant cathode for air-breathing microlaminar flow fuel cells. ChemElectroChem 2(9):1339–1345. https://doi.org/10.1002/celc.201500154
15. Xuan J, Wang H, Leung DYC, Leung MKH, Xu H, Zhang L, Shen Y (2013) Theoretical Graetz-Damköhler modeling of an air-breathing microfluidic fuel cell. J Power Sour 231:1–5. https://doi.org/10.1016/j.jpowsour.2012.12.090
16. Kjeang E, Djilali N, Sinton D (2009) Microfluidic fuel cells: A review. J Power Sour 186 (2):353–369
17. Kjeang E, Michel R, Harrington DA, Djilali N, Sinton D (2008) A microfluidic fuel cell with flow-through porous electrodes. J Am Chem Soc 130(12):4000–4006
18. Jayashree RS, Mitchell M, Natarajan D, Markoski LJ, Kenis PJA (2007) Microfluidic hydrogen fuel cell with a liquid electrolyte. Langmuir 23(13):6871–6874
19. Jayashree RS, Gancs L, Choban ER, Primak A, Natarajan D, Markoski LJ, Kenis PJA (2005) Air-breathing laminar flow-based microfluidic fuel cell. J Am Chem Soc 127(48):16758–16759
20. Hollinger AS, Maloney RJ, Markoski LJ, Kenis PJA (2008) Nanoporous separator to minimize fuel crossover in a direct methanol laminar flow fuel cell. In: 214th ECS Meeting, Abstract #701, © The Electrochemical Society, 2008
21. Hollinger AS, Maloney RJ, Jayashree RS, Natarajan D, Markoski LJ, Kenis PJA (2010) Nanoporous separator and low fuel concentration to minimize crossover in direct methanol laminar flow fuel cells. J Power Sour 195(11):3523–3528
22. Sun F, He H, Huo W (2015) Polymer separator and low fuel concentration to minimize crossover in microfluidic direct methanol fuel cells. Int J Energy Res 39(5):643–647

23. Ferrigno R, Stroock AD, Clark TD, Mayer M, Whitesides GM (2002) Membraneless vanadium redox fuel cell using laminar flow. J Am Chem Soc 124(44):12930–12931. https://doi.org/10.1021/ja020812q
24. Choban ER, Spendelow JS, Gancs L, Wieckowski A, Kenis PJA (2005) Membraneless laminar flow-based micro fuel cells operating in alkaline, acidic, and acidic/alkaline media. Electrochim Acta 50(27):5390–5398. https://doi.org/10.1016/j.electacta.2005.03.019
25. Cohen JL, Westly DA, Pechenik A, Abruña HD (2005) Fabrication and preliminary testing of a planar membraneless microchannel fuel cell. J Power Sour 139(1):96–105. https://doi.org/10.1016/j.jpowsour.2004.06.072
26. Cohen JL, Volpe DJ, Westly DA, Pechenik A, Abruña HD (2005) A dual electrolyte H_2/O_2 planar membraneless microchannel fuel cell system with open circuit potentials in excess of 1.4 V. Langmuir 21(8):3544–3550. https://doi.org/10.1021/la0479307
27. Chen F, Chang M-H, Hsu C-W (2007) Analysis of membraneless microfuel cell using decomposition of hydrogen peroxide in a Y-shaped microchannel. Electrochim Acta 52 (25):7270–7277. https://doi.org/10.1016/j.electacta.2007.05.072
28. Sung W, Choi J-W (2007) A membraneless microscale fuel cell using non-noble catalysts in alkaline solution. J Power Sour 172(1):198–208. https://doi.org/10.1016/j.jpowsour.2007.07.012
29. Hayes JR, Engstrom AM, Friesen C (2008) Orthogonal flow membraneless fuel cell. J Power Sour 183(1):257–259. https://doi.org/10.1016/j.jpowsour.2008.04.061
30. Jayashree RS, Yoon SK, Brushett FR, Lopez-Montesinos PO, Natarajan D, Markoski LJ, Kenis PJA (2010) On the performance of membraneless laminar flow-based fuel cells. J Power Sour 195(11):3569–3578
31. Salloum KS, Posner JD (2010) Counter flow membraneless microfluidic fuel cell. J Power Sour 195(19):6941–6944. https://doi.org/10.1016/j.jpowsour.2010.03.096
32. Morales-Acosta D, Rodríguez GH, Godinez LA, Arriaga LG (2010) Performance increase of microfluidic formic acid fuel cell using Pd/MWCNTs as catalyst. J Power Sour 195(7):1862–1865
33. Gago AS, Morales-Acosta D, Arriaga LG, Alonso-Vante N (2011) Carbon supported ruthenium chalcogenide as cathode catalyst in a microfluidic formic acid fuel cell. J Power Sour 196(3):1324–1328
34. López-Montesinos PO, Yossakda N, Schmidt A, Brushett FR, Pelton WE, Kenis PJA (2011) Design, fabrication, and characterization of a planar, silicon-based, monolithically integrated micro laminar flow fuel cell with a bridge-shaped microchannel cross-section. J Power Sour 196(10):4638–4645
35. Salloum KS, Posner JD (2011) A membraneless microfluidic fuel cell stack. J Power Sour 196(3):1229–1234
36. Ma J, Gago AS, Alonso-Vante N (2013) Performance study of platinum nanoparticles supported onto MWCNT in a formic acid microfluidic fuel cell system. J Electrochem Soc 160(8):F859–F866. https://doi.org/10.1149/2.101308jes
37. Hollinger AS, Kenis PJA (2013) Manufacturing all-polymer laminar flow-based fuel cells. J Power Sour 240:486–493. https://doi.org/10.1016/j.jpowsour.2013.04.053
38. Ma J, Habrioux A, Morais C, Alonso-Vante N (2014) Electronic modification of Pt via Ti and Se as tolerant cathodes in air-breathing methanol microfluidic fuel cells. Phys Chem Chem Phys 16(27):13820–13826. https://doi.org/10.1039/C3CP54564D
39. Hashemi SMH, Neuenschwander M, Hadikhani P, Modestino MA, Psaltis D (2017) Membrane-less micro fuel cell based on two-phase flow. J Power Sour 348:212–218. https://doi.org/10.1016/j.jpowsour.2017.02.079
40. Marschewski J, Ruch P, Ebejer N, Huerta Kanan O, Lhermitte G, Cabrol Q, Michel B, Poulikakos D (2017) On the mass transfer performance enhancement of membraneless redox flow cells with mixing promoters. Int J Heat Mass Transf 106:884–894. https://doi.org/10.1016/j.ijheatmasstransfer.2016.10.030
41. García-Rosado IJ, Uribe-Calderón J, Alonso-Vante N (2017) Nitrogen-doped reduced graphite oxide as a support for CoSe electrocatalyst for oxygen reduction reaction in alkaline media. J Electrochem Soc 164(6):F658–F666. https://doi.org/10.1149/2.1531706jes

42. Gochi-Ponce Y, Alonso-Nunez G, Alonso-Vante N (2006) Synthesis and electrochemical characterization of a novel platinum chalcogenide electrocatalyst with an enhanced tolerance to methanol in the oxygen reduction reaction. Electrochem Commun 8(9):1487–1491
43. Ma J, Canaff C, Alonso-Vante N (2013) The effect of tuning and origin of tolerance to organics of platinum catalytic centers modified by selenium. physica status solidi (a) 211 (9):2030–2034. https://doi.org/10.1002/pssa.201330148
44. Alonso-Vante N, Schubert B, Tributsch H (1989) Transition metal cluster materials for multi-electron transfer catalysis. Mater Chem Phys 22(3–4):281–307
45. Alonso-Vante N (2003) Chevrel phase and cluster-like chalcogenide materials. In: Vielstich W, Lamm A, Gasteiger H (eds) Handbook of fuel cells, vol 2. Wiley, Chichester, pp 534–543
46. Feng Y, Gago AS, Timperman L, Alonso-Vante N (2010) Chalcogenide metal centers for oxygen reduction reaction: activity and tolerance. Electrochim Acta 56(3):1009–1022
47. Montiel M, García-Rodríguez S, Hernández-Fernández P, Díaz R, Rojas S, Fierro JLG, Fatás E, Ocón P (2010) Relevance of the synthesis route of Se-modified Ru/C as methanol tolerant electrocatalysts for the oxygen reduction reaction. J Power Sour 195(9):2478–2487
48. Choi J-H, Johnston CM, Cao D, Babu PK, Zelenay P (2011) Se-modified Ru nanoparticles as ORR catalysts: Part 2: evaluation for use as DMFC cathodes. J Electroanal Chem 662(1):267–273. https://doi.org/10.1016/j.jelechem.2011.07.029
49. Neergat M, Gunasekar V, Singh RK (2011) Oxygen reduction reaction and peroxide generation on Ir, Rh, and their selenides—a comparison with Pt and RuSe. J Electrochem Soc 158(9):B1060–B1066. https://doi.org/10.1149/1.3604744
50. Gago AS, Luo Y, Alonso-Vante N (2017) Chalcogenide electrocatalysts for energy conversion fuel cell. In: Reference module in chemistry, molecular sciences and chemical engineering. Elsevier. https://doi.org/10.1016/B978-0-12-409547-2.13339-6
51. Feng YJ, He T, Alonso-Vante N (2010) Carbon-supported $CoSe_2$ nanoparticles for oxygen reduction reaction in acid medium. Fuel Cells 10(1):77–83
52. Feng Y, Alonso-Vante N (2012) Carbon-supported cubic $CoSe_2$ catalysts for oxygen reduction reaction in alkaline medium. Electrochim Acta 72:129–133. https://doi.org/10.1016/j.electacta.2012.04.003
53. Yu B, Jin J, Wu H, Wang S, Xia Q, Liu H (2017) Iron and nickel doped $CoSe_2$ as efficient non precious metal catalysts for oxygen reduction. Int J Hydrogen Energy 42(1):236–242. https://doi.org/10.1016/j.ijhydene.2016.10.052
54. Verjulio RW, Santander J, Ma J, Alonso-Vante N (2016) Selective $CoSe_2$/C cathode catalyst for passive air-breathing alkaline anion exchange membrane μ-direct methanol fuel cell (AEM-μDMFC). Int J Hydrogen Energy 41(43):19595–19600. https://doi.org/10.1016/j.ijhydene.2016.01.132
55. Gago AS, Arriaga LG, Gochi-Ponce Y, Feng YJ, Alonso-Vante N (2010) Oxygen reduction reaction selectivity of Ru_xSe_y in formic acid solutions. J Electroanal Chem 648(1):78–84
56. Nekooi P, Akbari M, Amini MK (2010) CoSe nanoparticles prepared by the microwave-assisted polyol method as an alcohol and formic acid tolerant oxygen reduction catalyst. Int J Hydrogen Energy 35(12):6392–6398. https://doi.org/10.1016/j.ijhydene.2010.03.134
57. Gago AS (2011) Développement de systèmes et matériaux pour la conversion de l'énergie chimique et lumineuse. University of Poitiers, Poitiers, France, PhD
58. Wippermann K, Richter B, Klafki K, Mergel J, Zehl G, Dorbandt I, Bogdanoff P, Fiechter S, Kaytakoglu S (2007) Carbon supported Ru-Se as methanol tolerant catalysts for DMFC cathodes. Part II: preparation and characterization of MEAs. J Appl Electrochem 37(12):1399–1411

59. Zehl G, Bogdanoff P, Dorbandt I, Fiechter S, Wippermann K, Hartnig C (2007) Carbon supported Ru–Se as methanol tolerant catalysts for DMFC cathodes. Part I: Preparation and characterization of catalysts. J Appl Electrochem 37(12):1475–1484
60. Colmenares L, Jusys Z, Behm RJ (2007) Activity, selectivity, and methanol tolerance of Se-modified Ru/C cathode catalysts. J Phys Chem C 111(3):1273–1283
61. Cheng H, Yuan W, Scott K (2007) Influence of thermal treatment on RuSe cathode materials for direct methanol fuel cells. Fuel Cells 7(1):16–20
62. Cheng H, Yuan W, Scott K, Browning DJ, Lakeman JB (2007) Evaluation of carbon-supported ruthenium-selenium-tungsten catalysts for direct methanol fuel cells. J Power Sour 172(2):597–603
63. Cheng H, Yuan W, Scott K, Browning DJ, Lakeman JB (2007) The catalytic activity and methanol tolerance of transition metal modified-ruthenium-selenium catalysts. Appl Catal B: Environ 75(3–4):221–228
64. Papageorgopoulos DC, Liu F, Conrad O (2007) A study of Rh_xS_y/C and $RuSe_x/C$ as methanol-tolerant oxygen reduction catalysts for mixed-reactant fuel cell applications. Electrochim Acta 52(15):4982–4986
65. Alonso-Vante N, Bogdanoff P, Tributsch H (2000) On the origin of the selectivity of oxygen reduction of ruthenium-containing electrocatalysts in methanol-containing electrolyte. J Catal 190(2):240–246
66. Boucher AC, Le Rhun V, Hahn F, Alonso-Vante N (2003) The CO-adsorbate electrooxidation on ruthenium cluster-like materials. J Electroanal Chem 554–555(1):379–384
67. Tritsaris GA, Nørskov JK, Rossmeisl J (2011) Trends in oxygen reduction and methanol activation on transition metal chalcogenides. Electrochim Acta 56(27):9783–9788
68. Mora-Hernández JM (2015) Desarrollo de electrocatalizadores a base paladio-calcogenuros de metales de transición y su evaluación electroquímica en reacciones redox de transferencia multi-electrónica para su utilización en pilas micro-fluíidicas a flujo laminar. PhD, Instituto Politecnico Nacional, CD Mexico, Mexico
69. Sreekuttan MU, Mora-Hernandez JM, Luo Y, Alonso-Vante N (2015) Substrate effects on the catalytic center of $CoSe_2$ for oxygen reduction reaction. ECS Trans 64(36):1–9. https://doi.org/10.1149/06436.0001ecst
70. Brushett FR, Zhou W-P, Jayashree RS, Kenis PJA (2009) Alkaline microfluidic hydrogen-oxygen fuel cell as a cathode characterization platform. J Electrochem Soc 156 (5):B565–B571
71. Brushett FR, Naughton MS, Ng JWD, Yin L, Kenis PJA (2012) Analysis of Pt/C electrode performance in a flowing-electrolyte alkaline fuel cell. Int J Hydrogen Energy 37(3):2559–2570. https://doi.org/10.1016/j.ijhydene.2011.10.078
72. Jhong H-R, Brushett FR, Yin L, Stevenson DM, Kenis PJA (2012) Combining structural and electrochemical analysis of electrodes using micro-computed tomography and a microfluidic fuel cell. J Electrochem Soc 159(3):B292–B298. https://doi.org/10.1149/2.033203jes
73. Luo Y, Calvillo L, Daiguebonne C, Daletou MK, Granozzi G, Alonso-Vante N (2016) A highly efficient and stable oxygen reduction reaction on Pt/CeOx/C electrocatalyst obtained via a sacrificial precursor based on a metal-organic framework. Appl Catal B: Environ 189:39–50. https://doi.org/10.1016/j.apcatb.2016.02.028
74. Luo Y, Kirchhoff B, Fantauzzi D, Calvillo L, Estudillo-Wong LA, Granozzi G, Jacob T, Alonso-Vante N (2018) Molybdenum-doping augments platinum-copper oxygen reduction electrocatalyst. Chemsuschem 11:193–201. https://doi.org/10.1002/cssc.201701822
75. Zhao T (2009) Micro fuel cells. principles and applications, 1st edn. Academic Press - Elsevier, Burlington, MA, USA; San Diego, California, USA; London UK
76. Kjeang E (2014) Microfluidic fuel cells and batteries. Springer International Publishing, Cham Heidelberg New York Dordrecht London

77. Nguyen N-T, Chan SH (2006) Micromachined polymer electrolyte membrane and direct methanol fuel cells—a review. J Micromech Microeng 16(4):R1
78. Yang Y, Ye D, Li J, Zhu X, Liao Q, Zhang B (2016) Microfluidic microbial fuel cells: from membrane to membrane free. J Power Sour 324:113–125. https://doi.org/10.1016/j.jpowsour.2016.05.078
79. Mora-Hernández J, Luo Y, Alonso-Vante N (2016) What can we learn in electrocatalysis, from nanoparticulated precious and/or non-precious catalytic centers interacting with their support? Catalysts 6(9):145
80. Mora-Hernández JM, Alonso-Vante N (2018) Membraneless micro fuel cell designs for portable applications. In: Ferreira-Aparicio P, Chaparro AM (eds) Portable hydrogen energy systems. Fuel cells and storage fundamentals and applications. Academic Press-Elsevier (ISBN 9780128131282)

Chapter 7
Outlook

Since the publication by Grove and Schönbein, in the mid-XIX century, of a gas-based energy converter device, nowadays known as fuel cell, platinum has been the electrode material of choice. Although this material is hampered by scarcity and hence high price, the material, as seen by the number of publications throughout the years, is still used a model to understand the physical–chemical basis of surface electrochemistry for many reactions of technological interest, as succinctly described in Chap. 2. Moreover, the scarcity and high price have not been a brake on the development of polymer electrolyte fuel cell (PEFCs), one of the most promising technologies for producing electricity, e.g., for transportation, as demonstrated by the Toyota-Mirai vehicle [1]. Herein, the platinum electrode working as cathode was modified to get a composition of PtCo to enhance its electrocatalytic, and stability properties. The strategy to optimize the platinum alloy synthesis has been pursued by performing the synthesis in non-aqueous medium to further increase the specific activity by a factor of 7 compared to commercial Pt/C, as reported by Shen and coworkers [2]. Despite the remarkable progress on precious metal centers, the present work has focused on the recent advancements in the synthesis, properties, and application of transition metal chalcogenides that enable efficient electrocatalysis of different reactions, such as ORR, OER, HER, and HOR, and tolerance to certain fuels. Otherwise, the pathways to oxygen and hydrogen electrochemical reactions which have not been dealt previously, this being the main reason to have written this book.

Even if the novel chalcogenide electrocatalysts discussed herein have still not reached the commercialization stage; the recent fast progress suggests a potential use of such materials in the near future. Specifically, the improvement in the rational design and synthesis of various nanostructured electrocatalysts with controllable size, shape, composition, surface morphology, and support has contributed to gain important insights into the nature of the catalytic activity and stability. Certainly, the upscaling synthesis and real-world evaluation may be the focus in the next step of materials' development. Among the transition metal chalcogenides, we

N. Alonso-Vante, *Chalcogenide Materials for Energy Conversion*, Nanostructure Science and Technology, https://doi.org/10.1007/978-3-319-89612-0_7

noted that two-dimensional materials are evolving as novel electrocatalysts for the HER, like MoS_2, with outstanding catalytic performances.

Regarding the catalysts' support, which affects the electrocatalysis performance of, e.g., HER, ORR, or OER, the most commonly used supporting substrates have been carbon-based materials that suffer corrosion. Chapter 5 delivers an additional appreciation of the modification of the catalytic center via the use of photons that generate electron–hole pairs on the support to better anchor the catalytic center, which could be extended to the most popular support: carbon. Therefore, this approach renders carbon one of the most interesting topics in materials science and research. Furthermore, progress of alkaline fuel cells, in microscale devices without membrane, promotes the development of metal chalcogenides (S, Se)-based catalysts for oxygen reduction reaction. Moreover, the ultimate need of reversible cells or batteries to sustain the energy economy will be the development of bifunctional electrodes, based on non-precious centers that enable the oxygen reduction reaction (ORR) and the oxygen evolution reaction (OER). This is one important challenge in the near future. The scope of the book does not allow for a full appreciation of all recent development in the field of chalcogenide materials. Moreover, it is advised to consult the respective original literature and reviews on special results and on further kinds of chalcogenide materials.

References

1. Yoshida T, Kojima K (2015) Toyota MIRAI fuel cell vehicle and progress toward a future hydrogen society. Electrochem Soc Interface 24(2):45–49. https://doi.org/10.1149/2.F03152if
2. Shen S, Li F, Luo L, Guo Y, Yan X, Ke C, Zhang J (2018) DMF-coordination assisted electrodeposition of highly active PtCo alloy catalysts for the oxygen reduction reaction. J Electrochem Soc 165(2):D43–D49. https://doi.org/10.1149/2.0471802jes

Index

N. Alonso-Vante, *Chalcogenide Materials for Energy Conversion*, Nanostructure Science and Technology, https://doi.org/10.1007/978-3-319-89612-0

The manufacturer's authorised representative in the EU is Springer Nature Customer Service Centre GmbH, Europaplatz 3, 69115 Heidelberg, Germany. If you have any concerns regarding our products, please contact ProductSafety@springernature.com

Printed and bound by CPI Group (UK) Ltd, Croydon, CR0 4YY
30/09/2026
02225221-0001